T0215460

# Data Science

Matthias Plaue

# Data Science

## An Introduction to Statistics and Machine Learning

 Springer

Matthias Plaue
MAPEGY GmbH
Berlin, Germany

ISBN 978-3-662-67881-7          ISBN 978-3-662-67882-4    (eBook)
https://doi.org/10.1007/978-3-662-67882-4

For Katja.

# Preface

**Preface to the original German edition.**

I wrote a large portion of this book in 2020, the year that saw the start of the COVID-19 pandemic. Data analysis was prominent in the media at the time, with geographic maps showing particularly affected areas (often brightly colored in red) and time series graphs illustrating the increase and decrease of confirmed cases or available hospital beds. Forecasts based on epidemiological models were made, and terms like "basic reproduction number" became common knowledge. These analyses and forecasts were rightfully used as the basis for significant political decisions, as they provide solid numerical evidence. However, due to their statistical nature, such analyses only allow for conclusions and actions to be taken with a high degree of uncertainty.

The world has been shaped by the advent of the information age in the twenty-first century. Significant advances in artificial intelligence are paving the way for its use in everyday life: search engines scour the World Wide Web for relevant content and attempt to capture the meaning of user input. Algorithms automatically sort vacation photos by image content, and "smart" household appliances are becoming more and more commonplace.

Whether we refer to statistics, big data, or machine learning, there can be little doubt that data science will continue to play a significant role in the coming decades. This book is designed to help its readers get started in this exciting and forward-looking field.

In my opinion, a successful study of data science requires two pillars. First, aspiring data scientists should gain practical experience by "playing" with data in a meaningful way, for example by programming their own models and performing their own parameter studies. Numerous tutorials, code examples, and datasets are available on the web for this purpose. The second important pillar is a comprehensive knowledge of methods and a deep understanding of the principles and ideas underlying data science methodology. This book helps arrive at such knowledge and understanding.

I would like to express my special thanks to my doctoral advisor and old partner in crime Mike Scherfner. Furthermore, I also thank Fred Hamprecht, Günter Bärwolff, Hartmut Schwandt, and Peter Walde for their guidance in my career as a data scientist. I am grateful to Alex Dirmeier for proofreading, and for his knowledgable and valuable comments. Finally, I would like to acknowledge the editors and staff of Springer who supported and made this project possible—my special thanks go to Andreas Rüdinger.

August 29, 2021

Berlin, Germany                                                                                     *Matthias Plaue*

### Preface to the revised, English edition.

This text is a translation of my German textbook, *Data Science – Grundlagen, Statistik und maschinelles Lernen*. The intention of publishing this translation is to make the content accessible to a wider, international readership. To that end, I have replaced some of the original book's application examples to make them more accessible to this audience.

As I continued working on the manuscript, it became clear that it would not merely be a translation but a complete revision of the original book. I corrected a number of errors that appeared in the German edition, and I would like to thank Andreas Zeh-Marschke and David Conti for identifying some of them. I have added many new references to the literature, some of which highlight more recent developments. This edition now includes exercises for readers to test their understanding of the subject matter (Chap. A).

I have also added a quick reference for the tools from linear algebra and calculus used throughout the book, making the book more self-contained (Chap. B). Other additions include subsections on stacked and grouped bar charts (Sect. 2.2.3.2), power transforms (Sect. 4.4.1.1), regularization (Sect. 6.1.2.1), and performance measures for regression (Sect. 6.1.3.1).

I thank Stefanie Adam and Andreas Rüdinger from Springer for their support, and I am grateful to Michael Wiley for proofreading the manuscript.

To prepare the manuscript, I found the machine translation service DeepL[1] to be helpful. Additionally, the virtual assistants ChatGPT[2] and Claude[3] were consulted to copy-edit some parts of it.

July 1, 2023

Berlin, Germany                                                                                     *Matthias Plaue*

---

[1] https://www.deepl.com/
[2] https://openai.com/product/chatgpt
[3] https://www.anthropic.com/product

# Contents

## Part III Machine learning

## Appendix

# Notation

| | |
|---|---|
| $a := b$ | assignment, definition; object $a$ is declared through object or formula $b$ |
| $\mathbb{Z}$ | set of integers $\{0, -1, 1, -2, 2, \ldots\}$ |
| $\mathbb{N}$ | set of natural numbers $\{0, 1, 2, \ldots\}$ |
| $\{0, 1, \ldots, D\}$ | set of the first $D + 1$ natural numbers |
| $\mathbb{R}$ | set of real numbers |
| $(x_1, \ldots, x_D)$ | $D$ tupel of elements (e.g., of real numbers) |
| $\mathbb{R}^D$ | set of all $D$ tupels of real numbers |
| $]a, b[, [a, b], ]a, b]$ | open, closed, half-open interval with endpoints $a, b \in \mathbb{R}$ |
| $]-\infty, b], [a, \infty[$ | improper intervals |
| $x \in A$ | $x$ is an element of the set $A$ |
| $\emptyset$ | empty set |
| $B \subseteq A$ | $B$ is a subset of $A$ |
| $B \subset A$ | $B$ is a proper subset of $A$ |
| $\bigcup_{i \in I} A_i$ | union with finite or countable index set $I$, e.g., $\bigcup_{i \in \{2,3,5\}} A_i = A_2 \cup A_3 \cup A_5$ |
| $f \colon X \to Y,$ $x \mapsto f(x)$ | $f$ is a map with domain $X$ and codomain $Y$, where each $x \in X$ maps to $f(x) \in Y$ |
| $g \circ f$ | composition of maps: $(g \circ f)(x) = g(f(x))$ |
| $f_\alpha(\,\cdot\,\|\theta; \beta)$ | map $x \mapsto f_\alpha(x\|\theta; \beta)$ from a family of maps, determined by parameters $\alpha, \beta, \theta$ |
| $\{x \in A \| \mathcal{B}[x]\}$ | set of all elements in $A$ that satisfy the condition $\mathcal{B}$ |

| | |
|---|---|
| $\|A\|$ | number of elements of a finite set $A$ |
| $\|B\|$ | area/volume/measure of a domain of integration $B \subseteq \mathbb{R}^D$ |
| $\|x\|$ | modulus of a real number $x$ |
| $\text{sgn}(x)$ | the sign function: $\text{sgn}(x) = -1$, if $x < 0$ etc. |
| $\lfloor x \rfloor, \lceil x \rceil$ | $x \in \mathbb{R}$ rounded down/up to the closest integer |
| $x \approx y$ | the value $x$ is approximately equal to $y$ (in some suitable sense) |
| $x \gg y$ | the value $x$ is much larger than $y$ (in some suitable sense) |
| $x \cong y$ | objects $x$ and $y$ are structurally identical in some suitable sense |
| $\sum_{k=1}^{K} x_k$ | sum of the form $x_1 + x_2 + \cdots + x_K$ |
| $\sum_{k \in I} x_k$ | sum with finite or countable index set $I$, e.g., $\sum_{k \in \{2,3,5\}} x_k = x_2 + x_3 + x_5$ |
| $\prod_{k=1}^{K} x_k$ | product of the form $x_1 \cdot x_2 \cdots x_K$ |
| $\langle x, y \rangle$ | scalar product of two vectors $x$ and $y$ |
| $\|x\|$ | length/norm of a vector $x$ |
| $f \propto g$ | the vectors/functions $f \neq 0$ and $g \neq 0$ are collinear/linearly dependent, i.e., there exists a scalar/constant $\lambda$, such that $f = \lambda g$ |
| $A^T$ | transposed matrix; the matrix $A$ after swapping columns with rows |
| $\det(A)$ | determinant of a square matrix $A$ |
| $\sphericalangle(x, y)$ | angle between vectors $x$ and $y$ |
| $\text{diag}(d_1, \ldots, d_K)$ | a diagonal matrix; square matrix with entries $d_1, \ldots, d_K$ on the diagonal, all other elements are zero |
| $\min A, \max A$ | minimum/maximum (smallest/largest element) of a (finite) set $A \subset \mathbb{R}$ |
| $\min_{n}\{x_n\}$ | minimum of a (finite) set or sequence of numbers $x_1, \ldots$ |
| $\inf A, \sup A$ | infimum/supremum (largest lower/upper bound) of a set $A \subseteq \mathbb{R}$ |
| $\inf_{\xi \in A}\{f(\xi)\}$ | infimum of a function $f$ with domain $A$ |
| $\lim_{n \to \infty} a_n$ | limit of a sequence $(a_n)_{n \in \mathbb{N}}$ |
| $\lim_{\xi \to \infty} f(\xi)$ | limit of a function $f$ |

| | |
|---|---|
| $\lim_{\xi \searrow u} f(\xi)$ | limit of a function $f$ from above |
| $\frac{\partial}{\partial x_2} f(x_1, x_2)$ | partial derivative of a (continuously differentiable) function $f$; alternative notation: $\partial_2 f(x_1, x_2)$ |
| $\operatorname{grad} f(x)$ | gradient of $f$; column vector of first derivatives $(\partial_i f)$ |
| $\operatorname{Hess} f(x)$ | Hesse matrix of a (twice continuously differentiable) function $f$: matrix of second derivatives $(\partial_i \partial_j f)$ |
| $\mathrm{D}f(x) = \frac{\mathrm{d}f}{\mathrm{d}x}(x)$ | Jacobian matrix of a (continuously differentiable, vector-valued) function $f$: matrix of first derivatives $(\partial_i f_j)$ |
| $\nabla_y f(x, y_1, y_2)$ | Nabla operator; partial gradient: $\nabla_y f = (\partial_2 f, \partial_3 f)^T$ |
| $\int_a^b f(x)\,\mathrm{d}x$ | integral of an (integrable) function with bounds $a, b \in \mathbb{R}$ |
| $\int_{-\infty}^b f(x)\,\mathrm{d}x$ | improper integral of a function |

# List of Figures

# List of Tables

# Introduction

According to the International Organization for Standardization (ISO) [1], **data** are "a reinterpretable representation of information in a formalized manner, suitable for communication, interpretation, or processing." Merriam–Webster [2] provides another characterization: data are "factual information (such as measurements or statistics) used as a basis for reasoning, discussion, or calculation."

The collection, processing, interpretation, and communication of data with the goal of obtaining robust and useful knowledge—usually with the aid of information technology—is the main function of **data science**.

In the empirical sciences, such as the natural sciences, the collection and analysis of data have long been an essential part of gaining knowledge. Modern physics, for example, would hardly be conceivable without the interaction of theory and experiment: the deviations from experimental data reveal the limits of theoretical models. One success of Newtonian celestial mechanics, for example, was the precise description and prediction of the motion of planets and other celestial bodies in the solar system. However, the theory could not explain the orbital perturbations of Mercury [3] that were measured to a high degree of accuracy in the nineteenth century by Urbain Le Verrier. These orbital perturbations could only be explained at a later time through the application of Einstein's theory of gravity, the general theory of relativity.

In a business context, the collection and analysis of data have also become increasingly important. Data can be seen as the raw material from which knowledge is produced, an economic good. Data analysis supports management in decision making and in understanding the impact that decisions have on the business (**business intelligence, business analytics**). For example, patent data can be used to show the innovation and networking activities of various organizations in a competitive environment, informing strategic decisions in innovation or technology management [4, 5].

Demographics are another discipline where data analysis is imperative. For example, statistical data are used to describe regional differences in human mobility behavior (cf. [6, 7]) and to provide actionable insights for urban planning.

Results from data analysis can be communicated to humans through language and through visual and graphical representations. The following diagram shows time series data on the air temperature measured at two meters above the ground in the district Tegel in Berlin, Germany for the year 2018 [8]:

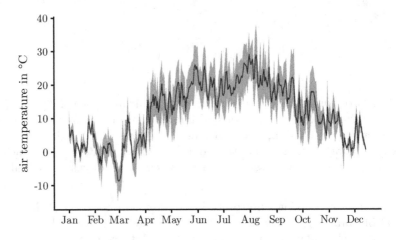

**Fig. 0.1.** *Line diagram of changes in temperature*

The gray ribbon represents the daily variation between the minimum and maximum temperatures. The following findings can easily be seen:

- The coldest days that year were in late February and early March.

- Obviously, it is colder in winter than in summer. However, the chart also illustrates how much more the daytime and nighttime temperatures differ in summer in comparison to the winter. Notice how the gray ribbon is much wider in the summer than in the winter.

The same information could not as easily or efficiently be determined from a table of measurements. The subfield of data analysis that focuses on graphical representation is called **data visualization**.

Another way of communicating data is to convert the data into acoustic non-linguistic signals, a process called **sonification**. However, this process is a much less common practice than graphical representation.

In summary:

> **Data analysis** aims to obtain relevant information and useful knowledge by systematic organization, statistical processing, and/or graphical (fur-

thermore: acoustic, audiovisual) representation of data (in this context also called **raw data**).

The following forms of data analysis are characterized by their respective objectives:

- **Descriptive statistics** are used to organize and summarize data.

- **Exploratory data analysis** finds hidden patterns in the data in order to formulate new hypotheses.

- **Inferential statistics** aim at describing observations by fitting them to statistical models and test the plausibility of hypotheses based on data.

When exploratory data analyses are performed on a large scale, they are also referred to as **data mining**. Possible descriptions of data mining are "the process of discovering interesting patterns from massive amounts of data" [9, Sec. 1.8] or "the practice of searching through large amounts of computerized data to find useful patterns or trends" [10].

Another central task of data science today is the development of algorithms for intelligent and autonomous systems using **machine learning**, so that these systems operate by rules that were not given explicitly but are mostly "learned" from training data.

This book describes concepts and methods from mathematics, statistics, and the computational sciences that are designed to accomplish the tasks described above. It is divided into three parts. The first part deals with aspects of **data organization**: the conceptual and logical structuring of data and how to ensure their quality. In that part, we also introduce the toolset of **descriptive statistics**, which aim to summarize and present essential characteristics of collected data.

The second part serves as an introduction to the mathematical field of **stochastics**, which includes probability theory and inferential statistics. These provide the conceptual and methodological foundation for reasoning under uncertainty.

In the last part of the book, we discuss aspects of statistical learning theory and present algorithmic methods for machine learning, which make substantial use of stochastics concepts.

This volume is an introductory text that presents essential topics and methods of data science. At the very end of the book, the reader will find a compilation of supplementary literature for further reading.

A prerequisite for the successful study of this book is having mathematical skills that include an understanding of linear algebra, calculus, and some aspects of multivariate calculus. These topics are typically taught during the first two to three semesters of an academic curriculum in science, technology, engineering, or mathematics (STEM). The appendix of this book provides a short summary and refresher.

"Practice makes perfect," they say—this holds especially true for the study of data science. This edition's appendix includes a number of exercises to test the reader's understanding of the subject matter. Additionally, the book contains numerous application examples based on data that are freely available. Readers are invited to reproduce these results using a programming language for statistical computing of their choice—for example, R [11] or Python [12]—and be creative, adding their own analyses and data-driven stories.

Many of the illustrations in this volume were generated using the data visualization package ggplot2 for R [13]. Manually layed out diagrams and graphs were drawn with the LaTeXpackage TikZ [14]. Other examples of program libraries for R that were used in the production of examples but not explicitly mentioned in the text include Cairo [15], cowplot [16], dplyr [17], extrafont [18], fastcluster [19], FNN [20], forcats [21], ggdendro [22] ggforce [23], ggrepel [24], Gmedian [25], gridExtra [26], igraph [27], latex2exp [28], lubridate [29], magick [30], mapproj [31], MASS [32], mlbench [33], mvtnorm [34, 35], neuralnet [36], nominatimlite [37], proxy [38], randomNames [39], reshape2 [40], rtweet [41], scales [42], sna [43], sp [44, 45], stringdist [46], stringr [47], tidyverse [48], tidytext [49], tsne [50], and usedist [51].

# References

[1]   ISO Central Secretary. *Information technology – Vocabulary*. Standard ISO/IEC 2382:2015. Genf, Schweiz: International Organization for Standardization, 2015, p. 2121272.

[2]   Merriam–Webster. *Data*. Accessed Apr. 10, 2022. URL: https://www. merriam-webster.com/dictionary/data.

[3]   Clifford M. Will. *Theory and Experiment in Gravitational Physics*. Cambridge University Press, Sept. 2018. DOI: 10.1017/9781316338612.

[4]   Holger Ernst. "Patent information for strategic technology management". In: *World Patent Information* 25.3 (Sept. 2003), pp. 233–242. DOI: 10.1016/s0172-2190(03)00077-2.

[5]   Peter Walde et al. "Erstellung von Technologie- und Wettbewerbsanalysen mithilfe von Big Data". In: *Wirtschaftsinformatik & Management* 5.2 (Feb. 2013), pp. 12–23. DOI: 10.1365/s35764-013-0274-7.

[6]   Robert Follmer and Dana Gruschwitz. *Mobility in Germany – short report (edition 4.0)*. Accessed Apr. 18, 2022. Bonn, Berlin, Sept. 2019. URL: http://www.mobilitaet-in-deutschland.de/pdf/MiD2017_ShortReport.pdf.

[7]   Fábio Duarte and Ricardo Álvarez. "The data politics of the urban age". In: *Palgrave Communications* 5.1 (May 2019). DOI: 10.1057/s41599-019-0264-3.

[8]   Deutscher Wetterdienst – Zentraler Vertrieb Klima und Umwelt. *Klimadaten Deutschland*. Accessed Apr. 1, 2020. Offenbach. URL: https://www.dwd.de/DE/leistungen/klimadatendeutschland/klimadatendeutschland.html.

[9]   Jiawei Han, Micheline Kamber, and Jian Pei. *Data Mining: Concepts and Techniques*. 3rd ed. Elsevier, 2012.

[10]   Merriam-Webster. *Data mining*. Accessed Apr. 10, 2022. URL: https://www.merriam-webster.com/dictionary/data%20mining.

[11]   R Core Team. *R: A Language and Environment for Statistical Computing*. R Foundation for Statistical Computing. 2020. URL: https://www.R-project.org/.

[12]   Guido van Rossum and Fred L. Drake. *Python 3 Reference Manual*. Scotts Valley, USA: CreateSpace, 2009.

[13]   Hadley Wickham. *ggplot2. Elegant Graphics for Data Analysis*. Springer, New York, 2009. DOI: 10.1007/978-0-387-98141-3.

[14]   Till Tantau. *The TikZ and PGF Packages. Manual for version 3.1.7*. Nov. 2020. URL: https://pgf-tikz.github.io/pgf/pgfmanual.pdf.

[15]   Simon Urbanek and Jeffrey Horner. *Cairo: R Graphics Device using Cairo Graphics Library for Creating High-Quality Bitmap (PNG, JPEG, TIFF), Vector (PDF, SVG, PostScript) and Display (X11 and Win32) Output*. R package. 2020. URL: https://CRAN.R-project.org/package=Cairo.

[16]   Claus O. Wilke. *cowplot: Streamlined Plot Theme and Plot Annotations for 'ggplot2'*. R package. 2020. URL: https://CRAN.R-project.org/package=cowplot.

[17]   Hadley Wickham et al. *dplyr: A Grammar of Data Manipulation*. R package. 2020. URL: https://CRAN.R-project.org/package=dplyr.

[18]  Winston Chang. *extrafont: Tools for using fonts*. R package. 2014. URL: https://CRAN.R-project.org/package=extrafont.

[19]  Daniel Müllner. "fastcluster: Fast Hierarchical, Agglomerative Clustering Routines for R and Python". In: *Journal of Statistical Software* 53.9 (2013), pp. 1–18. URL: http://www.jstatsoft.org/v53/i09/.

[20]  Alina Beygelzimer et al. *FNN: Fast Nearest Neighbor Search Algorithms and Applications*. R package. 2019. URL: https://CRAN.R-project.org/package=FNN.

[21]  Hadley Wickham. *forcats: Tools for Working with Categorical Variables (Factors)*. R package. 2020. URL: https://CRAN.R-project.org/package=forcats.

[22]  Andrie de Vries and Brian D. Ripley. *ggdendro: Create Dendrograms and Tree Diagrams Using 'ggplot2'*. R package. 2020. URL: https://CRAN.R-project.org/package=ggdendro.

[23]  Thomas Lin Pedersen. *ggforce: Accelerating 'ggplot2'*. R package. 2020. URL: https://CRAN.R-project.org/package=ggforce.

[24]  Kamil Slowikowski. *ggrepel: Automatically Position Non-Overlapping Text Labels with 'ggplot2'*. R package. 2020. URL: https://CRAN.R-project.org/package=ggrepel.

[25]  Herve Cardot. *Gmedian: Geometric Median, k-Median Clustering and Robust Median PCA*. R package. 2020. URL: https://CRAN.R-project.org/package=Gmedian.

[26]  Baptiste Auguie. *gridExtra: Miscellaneous Functions for 'Grid' Graphics*. R package. 2017. URL: https://CRAN.R-project.org/package=gridExtra.

[27]  Gabor Csardi and Tamas Nepusz. "The igraph software package for complex network research". In: *InterJournal* Complex Systems (2006), p. 1695. URL: https://igraph.org.

[28]  Stefano Meschiari. *latex2exp: Use LaTeX Expressions in Plots*. R package. 2015. URL: https://CRAN.R-project.org/package=latex2exp.

[29]  Garrett Grolemund and Hadley Wickham. "Dates and Times Made Easy with lubridate". In: *Journal of Statistical Software* 40.3 (2011), pp. 1–25. URL: https://www.jstatsoft.org/v40/i03/.

[30]  Jeroen Ooms. *magick: Advanced Graphics and Image-Processing in R*. R package. 2020. URL: https://CRAN.R-project.org/package=magick.

[31]  Doug McIlroy et al. *mapproj: Map Projections*. R package. 2020. URL: https://CRAN.R-project.org/package=mapproj.

[32]  W. N. Venables and B. D. Ripley. *Modern Applied Statistics with S*. 4th ed. Springer, New York, 2002. DOI: 10.1007/978-0-387-21706-2.

[33]  Friedrich Leisch and Evgenia Dimitriadou. *mlbench: Machine Learning Benchmark Problems*. R package. 2010. URL: https://CRAN.R-project.org/package=mlbench.

[34]  Alan Genz et al. *mvtnorm: Multivariate Normal and t Distributions*. R package. 2020. URL: https://CRAN.R-project.org/package=mvtnorm.

[35]  Alan Genz and Frank Bretz. *Computation of Multivariate Normal and t Probabilities*. Lecture Notes in Statistics. Springer, Berlin, Heidelberg, 2009.

[36] Stefan Fritsch, Frauke Günther, and Marvin N. Wright. *neuralnet: Training of Neural Networks*. R package. 2019. URL: https://CRAN.R-project.org/package=neuralnet.

[37] Diego Hernangómez. *nominatimlite: Interface with 'Nominatim' API Service*. R package. 2022. DOI: 10.5281/zenodo.5113195. URL: https://dieghernan.github.io/nominatimlite/.

[38] David Meyer and Christian Buchta. *proxy: Distance and Similarity Measures*. R package. 2020. URL: https://CRAN.R-project.org/package=proxy.

[39] Damian W. Betebenner. *randomNames: Function for Generating Random Names and a Dataset*. R package. 2019. URL: https://cran.r-project.org/package=randomNames.

[40] Hadley Wickham. "Reshaping Data with the reshape Package". In: *Journal of Statistical Software* 21.12 (2007), pp. 1–20. URL: http://www.jstatsoft.org/v21/i12/.

[41] Michael W. Kearney. "rtweet: Collecting and analyzing Twitter data". In: *Journal of Open Source Software* 4.42 (2019). R package, p. 1829. DOI: 10.21105/joss.01829.

[42] Hadley Wickham and Dana Seidel. *scales: Scale Functions for Visualization*. R package. 2020. URL: https://CRAN.R-project.org/package=scales.

[43] Carter T. Butts. *sna: Tools for Social Network Analysis*. R package. 2020. URL: https://CRAN.R-project.org/package=sna.

[44] Edzer J. Pebesma and Roger S. Bivand. "Classes and methods for spatial data in R". In: *R News* 5.2 (Nov. 2005), pp. 9–13. URL: https://CRAN.R-project.org/doc/Rnews/.

[45] Roger S. Bivand, Edzer Pebesma, and Virgilio Gomez-Rubio. *Applied spatial data analysis with R*. 2nd ed. Springer, New York, 2013. URL: https://asdar-book.org/.

[46] Mark P. J. van der Loo. "The stringdist package for approximate string matching". In: *The R Journal* 6 (1 2014), pp. 111–122. URL: https://CRAN.R-project.org/package=stringdist.

[47] Hadley Wickham. *stringr: Simple, Consistent Wrappers for Common String Operations*. R package. 2019. URL: https://CRAN.R-project.org/package=stringr.

[48] Hadley Wickham et al. "Welcome to the tidyverse". In: *Journal of Open Source Software* 4.43 (2019), p. 1686. DOI: 10.21105/joss.01686.

[49] Julia Silge and David Robinson. "tidytext: Text Mining and Analysis Using Tidy Data Principles in R". In: *JOSS* 1.3 (2016). DOI: 10.21105/joss.00037.

[50] Justin Donaldson. *tsne: t-Distributed Stochastic Neighbor Embedding for R (t-SNE)*. R package. 2016. URL: https://CRAN.R-project.org/package=tsne.

[51] Kyle Bittinger. *usedist: Distance Matrix Utilities*. R package. 2020. URL: https://CRAN.R-project.org/package=usedist.

# Part I

## Basics

# Elements of data organization

We use the term **database** to refer to all data captured, stored, organized, and made available for access and processing. A large collection of data is often **heterogeneous**, as it is the result of **data integration**, the merging of data from different and diverse **data sources**. When planning to build a large and/or heterogeneous database (buzzword: **big data**) while ensuring the quality of the collected data, special challenges arise that we address in this chapter.

First, we describe approaches to modeling data. The primary goal of data modeling is to capture the meaning of the data and to structure it to best serve the use case at hand.

Three stages or milestones of the data modeling process can be distinguished, each of which serves to answer the following questions [1]:

1. **Conceptual data model:** What are the relevant objects of the knowledge domain that the data describe? What properties can we associate with those objects, and how do they relate to each other?

2. **Logical data model:** How are the data structured (i.e., how are they arranged, grouped, combined, or compared)?

3. **Physical data model:** What are the technical specifications of the hardware and software used to capture and store the data?

The conceptual data model is a systematic description of the **semantics** of the data in the context of the knowledge domain, while the logical data model focuses on formal structure. In practice, the key difference between these stages lies in the level of detail: the conceptual data model aims at a broad, global description of the underlying concepts, while the logical data model usually provides a greater level of detail. It is not uncommon for both stages to be combined.

Physical data modeling focuses on the technical challenges associated with, for example, the collection and validation, performant processing, and scalable

© Springer-Verlag GmbH Germany, part of Springer Nature 2023
M. Plaue, *Data Science*, https://doi.org/10.1007/978-3-662-67882-4_1

storage of data using a specific **database management system** (DBMS). We do not elaborate on these challenges of **data engineering** in this book.

Hereafter, we present criteria and procedures for measuring and ensuring the quality of data.

## 1.1 Conceptual data models

In order to be a useful resource, rather than a meaningless collection of numerical values, strings, etc., data must be connected to the real world and represent information about the domain of interest.

> Data represent information about individually identifiable real or abstract objects, in this context called **entities** or **information objects**.

Entities could be, for example, persons in a customer database, and data that describe each entity include name and place of residence.

If the entities described by the data are themselves data, then the descriptive data are also referred to as **metadata**. The **data documentation** (more on this topic in Sect. 1.3.1) represents an important example of metadata.

A **conceptual data model** is a systematic representation of the entities, their relevant properties, and their relationships with each other. Such a model can be helpful, for example, in planning a data analysis project to determine information needs.

### 1.1.1 Entity–relationship models

A frequently used systematic representation of a data model is the **entity–relationship model** (ERM) [2]. First, note that similar entities can be grouped into **entity types** by way of abstraction. For example, in a database that contains data about customers of a retail chain, each customer may be regarded as an **instance** of the entity type *customer*.

After determining which entity types to use in the data model, each entity can be assigned specific **attributes**, the properties that define an entity type. For example, *first name*, *family name*, and *address* are attributes that may be useful for the entity type *customer*.

A specific assignment of an attribute is an **attribute value**. Possible attribute values for the attribute *first name* of a customer may be, for example, "Anna" or "Carl." In the context of statistics or machine learning, attributes are also known as *variables* or *features*, respectively.

Given a knowledge domain, there is no standard way to model the underlying data. Any distinction between attributes and entities must make sense for the use case at hand. For example, a customer's place of residence might be understood as an attribute or it might be interpreted as an entity type on its own, represented by the address.

Finally, entities are connected by **relationships**. For example, when customers of a retail company make their purchases in the company's physical stores, the customers can be linked to those stores by a relationship type *shops at*.

The data model described can be summarized in an **entity–relationship diagram** like the following:

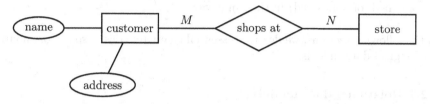

**Fig. 1.1.** *Entity–relationship diagram*

The above graphical representation of an ERM is also called **Chen notation**. In Chen notation, entity types are represented by rectangles, attributes by oval graphical elements, and relationship types are diamond-shaped.

The descriptors $M$, $N$ in the above diagram denote the **maximum cardinality** of the relationship type, as described below between two entity types $X$ and $Y$:

- **1:1 relationship:** Given a single entity of type $Y$, at most one entity of type $X$ is related to it, and vice versa. For example, every motor vehicle is related to one and only one license plate (with the odd exception of changeable/seasonal license plates).

- **1:N relationship:** Given a single entity of type $Y$, at most one entity of type $X$ is related to it. Conversely though, an entity of type $X$ may be related to multiple entities of type $Y$. For example, a person has only a single (primary) place of residence, but several persons may live at the same address.

- **M:N relationship:** There is no numerical limit on the number of entities that are related to each other. An example is the relationship type *customer x shops at store y*: a store usually has more than one customer, and one customer may make purchases in multiple stores.

## 1.2 Logical data models

A conceptual data model structures a knowledge domain—an entity–relationship model achieves this goal by definition of entities, attributes, and relationships between the entities. These real-world objects are associated to data that represent them. For example, a person's first name corresponds to a particular string of characters, and their place of residence can be stored as latitude and longitude coordinates in the form of a pair of floating point numbers (e.g. 52.512652, 13.316714). In order to process data with information technology, it should be structured in a suitable manner. A **logical data model** structurally mirrors the conceptual data model.

We can understand a logical data model as an organization of smallest units:

> A **data item** is the smallest organizational unit of data. A **dataset** is a collection of data items that are structured (i.e., arranged, grouped, combined, or compared) in the same way.

In the following sections, we describe essential characteristics of some commonly used logical data models.

### 1.2.1 Relational data models

A very common way of structuring data items is to arrange and group them in tabular form. Here is an example of a dataset of fictitious customer data:

| ID | last name | first name | street | str. no. | zip code | city |
|----|-----------|------------|--------------|----------|----------|-----------------|
| 1  | Smith     | Anna       | First Street | 1        | 12345    | Model City      |
| 2  | Miller    | Robert     | Second place | 3        | 54321    | Example Village |
| 3  | Miller    | Carl       | Second place | 3        | 54321    | Example Village |

**Table 1.1.** *Customer data*

Each row of such a data table may be called a **data record** or **data tuple**.

The **primary key** attribute, *ID* in the first column of the table above, uniquely identifies each entity that is represented and described by a data record. It may be as simple as a consecutive sequence of positive integers, as shown in the example.

> Any given entity–relationship model can be mapped onto a **relational data model** by applying the concepts of entities, attributes, and relationships. In this context, a data table is also referred to as a **relation**.
>
> **Entities:** Each entity type corresponds to a relation. Each data record corresponds to an entity (in the above example, a person/customer).
>
> **Attributes:** Each column corresponds to an attribute. The attribute values in a single row constitute a data record/tuple.
>
> **Relationships:** Each relationship type corresponds to a relation. Each data record represents a relationship by listing the values of the primary keys of the related entities.

To illustrate how this framework can be applied, let us once more consider the hypothetical retail company's database, which may also include the below example data about stores and customers.

| ID | name | city |
|----|------|------|
| 1 | Good Buy | Example Village |
| 2 | Cheap Buy | Model City |

The relation type *shops at* can be stored as shown by the following relation:

| customer ID | store ID |
|-------------|----------|
| 1 | 2 |
| 2 | 1 |
| 2 | 2 |
| 3 | 1 |

The customer with primary key "1" (= Smith, Anna) shops at the store with primary key "2" (= Cheap Buy), and so on.

Relations can be processed by operations that make up the so-called **relational algebra**. To explain these operations, it is helpful to think of a relation as a set of tuples. For a quick reminder on the mathematical concept of a set, and set operations like intersection and union, we refer to the appendix (Sect. B.1). Thus, a relation $R$ with $D = D(R)$ columns/attributes that yield values taken from the sets $R_1, \ldots, R_D$, and $N$ rows/data records/tuples can be written as follows:

$$R = \left\{ \left( r^1{}_1, \ldots, r^1{}_D \right), \left( r^2{}_1, \ldots, r^2{}_D \right), \ldots, \left( r^N{}_1, \ldots, r^N{}_D \right) \right\}$$

Here, $r^n{}_d \in R_d$ for all $d \in \{1, \ldots, D\}$ and $n \in \{1, \ldots, N\}$ is the entry in the $n$-th row and the $d$-th column of the relation. In the following, the row index will not be important, so we write more briefly: $R = \{(r_1, \ldots, r_D)\}$.

Two relations, $R$ and $S$, are called **type-compatible** if they are described by identical attributes: the number of attributes $D$ of both relations is identical and it holds $R_1 = S_1, R_2 = S_2, \ldots, R_D = S_D$.

For type-compatible relations $R$ and $S$, the usual set operations are well-defined: **union** $R \cup S$, **intersection** $R \cap S$, **(set) difference** $R \setminus S$.

For arbitrary relations that are not necessarily type-compatible, the **Cartesian product** is also well-defined:

$$R \times S = \{(r_1, \ldots, r_{D(R)}, s_1, \ldots, s_{D(S)})\}$$

We can perform a **projection** onto certain attributes or columns as follows:

$$R = \{(r_1, \ldots, r_D)\} \mapsto \{(r_{\iota(1)}, \ldots, r_{\iota(K)})\} = \pi_{\iota(1), \ldots, \iota(K)}(R)$$

where $\iota \colon \{1, \ldots, K\} \to \{1, \ldots, D\}$, $K \leq D$.

Given some condition, $\mathcal{B}$, imposed on the attribute values, we can collect the data records that satisfy that condition—this operation is called a **selection**:

$$R[\mathcal{B}] = \{r \in R | \mathcal{B}[r]\}$$

The **join** operation is the composition of Cartesian product and selection:

$$R \bowtie_{\mathcal{B}} S = (R \times S)[\mathcal{B}]$$

Let us give a concrete example for applying these operations. Consider again the relations representing the entity types *customer* ($=: A$) and *stores* ($=: B$) as well as the relation type *shops at* ($=: R$). We want to construct a new relation that lists the last name and city of residence of those customers who shop at stores in "Model City" ($=: x$).

This can, for example, be achieved as follows:

1. Join the individual relations via their primary keys. We obtain the new relation $A \bowtie_{a_1=r_1} R \bowtie_{r_2=b_1} B$.

2. Selection by the desired stores: $A \bowtie_{a_1=r_1} R \bowtie_{r_2=b_1} B[b_3 = x]$.

3. By projecting on the attributes *last name* and *city* of the individuals (2nd and 7th column, respectively) we finally obtain:

$$\pi_{2,7} \left( A \bowtie_{a_1=r_1} R \bowtie_{r_2=b_1} B[b_3 = x] \right)$$

In practice, a standard for implementing these operations is the widely used database language SQL (Structured Query Language). SQL has a simple and explainable syntax. A database query written in SQL for the above example can look something like this:

```
SELECT
        c.last_name,
        c.city
FROM    customers c
JOIN    buys_at r ON c.id = r.id_customer
JOIN    stores s ON s.id = r.id_store
WHERE   s.city = 'Model City';
```

SQL also enables the implementation of important operations that have not been presented so far. For example, records can be **grouped** and their attribute values **aggregated**. A database query to determine the total number of customers for each store might look something like this:

```
SELECT
        s.id, s.name, s.city,
        COUNT(DISTINCT c.id) AS number_customers
FROM    customers c
```

```
JOIN    buys_at r ON c.id = r.id_customer
JOIN    stores s ON s.id = r.id_store
GROUP BY s.id;
```

Database management systems that implement relational data models include MySQL [3], MariaDB [4], and PostgreSQL [5].

### 1.2.2 Graph-based data models

Mathematically, relational data models represent data as sets of data tuples, and these sets can be combined by operations defined by the relational algebra. Graph-based data models on the other hand are based on the following mathematical concept:

A **directed graph** $G$ is a pair $G = (V, E)$ that consists of a finite set $V$ and a set $E \subseteq V \times V$.

The elements of $V$ are called **nodes** or **vertices**, the elements of $E$ are called **directed edges**.

The above meaning of the term "graph" must not be confused with that of a function graph. We can think of a directed edge $(u, v) \in E$ as connecting the **start node** $u \in V$ to the **end node** $v \in V$.

Interpreting edges as directed connections between nodes allows a directed graph to be represented as a **node–link diagram**, as shown below.

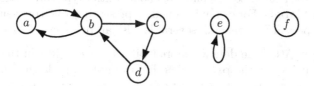

**Fig. 1.2.** *Node–link diagram*

The figure shows the directed graph $G = (V, E)$ with vertex set $V = \{a, \ldots, f\}$ and edges $E = \{(a, b), (b, a), (b, c), (c, d), (d, b), (e, e)\}$.

An **undirected graph** differs from a directed graph in that the edges do not exhibit a preferred direction. In other words, there is no distinction being made between start and end nodes.

In data modeling practice, the term "graph" is usually understood to mean multigraph. In a multigraph, nodes may be connected by more than one edge.

A **directed multigraph** $G = (V, E)$ consists of a finite set of vertices $V$ and a set of enumerated directed edges

$$E \subseteq \{(u, v, n) | u, v \in V, n \in \{1, 2, 3, \ldots\}\}.$$

More precisely, "enumerated" means: for all $u, v \in V$ and $n \in \mathbb{N}$, $n > 1$ the following holds: if $(u, v, n)$ is an edge, then $(u, v, n-1)$ is also an edge.

Instead of denoting the edge of a multigraph by $(u, v, n)$, we can also write $(u, v)_n$ to more clearly distinguish the edge number from start and end nodes notationally. We illustrate by example:

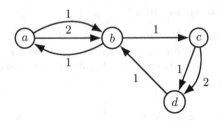

**Fig. 1.3.** *Multigraph*

The figure above shows a node–link diagram of the multigraph with nodes $V = \{a, \ldots, e\}$ and edges:

$$E = \{(a, b)_1, (a, b)_2, (b, a)_1, (b, c)_1, (c, d)_1, (c, d)_2, (d, b)_1\}$$

An entity–relationship model can be mapped onto a logical structure known as a **property graph** as follows [6, Chap. 3]:

**Entities:** Each entity corresponds to a node. Entity types are distinguished by so-called **labels.** For example, a node representing a person would carry the label "person." In general, several labels can be assigned to a node.

**Attributes:** Attribute data are stored with nodes or edges in the form of key–value pairs, i.e. the **properties.** For example, a node that is labeled as a person could have the property ("first_name" : "Carl").

**Relationships:** Each relationship between two entities corresponds to an edge between the corresponding nodes. Relationship types are distinguished from each other by labels.

An example of a graph-based database management system is neo4j [7]. A query using the Cypher database language in neo4j might look like the following:

```
MATCH    (c:customer)-[:shops_at]->(s:store)
WHERE    s.city = 'Model City'
RETURN   c.name, c.city
```

While relationally structured data are best represented as tables in order to be read and processed by humans, a node–link diagram such as the following is suitable for visualizing a property graph.

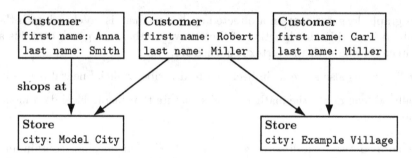

Fig. 1.4. *Property graph*

### 1.2.3 Hierarchical data models

Hierarchical data models are built on the mathematical concept of a tree. Trees are graphs that have a special shape, starting from a particular node—the "root"—there is only one direction in which we can walk along the edges, toward the "leaves." Before we can formally define trees, we first need to explain some additional terminology from graph theory.

Let $G = (V, E)$ be a directed graph. A **directed walk** with a start node $u \in V$ and an end node $w \in V$ of length $N \geq 1$ is an alternating sequence of nodes and edges $u = v_0, e_1, v_1, e_2, \ldots, v_{N-1}, e_N, v_N = w$ such that $e_n = (v_{n-1}, v_n)$ for all $n \in \{1, \ldots, N\}$.

For an **undirected walk**, $e_n = (v_{n-1}, v_n)$ or $e_n = (v_n, v_{n-1})$ may hold. In other words, an undirected walk may also make steps against the direction of the edges.

If all of the nodes $u, v_1, \ldots, v_N, w$ of a directed/undirected walk are pairwise distinct, it may also be called a directed/undirected **path**.

A directed/undirected **cycle** is a directed/undirected walk where the start and end nodes are identical, $u = w$, but all of the remaining path nodes, $v_1, \ldots, v_N$, are pairwise distinct.

Graphs that do not contain a directed cycle are called **acyclic**.

The following graph is an example of an acyclic graph:

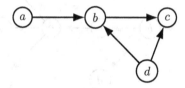

Fig. 1.5. *Acyclic graph with undirected cycle*

This graph does not contain a directed cycle—so there is no closed path that can be traversed in the direction of the arrows. Rather, the graph contains an undirected cycle with the vertex sequence $b, c, d, b$.

Acyclic graphs also serve us in Sect. 6.4 to describe artificial neural networks.

Finally, we can define the mathematical structure that hierarchical data models are based on.

> A **rooted tree** is a directed graph $G = (V, E)$ that does not contain an undirected cycle, together with a distinguished node $r \in V$, the **root** of $G$, such that for all $v \in V$ with $v \neq r$ there exists a directed path from $r$ to $v$.
>
> The **leaves** of a tree are the nodes that are different from the root and have only one neighboring node.

The following node–link diagram shows an example of a tree with root $a$:

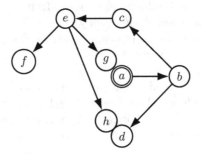

**Fig. 1.6.** *Rooted tree*

Each node can be assigned a hierarchical level by the (uniquely determined) length of the path from the root node. On the "zeroth" level, there is only the root node $a$, on the first level the node $b$, the second level consists of the nodes $c$ and $d$, and so on. The leaves are given by the nodes $d$, $f$, $g$ and $h$.

The following diagram shows the same tree again, but in this representation the hierarchical structure is more apparent since nodes at the same hierarchical level are shown in the same column:

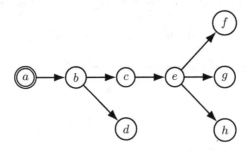

**Fig. 1.7.** *Levels of hierarchy in a rooted tree*

Trees can be used to represent the results of a hierarchical cluster analysis, which we will discuss in Sect. 7.3.2.

**Example.** The following is a small excerpt of a so-called Extensible Markup Language (XML) file submitted by an RSS feed [8]:

```
<?xml version="1.0" encoding="UTF-8"?>
<rss xmlns:dc="http://purl.org/dc/elements/1.1/"
xmlns:content="http://purl.org/rss/1.0/modules/content/"
xmlns:atom="http://www.w3.org/2005/Atom" version="2.0"
xmlns:cc="http://cyber.law.harvard.edu/rss/
creativeCommonsRssModule.html">
    <channel>
        <title><![CDATA[Towards Data Science - Medium]]></title>
        <description><![CDATA[A Medium publication sharing
        concepts, ideas, and codes. - Medium]]></description>
        <link>https://towardsdatascience.com?source=rss
        ----7f60cf5620c9---4</link>
        <item>
            <title><![CDATA[Why Psychologists Can Be Great
            Data Scientists]]></title>
            <guid isPermaLink="false">https://medium.com/p/
            970552b5223</guid>
            <category><![CDATA[careers]]></category>
            <category><![CDATA[machine-learning]]></category>
            <dc:creator><![CDATA[Maarten Grootendorst]]>
            </dc:creator>
        <pubDate>Fri, 18 Sep 2020 14:00:02 GMT</pubDate>
            <atom:updated>2020-09-18T14:00:02.568Z</atom:updated>
        </item>
        <item>
            ...
        </item>
    </channel>
</rss>
```

XML files are examples of hierarchically structured data. Each node of the tree is referred to as an **element** in this context. Each element and the elements of the hierarchy levels below that element are enclosed by tags of the form <element> and </element>. In this example, the root is represented by the tags <rss>...</rss>, which enclose all other elements.

## 1.3 Data quality

The principal purpose of data science is to generate trusted and useful knowledge from data. A high-quality dataset is a prerequisite for fulfilling this purpose.

### 1.3.1 Data quality dimensions

The Data Management Association UK proposes an assessment of the quality of data based on six key characteristics that they call data quality dimensions [9]: **completeness, consistency, validity, uniqueness, accuracy**, and **timeliness**. Tab. 1.6 lists these data quality dimensions, with each one explained in the table in the form of a probing question. The table also provides examples of typical **data defects. Data profiling** is the practice of collecting metrics in order to assess data quality. The last column of the table lists such metrics, which measure the severity or frequency of data defects.

> **Data quality** measures the extent to which a dataset can be used as intended. Consequently, the significance of each data quality dimension and the priority with which data defects need to be managed depend on the use case at hand.

For example, the validity of data is important for automated processing. Invalid inputs, for example when inputs are outside of the domain of the system's functions, can lead to software errors if they are not caught by some form of exception handling. In general, the accuracy of data plays only a minor role in the automated processing stage. On the other hand, accuracy is a vital feature for information that is presented to and evaluated by a human.

Completeness of metadata is also a relevant indicator of high data quality, especially when it comes to **data documentation**. Good data documentation ensures reproducibility of results and transparency of the analysis process. Data documentation may include information such as:

- the context of data collection, underlying hypotheses, and goals of the associated data analysis project;
- the semantics of the data, a conceptual data model, e.g., an entity–relationship diagram;
- the logical data model; a description of attributes, properties, etc. including their range of valid values, units of measurement, etc.; in the context of survey research, the **codebook** [10, Chap. 5];
- an evaluation of data quality, results of data profiling, or a description of any data cleaning measures performed on the dataset (see next section);
- the location, time, and modality of data collection: methods, software, measurement apparatus used, data sources tapped into, etc.;
- information about **data versioning**, differences between different versions;
- technical information, such as interfaces for access or the physical size of files;
- terms of use, data protection measures.

# 1.4 Data cleaning

In the following section, we present a few methods to identify and manage data defects. In general, it is highly recommended to keep a historical record of such **data preprocessing** tasks, which may include maintaining the original raw data with the database or a backup thereof. Such a historical record is called **data provenance**, and it helps with tracking down the cause of data quality issues.

### 1.4.1 Validation

Formally or syntactically incorrect, invalid attribute values can, under certain circumstances, lead to malfunctions of the data processing software. Implausible values outside of a permissible range may also skew the results of data analysis.

Invalid attribute values can be identified with the help of validation rules. On the one hand, such rules can refer to the syntax. For example, a birth date can be required to be stored in the format YYYY-MM-DD, where YYYY is a four-digit year, MM is a two-digit month (with a leading zero if applicable), and TT is a two-digit number that represents the day. According to this validation rule, 2010-12-01 would be a valid date, but 10-12-01 or 2010-12-1 would not be.

On the other hand, values can be recognized as invalid if they lie outside of a specified valid range. Thus, 2010-23-01 would be a valid date according to the above validation rule—but not if 1 <= MM <= 12 was also required.

Additionally, necessary conditions for the accuracy of the attribute value can be checked, i.e., their plausibility. Thus, birth dates can be marked as invalid when they lie in the future at the time of the creation of the dataset, or also if they are too far in the past.

In many cases, invalid or implausible attribute values are handled by deleting them. For other cases, a transformation can be performed to correct the data defect. For example, dates of the form YYYY-MM-D can be transformed to the valid syntax YYYY-MM-DD by inserting a leading zero: JJJJ-MM-0D. Of course, this procedure is only recommendable if the data's provenance and the data generating process imply that the result obtained is not only valid but also accurate.

We summarize:

> Invalid or implausible attribute values can be identified by checking against **validation** and/or **plausibility rules**. Removing invalid values or converting them into a valid format may be part of such a **data validation** process.

For the implementation of data validation processes, so-called **regular expressions** are often helpful. With the help of regular expressions, syntactic patterns in strings can be searched and replaced. They are included with many programming languages and standard libraries [11].

### 1.4.2 Standardization

The main goal of data integration is to combine data from different data sources. Significant challenges arise from the heterogeneity of the sources, such as varying data models or different technical means to store or send the data (e.g., different file formats or interfaces). Consequently, an important aspect of successful data integration is **data harmonization**, which aims to reduce heterogeneity.

An important subtask of harmonizing a dataset is to standardize the data so that the data are available in a uniform syntax and with the same units of measurement. Here is a fictitious example of two datasets with different syntax or unit of measurement:

| | | |
|---:|---|---|
| **name** | Rose Smith | Smith, Rose |
| **phone no.** | +498932168 | +4989-321-68 |
| **date of birth** | 1981-12-07 | 12/07/1981 |
| **organization** | Example Incorporated | Example Inc. |
| **body height** | 180 cm | 5' 11" |

**Table 1.2.** *Heterogeneous, non-standardized data*

In order to standardize the dates, we can choose the target syntax YYYY-MM-DD, where the letters stand for year, month, and day. Other date formats such as MM/DD/YYYY can be mapped onto this syntax in the obvious way. Similarly, data that represent lengths or distances can be converted to the same unit of measurement. For example, all body heights across data sources can be given in centimeters.

> **Data standardization** aims to align a heterogeneous syntax or units of measurement. For this purpose, attribute values are converted into a specified target syntax and common units of measurement.

### 1.4.3 Imputation

Missing attribute values in a dataset can occur for various reasons. In the case of a survey, a respondent may have refused to provide information. In the case of a physical measurement, the measuring device may have been impaired in its functionality. Missing values will also occur if invalid values were discarded during a data validation process.

> **Data imputation** is the process of completing data by replacing missing attribute values with synthetic but plausible values.

Missing attribute values can lead to a bias in the results of data analysis. The primary goal of data imputation is not to determine the correct attribute values but to supplement the data in order to improve the results.

An alternative to imputation is the **complete case analysis**, a simple and commonly used procedure. With this approach, data tuples that are missing relevant values are excluded from the analysis.

### 1.4.3.1 Imputation with measures of central tendency

Even when some information objects have no value assigned to a certain attribute, a substitute value can be determined on the basis of the remaining existing values of the same attribute. The arithmetic mean is a popular choice for computing such a substitute value. In Sect. 2.3, we introduce other measures of central tendency, like the mode or median. The missing attribute values can be replaced by this substitute value. This procedure is referred to as **mean substitution**, especially when using the arithmetic mean.

**Example.** The following table represents a small sample of results from a 2018 telephone survey of U.S. citizens provided by the U.S. Centers for Disease Control and Prevention (CDC) [12]:

| sex | age in years | height in cm | weight in kg |
|---|---|---|---|
| male | 35 | 170 | 98 |
| male | 66 | **NA** | **NA** |
| female | 67 | 155 | **NA** |
| ⋮ | ⋮ | ⋮ | ⋮ |

**Table 1.3.** *Missing attribute values*

Here, the attribute values "NA" represent missing data. The missing data in this case is because the respondent provided no data, invalid data, or insufficient data.

The mean body height and mean body weight, determined from the entire dataset as provided by the CDC, are 170 cm and 82 kg, respectively. Thus, the mean-substituted dataset is the following:

| sex | age in years | height in cm | weight in kg |
|---|---|---|---|
| male | 35 | 170 | 98 |
| male | 66 | **170** | **82** |
| female | 67 | 155 | **82** |

**Table 1.4.** *Mean substitution*

The dataset can also be partitioned into convenient **imputation classes** in order to apply a mean substitution to each of them. In the above example, instead of averaging across the entire dataset, missing data can instead be replaced by the mean value with respect to each sex. Taking into account that male respondents typically have larger body weight and body height, the imputed data table becomes:

| sex | age in years | height in cm | weight in kg |
|---|---|---|---|
| male | 35 | 170 | 98 |
| male | 66 | **178** | **90** |
| female | 67 | 155 | **75** |

**Table 1.5.** *Use of imputation classes*

### 1.4.3.2 Imputation via regression and classification

In the chapters on inferential statistics and machine learning (Chap. 4 and 6), we will present various procedures by which a missing or invalid value of an attribute can be predicted by the remaining attribute values of the same entity. These procedures can also be used for data imputation. In fact, mean substitution may be thought of as a simple regression procedure where only a single parameter is inferred from the data. Nearest-neighbor imputation ([13], see Section 6.3.2) replaces the missing attribute values with values of otherwise similar data tuples of the same dataset.

### 1.4.4 Augmentation

In some cases, it may make sense or even be necessary to expand the dataset with new data records if coverage is insufficient for the use case at hand.

Data coverage can be increased by tapping into new data sources and collecting more data. However, especially in the field of machine learning, **synthetic data** is also used to provide additional data for training algorithms.

> **Data augmentation** refers to an extension of the dataset by artificially generated data records. Typically, these data records are generated by a suitable transformation of already existing data records.

Similar to imputation, augmentation is not about adding accurate data but instead aims to avoid bias in the results due to "lack of data." It can be used to improve the robustness of a machine learning model that was trained with the data, and it is used especially to avoid so-called overfitting (see Sect. 6.1.2).

- In image or video analysis, already available images can be subjected to various image transformations to generate new data [14], such as with geometric transformations, like shear, zoom, reflection, or rotation. Additionally, image sharpness, contrast, brightness, or color temperature can be changed to create additional images.

- In text analysis, possible techniques [15] include the replacement of words with synonyms [16] or back translation [17].

### 1.4.5 Deduplication

A particular challenge for data harmonization is posed by **unstructured data** that does not follow a fixed syntax. For example, the date may have been entered manually into a free text field so that the strings `December 7, 1981`, `7 December 81`, `07 Dec. 1981`, etc. all represent the same date. They might even contain typographic errors, like `Decebmer 7, 1981`. In this case, more complex rules must be developed for an extraction of month, day, and year.

Organization names can be especially difficult to harmonize. The following list comes from the patent database PATSTAT [18, 19] and shows a small selection of spellings with which the German multinational conglomerate corporation Siemens can appear as the applicant of a patent:

`SIEMENES AKTIENGESELLSCHAFT; Siemens; Siemens A.G.; Siemens AG;`
`SIEMENS AG (DE); SIEMENS AG, 8000 MUNICH, DE;`
`SIEMENS AG, 80333 MUNICH, DE; Siemens Akteingesellschaft;`
`Siemens Aktiengellschaft; Siemens Aktiengesellscahft;`
`Siemens Aktiengesesellschaft; SIMENS AKTSIENGEZELL'SHAFT`

These examples exhibit the many causes for such heterogeneity: typographic errors, abbreviated spellings, address information included with the name, or transcription.

Even determining the number of patents registered by Siemens thus becomes a non-trivial task, since the organization is not represented in a unique way.

> **Data deduplication** is the task of identifying the data records that describe a single entity in order to merge them into a single representative data record.

At their core, many commonly used automated duplicate detection and data fusion procedures are based on so-called cluster analysis. In cluster analysis, sufficiently similar information objects are grouped together. More about cluster analysis can be found in Sect. 7.3. The terms **object identification**, **entity resolution**, and **record linkage** are also commonly used to describe the task of data deduplication.

### 1.4.5.1 Distance and similarity measures for strings

One approach to data deduplication is to fuse data records that have, in an appropriate sense, a sufficiently large similarity or small "distance," respectively. In Sect. 5.2, we will introduce a few distance and similarity measures that can be used to compare data tuples. At this point, we would like to introduce such metrics for comparing lists of symbols, or strings, which may be particularly useful for detecting duplicates in lists of organization names or person names. Metrics of this type are also used in other contexts, such as in information retrieval for **fuzzy search** or in bioinformatics for comparing DNA sequences [20].

A very simple but also rather crude definition for the distance between two strings $a$ and $b$ is the **discrete metric**:

$$\delta(a, b) = \begin{cases} 0 & \text{if } a = b \\ 1 & \text{if } a \neq b \end{cases}$$

This distance definition is universal in that it can be applied to other data types. However, it obviously does not allow gradual comparison between objects. The strings $a = \text{SIEMENS}$ and $b = \text{IEMENS}$ are "just as different" under the discrete metric as the strings $a = \text{SIEMENS}$ and $c = \text{GRUNDIG}$. Intuitively, however, it seems clear that the strings $a$ and $b$ have a greater similarity than the strings $a$ and $c$. A meaningful distance measure should reflect this fact numerically.

To be able to define less coarse distance measures than the discrete metric, we first note the different ways in which strings can be transformed, or edited.

The following elementary **edit operations** can be applied to strings:

**Delete a character:** SIEMENS ↦ SIMENS

**Insert a character:** SIEMENS ↦ SIEMENES

**Replace a character:** SIEMENS ↦ SOEMENS

**Swap two adjacent characters:** SIEMENS ↦ SEIMENS

Given any two strings, one can be transformed into the other by a suitable sequence of the above edit operations. The basic idea of an **edit distance** is to determine the number of elementary edit operations necessary to convert two strings into each other. The more operations necessary, the more different the strings can be considered to be.

Let $a$ and $b$ be strings.

The **Levenshtein distance** between $a$ and $b$ is given by the minimum number of insert, delete, and replace operations required to convert $a$ into $b$.

The **Damerau–Levenshtein distance** between $a$ and $b$ is given by the minimum number of swap, insert, delete, and replace operations required to convert $a$ into $b$.

For example, the Levenshtein distance between $a = \text{SIEMENS}$ and $b = \text{SOEMEN}$ is two, since at least one replace and one delete operation is necessary to get $b$

from $a$. Conversely, at least one replace and one insert operation is necessary to get $a$ from $b$.

We denote the lengths of the strings by $|a|$ and $|b|$, respectively, e.g. $|\text{xyYz}| = 4$. Let $\text{lev}_{i,j}(a,b)$ be the Levenshtein distance between the first $i$ characters of $a$ and the first $j$ characters of $b$. In order to calculate the full Levenshtein distance, $\text{lev}(a,b) = \text{lev}_{|a|,|b|}(a,b)$, the following recursive algorithm can be used [21, Theorem 2]:

$$\text{lev}_{i,j}(a,b) = \begin{cases} \max\{i,j\} & \text{if } \min\{i,j\} = 0, \\ \min \begin{cases} \text{lev}_{i-1,j}(a,b) + 1 \\ \text{lev}_{i,j-1}(a,b) + 1 \\ \text{lev}_{i-1,j-1}(a,b) + \delta(a_i,b_j) \end{cases} & \text{otherwise} \end{cases}$$

Here, $a_i$ is the character at the $i$-th position of $a$, and $b_j$ is the character at the $j$-th position of $b$, and:

$$\delta(a_i,b_j) = \begin{cases} 0 & \text{if } a_i = b_j \\ 1 & \text{if } a_i \neq b_j \end{cases}$$

We have $0 \leq \text{lev}(a,b) \leq \max\{|a|,|b|\}$, so the **normalized Levenshtein distance** (with value at most one) can be defined as follows:

$$\text{lev}_{\text{norm}}(a,b) = \frac{\text{lev}(a,b)}{\max\{|a|,|b|\}}$$

The number $1 - \text{lev}_{\text{norm}}(a,b)$ can then be understood as a measure of the similarity of the two strings. Identical strings have a maximum possible Levenshtein similarity of one, while the "maximally different" strings have a vanishing similarity measure.

The Damerau–Levenshtein distance is always at most as large as the Levenshtein distance, since every swap operation is the product of two replacement operations, e.g., SIEMENS $\mapsto$ SIIMENS $\mapsto$ SEIMENS.

While Levenshtein distance or similarity measures are used for a variety of different applications, the similarity measures described below were developed specifically for comparing person names [22].

For two strings, $a$ and $b$, of length $|a|$ and $|b|$, respectively, we first perform the following construction: Let $a \cap b = a_{i_1} \cdots a_{i_m}$ with $i_1 < \cdots < i_m$ be the string that emerges from $a$ by stepwise selection of characters $a_i$, starting from the beginning of the string for which there is a corresponding character $b_j$, with $a_i = b_j$ and $|j - i| \leq \frac{1}{2}\max\{|a|,|b|\} - 1$. In case of repeating characters within a string, a one-to-one correspondence shall be established by selecting $b_j$ starting from the beginning of the string.

Similarly, we define $b \cap a$. Necessarily, $b \cap a$ consists of the same characters as $a \cap b$, those characters contained in both $a$ and $b$ that are not too far apart. In general, however, $a \cap b$ and $b \cap a$ may differ in the order of characters.

For two strings, $a$ and $b$, let $m = |a \cap b| = |b \cap a|$, and let $t$ be the minimum number of permutations of not necessarily adjacent characters required to convert $a \cap b$ to $b \cap a$.

The **Jaro similarity** between $a$ and $b$ is then given as follows:

$$\text{jaro}(a, b) = \begin{cases} 0 & \text{if } m = 0 \\ \frac{1}{3} \cdot \left( \frac{m}{|a|} + \frac{m}{|b|} + \frac{m-t}{m} \right) & \text{otherwise} \end{cases}$$

As a concrete example, we want to determine the Jaro similarity between the strings $a = $ SIEMENS and $b = $ SIMESN. We have $\frac{1}{2} \max\{|a|, |b|\} - 1 = \frac{1}{2} \max\{7, 6\} - 1 = 2.5$, so only characters with a distance of at most two are put into correspondence. This gives $a \cap b = $ SIEMNS and $b \cap a = $ SIMESN, consequently $m = 6$ and $t = 2$. Finally:

$$\text{jaro}(a, b) = \frac{1}{3} \cdot \left( \frac{6}{6} + \frac{6}{7} + \frac{6-2}{6} \right) \approx 0.84$$

Let $a$ and $b$ be strings of length $|a|$ and $|b|$, and $L \in \mathbb{N}, p \in \mathbb{R}$ with $L \geq 1$ and $0 \leq p \leq \frac{1}{L}$ are fixed parameters. For the case $L = 0$, the choice of $p$ is arbitrary.

Let $l = l(a, b) \in \mathbb{N}$ be the largest number, such that $l \leq \min\{|a|, |b|, L\}$ holds and the first $l$ characters of $a$ and $b$ match as strings: $a_1 \cdots a_l = b_1 \cdots b_l$.

Then, the **Jaro–Winkler similarity** is given as follows:

$$\text{jw}_{p,L}(a, b) = \text{jaro}(a, b) + l(a, b) \cdot p \cdot (1 - \text{jaro}(a, b))$$

Typical choices for the parameters of the Jaro–Winkler similarity are $L = 4$ and $p = 0.1$. The basic idea of the Winkler correction is to give more weight to the matching of the first $L$ characters. For example, the string $a = $ SIEMENS has equal Jaro similarity with the strings $b = $ SIEMENX and $c = $ SXEMENS:

$$\text{jaro}(a, b) = \text{jaro}(a, c) \approx 0.90$$

However, the Jaro–Winkler similarities are different, since $l(a, b) = 4$ and $l(a, c) = 1$:

$$\text{jw}_{0.1,4}(a, b) = 0.90 + 4 \cdot 0.1 \cdot (1 - 0.90) = 0.94$$
$$\text{jw}_{0.1,4}(a, c) = 0.90 + 1 \cdot 0.1 \cdot (1 - 0.90) = 0.91$$

Finally, we sketch how distance measures between strings can be used for deduplication, again drawing on the example of a list of company names. First of all, descriptors of the legal entity type of the company, such as "Inc."

or "Aktiengesellschaft," can be removed or normalized. For example, regular expressions can be used to take into account frequently occurring typos and misspellings. This preprocessing may lead to a list of organization names like the following:

SIEMENES; Siemens; SIMENS; Bosch; BOSH

We further assume that for this particular dataset, it is not relevant whether the name is written in upper or lower case letters. Therefore, we can replace all characters with the corresponding capitalized letters.

A pairwise comparison of these **harmonized names** via the normalized Levenshtein similarity leads to the following matrix:

|          | SIEMENES | SIEMENS | SIMENS | BOSCH | BOSH |
|----------|----------|---------|--------|-------|------|
| SIEMENES | 1.00     | 0.88    | 0.75   | 0.00  | 0.00 |
| SIEMENS  | 0.88     | 1.00    | 0.86   | 0.00  | 0.00 |
| SIMENS   | 0.75     | 0.86    | 1.00   | 0.00  | 0.00 |
| BOSCH    | 0.00     | 0.00    | 0.00   | 1.00  | 0.80 |
| BOSH     | 0.00     | 0.00    | 0.00   | 0.80  | 1.00 |

We can choose to merge all entities with a Levenshtein similarity of at least 0.80, for example. According to that rule, we obtain the pairs (SIEMENES, SIEMENS), (SIEMENS, SIMENS), and (BOSCH, BOSH). After obtaining the pairs, we can then merge the corresponding data records.

SIEMENES and SIMENS are also matched, due to their common similarity to SIEMENS, even though the pairing does not exceed the similarity threshold, yielding the group of data records $S = \{$SIEMENES, SIEMENS, SIMENS$\}$. After the merging process, there is still the problem of providing unique information with the final data record. In particular, we would like to pick an organization name to include. If all names can be considered equally reliable information, we can determine a **medoid** of the group. A medoid (see also Sect. 5.3.1) is an element, $s \in S$, that maximizes the sum over the similarities, thus minimizing the sum over the distances $\sum_{t \in S} \text{lev}_{\text{norm}}(s, t)$. In this case, the medoid provides the correct name SIEMENS.

In addition to names, other information from the data records can be used for deduplication, such as the address/location of the headquarters or the year that the organization was founded in, if available. A general methodological framework for statistical data deduplication is the **Fellegi–Sunter model**, formulated in 1969 [23, 24]. This model is equivalent to the naive Bayes classification procedure that we discuss in Sect. 6.3.3. The Fellegi–Sunter model and other machine learning techniques are increasingly used today for the purpose of record linkage [25].

| | description | data defect example | data profiling example |
|---|---|---|---|
| **completeness** | Have all relevant data been collected and stored? | Some persons in the customer database are missing address information. | number, proportion, or percentage of customers with missing address information |
| **consistency** | Are redundantly stored data consistent, or does the information contradict itself? | Customer addresses stored at a branch location are different from those stored at the central location. | proportion of customers with inconsistent addresses |
| **validity** | Are the data formally correct? That is, are they syntactically correct, of the correct type, within the valid range of values? | Customer addresses contain postal codes with invalid characters, such as letters, e.g., "123Q5" | proportion of customers with invalid postal code, calculated by matching against: <br>• a complete and correct list of postal codes, or <br>• a fixed pattern, for example: exactly five digits, no letters |
| **uniqueness** | Can the data associated with any given entity be uniquely identified? | Several different primary keys have been assigned to the same person. | identify potential duplicates, see Sect. 1.4.5; proportion of potential duplicates |
| **accuracy** | Do the data reflect the truth, or are they at odds with reality? | For some customers, the name or address are incorrect. | manually check the accuracy of the data (in a sample), for example by telephone survey; proportion of customers with incorrect information |
| **timeliness** | Do the data reflect the current status of the information, or are they outdated? | Some customers have moved but their addresses have not been updated. | proportion of customers whose address data have not been updated for some time |

**Table 1.6.** *Data quality dimensions*

# References

[1] "Interim Report: ANSI/X3/SPARC Study Group on Data Base Management Systems 75-02-08". In: *Bulletin of ACM SIGMOD* 7.2 (1975). Ed. by Thomas B. Steel, Jr.

[2] Peter Pin-Shan Chen. "The entity-relationship model—toward a unified view of data". In: *ACM Transactions on Database Systems* 1.1 (Mar. 1976), pp. 9–36. DOI: 10.1145/320434.320440.

[3] Michael Widenius, Davis Axmark, and Paul DuBois. *MySQL Reference Manual*. 1st ed. Sebastopol, USA: O'Reilly & Associates, 2002.

[4] MariaDB Corporation. *MariaDB Server Documentation*. Accessed Aug. 29, 2021. URL: https://mariadb.com/kb/en/documentation/.

[5] PostgreSQL Development Team. *PostgreSQL Documentation*. Accessed July 10, 2020. URL: https://www.postgresql.org/docs/.

[6] Ian Robinson, Jim Webber, and Emil Eifrem. *Graph Databases*. 2nd ed. Sebastopol, USA: O'Reilly, 2015.

[7] Neo4j Team. *The Neo4j Operations Manual v4.1*. Accessed July 11, 2020. URL: https://neo4j.com/docs/pdf/neo4j-operations-manual-4.1.pdf.

[8] *RSS feed from Towards Data Science*. Accessed Sep. 18, 2020. URL: https://towardsdatascience.com/feed.

[9] Nicola Askham et al. *The Six Primary Dimensions for Data Quality Assessment*. Tech. rep. Data Management Association UK, Oct. 2013.

[10] Mark Litwin. *How to Measure Survey Reliability and Validity*. SAGE Publications, Inc., 1995. DOI: 10.4135/9781483348957.

[11] Jeffrey E. F. Friedl. *Reguläre Ausdrücke*. 3rd ed. O'Reilly, Oct. 2012. ISBN: 978-3-897-21720-1.

[12] CDC Population Health Surveillance Branch. *Behavioral Risk Factor Surveillance System (BRFSS) Survey Data 2018*. Accessed Feb. 1, 2020. URL: https://www.cdc.gov/brfss/.

[13] Lorenzo Beretta and Alessandro Santaniello. "Nearest neighbor imputation algorithms: a critical evaluation". In: *BMC Med Inform Decis Mak* 74.16 (2016). DOI: 10.1186/s12911-016-0318-z.

[14] Agnieszka Mikołajczyk and Michał Grochowski. "Data augmentation for improving deep learning in image classification problem". In: *2018 International Interdisciplinary PhD Workshop (IIPhDW)*. IEEE, May 2018. DOI: 10.1109/iiphdw.2018.8388338.

[15] Steven Y. Feng et al. *A Survey of Data Augmentation Approaches for NLP*. Dec. 2021. arXiv:2105.03075v5.

[16] Xiang Zhang, Junbo Zhao, and Yann LeCun. "Character-Level Convolutional Networks for Text Classification". In: *28th International Conference on Neural Information Processing Systems, Montreal, Canada*. Vol. 1. NIPS'15. MIT Press, 2015, pp. 649–657.

[17] Rico Sennrich, Barry Haddow, and Alexandra Birch. "Improving Neural Machine Translation Models with Monolingual Data". In: *54th Annual Meeting of the Association for Computational Linguistics*. Vol. 1. Berlin,

Germany: Association for Computational Linguistics, Aug. 2016, pp. 86–96. DOI: 10.18653/v1/P16-1009.

[18]    Gaétan de Rassenfosse, Hélène Dernis, and Geert Boedt. "An Introduction to the Patstat Database with Example Queries". In: *Australian Economic Review* 47.3 (Sept. 2014), pp. 395–408. DOI: 10.1111/1467-8462.12073.

[19]    European Patent Office. *PATSTAT Global - 2018 Autumn Edition (sample data)*. Accessed Sep. 20, 2020. URL: https://www.epo.org/searching-for-patents/business/patstat.html.

[20]    Bonnie Berger, Michael S. Waterman, and Yun William Yu. "Levenshtein Distance, Sequence Comparison and Biological Database Search". In: *IEEE Transactions on Information Theory* (2020). Early Access. DOI: 10.1109/tit.2020.2996543.

[21]    Robert A. Wagner and Michael J. Fischer. "The String-to-String Correction Problem". In: *Journal of the ACM* 21.1 (Jan. 1974), pp. 168–173. DOI: 10.1145/321796.321811.

[22]    William E. Winkler. *Overview of Record Linkage and Current Research Directions*. Tech. rep. Washington: U.S. Census Bureau, Statistical Research Division, 2006.

[23]    H. B. Newcombe et al. "Automatic Linkage of Vital Records: Computers can be used to extract 'follow-up' statistics of families from files of routine records". In: *Science* 130.3381 (Oct. 1959), pp. 954–959. DOI: 10.1126/science.130.3381.954.

[24]    Ivan P. Fellegi and Alan B. Sunter. "A Theory for Record Linkage". In: *Journal of the American Statistical Association* 64.328 (1969), pp. 1183–1210. DOI: 10.1080/01621459.1969.10501049.

[25]    D. Randall Wilson. "Beyond probabilistic record linkage: Using neural networks and complex features to improve genealogical record linkage". In: *2011 International Joint Conference on Neural Networks, San Diego, USA*. IEEE, July 2011. DOI: 10.1109/ijcnn.2011.6033192.

# 2

# Descriptive statistics

Through our everyday experience, we have an intuitive understanding of what a typical body height is for people in the population. In much of the world, adult humans are typically between 1.60 m and 1.80 m tall, while people taller than two meters are rare to meet.

By providing a **frequency distribution** of body height, this intuited fact can be backed up with numerical evidence:

| body height $l$ | number of people/ absolute frequency | relative frequency |
|:---:|:---:|:---:|
| $l < 1.60\,\mathrm{m}$ | 81,199 | 19% |
| $1.60\,\mathrm{m} \leq l < 1.80\,\mathrm{m}$ | 260,433 | 62% |
| $1.80\,\mathrm{m} \leq l < 2.00\,\mathrm{m}$ | 77,462 | 18% |
| $2.00\,\mathrm{m} \leq l$ | 770 | <1% |

**Table 2.1.** *Frequency of human body height*

These figures are based on a dataset collected by the U.S. Centers for Disease Control and Prevention (CDC) that lists, among other attributes, the height of more than 340,000 individuals [1].

An inspection of this **frequency table** shows that, in fact, more than half of the people interviewed for the survey reported their height to be between 1.60 m and 1.80 m. Highlighting key characteristics of frequency distributions via numeric measures and graphical representation is a primary goal of **descriptive statistics**.

## 2.1 Samples

When a dataset is the subject of a statistical study, it is also referred to as a **sample**. Here is a small excerpt from the CDC sample:

© Springer-Verlag GmbH Germany, part of Springer Nature 2023
M. Plaue, *Data Science*, https://doi.org/10.1007/978-3-662-67882-4_2

| sex | income category | body height in m | body weight in kg |
|---|---|---|---|
| female | 6 | 1.62 | 59 |
| female | 4 | 1.66 | 91 |
| female | 3 | 1.47 | 64 |
| male | 3 | 1.79 | 86 |
| ⋮ | ⋮ | ⋮ | ⋮ |

**Table 2.2.** *CDC sample*

The income category takes discrete values from 1 to 8 and denotes a specific interval of annual household income in ascending order. For example, category 3 represents an income of $15,000 or more but less than $20,000 (income in USD).

In the last chapter, we already identified the columns of such a data table as *attributes*. In the context of statistics, they are also called **variables**. In machine learning, we will often call them **features**, especially when the variables' values are used to inform the algorithm about essential characteristics of the associated entities in order to compare them. All possible values that a variable/feature may take constitute the **domain/feature space**.

Variables that may take only a finite number of values which have no natural, meaningful order are called **nominal variables**. For example, a person's sex is a nominal variable.

Variables that can produce a finite number of possible outcomes which *can* be ordered in some natural and meaningful way are called **ordinal variables**. For example, the income category is an ordinal variable because a person placed in income category 6 earns more than a person in income category 3.

A variable that is either nominal or ordinal is also known as **categorical** or **qualitative**.

Variables that allow for a meaningful comparison of their values as differences, ratios, etc. are called **numeric** or **quantitative**. For example, height is a quantitative variable because statements such as "this person is 10 cm taller than that person" are meaningful.

The rows of the tabulated sample represent **statistical units**, **entities**, or **objects**. In machine learning, the term **instance** is commonly used. We can also refer to them more concretely: in this case, for example, we can speak of "persons." The number of statistical units is called the **sample size**.

A column represents a series of **measurements** of a single variable, or in other terms the **observations** of a feature.

We refer to the set $\Omega$ of all existing statistical units as the **population**—i.e., not only those covered by the sample. In the CDC survey, the population is all U.S. adults and their households. Each subset $\Omega_0 \subseteq \Omega$ of the population (i.e., including the sample) is a **subpopulation**. If the subpopulation is specifically a group of people—for example, in the context of a demographic study—it is also referred to as a **cohort**.

Each subpopulation uniquely defines a **dichotomous** or **binary variable**, i.e., a variable with domain $\{0, 1\}$: a statistical unit is assigned the value one if it belongs to the subpopulation, and a value of zero if it does not belong to the subpopulation. Conversely, each binary variable uniquely defines a subpopulation: all those statistical units for which we observe the value 1. In summary, there is a one-to-one correspondence between subpopulations and binary variables.

All of the above are examples for **univariate variables**: each observation corresponds to only a single value at a time. However, we can also combine several univariate variables into one **multivariate variable** to yield a sequence of **data points** or **feature vectors**. For example, the feature vector (height in m, body weight in kg) can be considered for each person. Such a multivariate variable or feature with exactly two entries is also referred to as **bivariate**.

In summary, the variables in the CDC dataset can be categorized as follows:

| variable/feature | domain/feature space | data type |
|---|---|---|
| sex | $\{\text{female}, \text{male}\}$ | nominal |
| income category | $\{1, 2, \ldots, 8\}$ | ordinal |
| body height | $]0, \infty[ \subset \mathbb{R}$ | numeric |
| (body height, body weight) | $]0, \infty[ \times ]0, \infty[$ | numeric, bivariate |

**Table 2.3.** *Data types*

Finally, subpopulations can be defined by imposing conditions on variables. Examples would be the cohort of males $\Omega_{\text{male}} = \{\omega | \text{sex}(\omega) = \text{male}\}$ or the cohort of individuals with a body mass index (BMI) between $25\,\text{kg}/\text{m}^2$ and $30\,\text{kg}/\text{m}^2$:

$$\Omega_{\text{pre-obese}} = \left\{ \omega \,\middle|\, 25 \leq \frac{\text{weight}(\omega)}{\text{height}(\omega)^2} < 30 \right\}$$

## 2.2 Statistical charts

In the introduction, we had already introduced a form of statistical chart, the **line chart**, or **line graph** (Fig. 0.1). It is similar to the scatter plot discussed in one of the following sections, with successive data points connected by a (straight) line to highlight an increase or decrease in the ordinate value. This type of chart is particularly suitable for plotting **time series**, that is, numeric observations ordered over time.

The use of statistical charts is well suited for **exploratory data analysis**. In this context, they are used to uncover patterns and relationships to formulate hypotheses that can be tested against the data.

### 2.2.1 Bar charts and histograms

In order to make us—quite literally—a picture of a given frequency distribution, we can represent it as a **bar chart** in the case of a categorical feature, or a **histogram** in case of a numeric variable. In Fig. 2.7, a few bar charts and histograms are shown for features exhibited by the CDC dataset.

For qualitative variables, such as sex or income category, the height of each rectangle indicates the relative frequency that the respective value occurs with. The absolute frequency can also be plotted; this simply corresponds to a global scaling of all column heights by the sample size. The widths of the rectangles have no particular meaning.

For quantitative variables such as body height or body weight, the domain must be divided into intervals, called **bins** or **buckets**, in order to count the observations within each bin and thus determine the frequency. The **bin width** is represented by the width of each rectangle in the chart. For example, the bin width for the histogram of the body height in the example charts is exactly 2.54 cm. This value corresponds to one inch, the U.S. customary unit of length that the body height had been measured in originally. The frequency is represented by the *area* of each rectangle, and the height of each rectangle is given by the frequency divided by the bin width. In general, although not commonly found, a histogram may have variable bin width.

**Example.** We notice that the histogram of the body weight reported with the CDC survey shows gaps at regular intervals. A "zoom" into the histogram, given in the original units of measurement provides more detail:

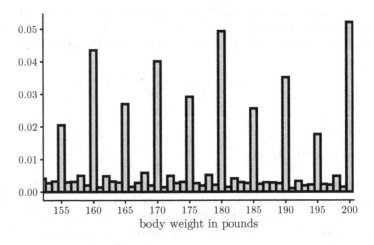

**Fig. 2.1.** *Frequency distribution of self-reported body weight*

The data were collected via a telephone survey among U.S. citizens. Participants rarely reported their weight exactly to the pound, but instead they often used rounded values. This behavior gives us an example of **response bias**, noticeable by how the body weights divisible by five are overrepresented.

The example shows that we can use bar charts and histograms to identify patterns in the distribution of values. Characteristics that can be inferred are:

- Which values of a categorical variable are the most common, and which values of a numeric variable can be considered "typical"?

- Do all values occur equally often, i.e., do they show a large variation or dispersion—or are they concentrated around a "typical" value?

Statistical measures that gauge a central tendency or dispersion in the distribution of values help us ground such obervations on numeric evidence. These will be described in the later sections.

### 2.2.2 Scatter plots

Variables often depend on each other, i.e., there may be functional relationships between them. Such association between two numeric variables can be represented graphically by **scatter plots**. These plots may illustrate an already known association or be used to explore the data for previously unknown relationships between features.

In a scatter plot, the *paired* variable values—i.e., values that belong to the same statistical unit—are plotted against each other as data points in the coordinate plane.

The following scatter plot shows height and weight for 150 randomly selected male respondents from the CDC survey.

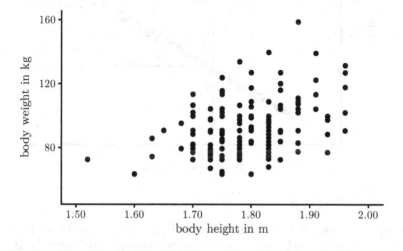

**Fig. 2.2.** *Scatter plot of body height vs. body weight*

As expected, there is a general tendency for taller people to also have higher weight. Nevertheless, even amongst those persons with the same height, there is a large variation in body weight. Therefore, height is not the only variable that factors into a person's weight, a finding that is consistent with our everyday experience:

**Example.** Let's use the CDC dataset to compare the body mass index and the Broca index. These measures serve as a rough indication of what can be considered a healthy body weight. BMI is calculated using the following formula:

$$\text{BMI} = \frac{\text{body weight}}{(\text{body height})^2}$$

Usually, BMI is expressed in units of kilograms per square meter. Adults with a BMI of more than $30\,\frac{\text{kg}}{\text{m}^2}$ are generally considered obese, according to the World Health Organization [2, Sect. 2.3].

*Normal weight* according to Broca is calculated as follows:

normal weight according to Broca in kg = body height in cm − 100

The Broca index is the difference of the actual body weight from this reference point, given as a percentage. According to this metric, adults with a deviation above +20% from the normal weight can be considered obese.

The following chart shows a plot of BMI against Broca's index for the male respondents in the CDC survey selected above.

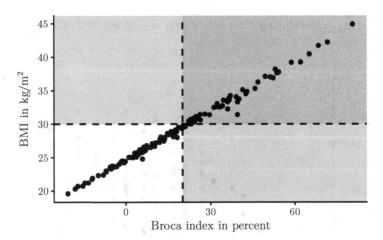

**Fig. 2.3.** *Scatter plot of Broca vs. body mass index*

The gray areas mark the "critical" regions BMI $> 30\,\frac{\text{kg}}{\text{m}^2}$ and Broca index $>$ 20%, respectively. Both metrics are associated strongly: although the Broca index has become "out of fashion" compared to the BMI, both measures come to very similar conclusions. The definitions for obesity are also compatible with each other: there are only very few observations in the light gray areas at the top left or bottom right that would indicate an inconsistent assessment.

In addition to graphical representation, we can compute measures of association to gauge whether statistical variables are related. These measures will be discussed in Sect. 2.5.

Furthermore, associations can be described by statistical models. For example, the dependence of the Broca index on the body mass index, and vice versa, appears to be linear: the data points do not lie exactly but are approximately on a straight line. The optimal position and orientation of such a straight line can be determined by a regression analysis, see Sect. 4.5.1.

### 2.2.3 Pie charts, grouped and stacked bar charts, heatmaps

### 2.2.3.1 Pie chart

One popular method used to represent proportions is the **pie chart**, or **circle chart**. In a pie chart, proportions are represented as sectors of a circle, similar to pie or pizza slices. Although it is widely used in popular publications, whether or not the pie chart is an effective form of visualization is considered controversial among experts [3]. If possible, a pie chart should only be used if it does not contain too many pieces. The use of three-dimensional pie charts should always be refrained from.

A variant of the pie chart is the **doughnut chart**, which is basically a single stacked bar (see the next section) bent to a circle. The following doughnut chart shows the outcome of the 2021 German federal election [4]:

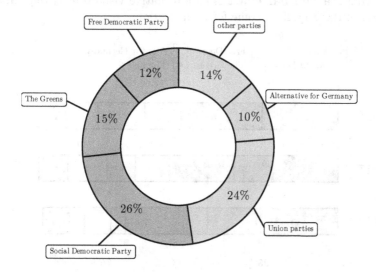

**Fig. 2.4.** *Doughnut chart*

### 2.2.3.2 Grouped and stacked bar charts

In order to compare distributions across (sub-)populations, we can show their bar charts in a single figure. The following **grouped bar chart** shows the outcome of consecutive German federal elections [4, 5, 6]:

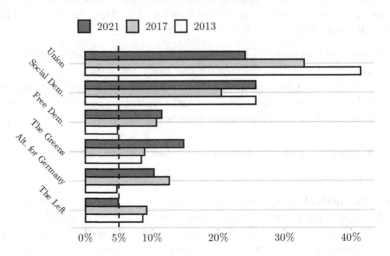

**Fig. 2.5.** *Grouped bar chart*

For each party, we can readily compare how the share of votes changed over time. Nevertheless, we are also able to gauge which parties received a larger share in general. The vertical dashed line indicates the five-percent electoral threshold, which must be met for a party to hold office in the German parliament.

The following **stacked bar chart** is an alternative visualization that allows comparison of the overall outcome for each election:

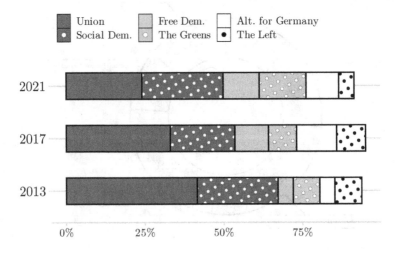

**Fig. 2.6.** *Stacked bar chart*

### 2.2.3.3 Heatmap

A **heatmap** is used to visualize key figures arranged in a matrix, where each numeric value corresponds to a color-coded tile. Fig. 2.8 shows on the left a heatmap diagram based on data from the German 2018 ALLBUS survey [7]. The chart compares the preference for a political party in Germany with responses to the following question: "What do you think of the following statement: the influx of refugees should be stopped." The color saturation of the tiles indicates the proportion of people with a given party preference, i.e., each row represents the distribution of the frequency of responses by people with the given party preference. Consequently, the distributions in the respective cohorts can be compared easily.

Another variant of the heatmap is the **choropleth map**, in which geographic areas are colored. Such a map is shown in Fig. 2.8 on the right. It shows the percentage of ALLBUS respondents who tend to agree or disagree that the influx of refugees should be stopped for each German federal state.

## 2.3 Measures of central tendency

Measures of central tendency answer the question: which measurements or observations occurring in a sample can be considered "typical"?

Tab. 2.1 at the beginning of the chapter shows that most people, 62%, have a height between 160 and 180 cm, which is a larger proportion than in any other of the given height cohorts. Therefore, the range of values between 160 and 180 cm can be considered typical because people with heights in this range are most frequently encountered. In other words, the frequency distribution assumes a maximum number of observations at that range. Such a maximum is also called a **mode** of the distribution. In practice, we often speak of *the* mode, even though the mode need not be uniquely determined. Firstly, variable values can occur with exactly the same or nearly the same frequency. Secondly, it may happen that a histogram has several *local* maxima, in which case the distribution is called **multimodal**, otherwise **unimodal**. Fig. 2.10 shows the histogram of a multimodal distribution.

For categorical variables, determining the mode simply requires counting observations within each category and selecting the one that occurs most frequently. For numeric variables, the mode depends on the division of the variable's domain into bins that is used to determine the histogram or frequency distribution. The mode can also be determined via smooth approximations to the histogram, like kernel density estimation (see Sect. 4.4.3).

### 2.3.1 Arithmetic mean and sample median

Besides the mode, the following measures of central tendency are among the most commonly used.

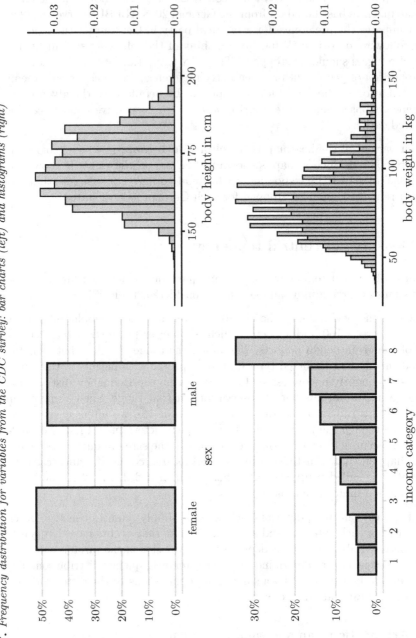

**Fig. 2.7.** *Frequency distribution for variables from the CDC survey: bar charts (left) and histograms (right)*

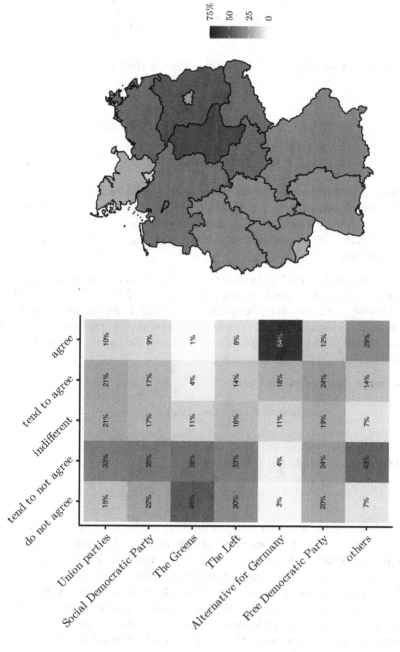

**Fig. 2.8.** Choropleth map of Germany (right): percentage of persons that tend to agree or agree that the influx of refugees should be stopped; heatmap (left): political party preference vs. opinion on the necessity of migration control

For a sequence of numeric observations $x = (x_1, x_2, \ldots, x_N) \in \mathbb{R}^N$, the **arithmetic mean**, or **arithmetic average**, is defined as follows:

$$\bar{x}_{\text{arithm}} = \frac{1}{N}(x_1 + x_2 + \cdots + x_N) = \frac{1}{N}\sum_{n=1}^{N} x_n$$

Now assume that the observations have been sorted in ascending order: $x_1 \leq x_2 \leq \cdots \leq x_N$. The **sample median** of $x$ is defined as the midpoint of this sorted sequence:

$$\bar{x}_{\text{median}} = \begin{cases} x_{\frac{N+1}{2}} & \text{if } N \text{ is odd} \\ \frac{1}{2}\cdot\left(x_{\frac{N}{2}} + x_{\frac{N}{2}+1}\right) & \text{if } N \text{ is even} \end{cases}$$

The sample median can also be calculated for ordinal variables. However, in general, there is no longer a meaningful convention with which to enforce a unique result in the case of an even number of observations. For example, for a sample $x = (1, 5, 6, 8)$ of income categories, both $\bar{x}_{\text{median}} = 5$ and $\bar{x}_{\text{median}} = 6$ are valid medians.

In practice, the sample median in this example could still be given as $\bar{x}_{\text{median}} = 5.5$, even though no fractional income categories are defined. Even computing the arithmetic mean for—strictly speaking—ordinal features can be useful in some cases. For example, in the course of a game of dice, the average die roll might be given by 3.14—although, of course, a real die has no side that would correspond to such a fractional number.

For the sample median we also write:

$$\text{median}(x) = \underset{n\in\{1,\ldots,N\}}{\text{median}}(x_n) = \bar{x}_{\text{median}}$$

For the arithmetic mean, we often omit the subscript: $\bar{x} = \bar{x}_{\text{arithm}}$. Other possible notations include the following:

$$\mu(x) = \langle x \rangle = \langle x_n \rangle_{n\in\{1,\ldots,N\}} = \bar{x}_{\text{arithm}}$$

The sample median is also called the **empirical median**, while the arithmetic mean can also be called the **empirical mean**, **sample mean**, or **empirical expected value**. The labels "sample" or "empirical" indicate that these measures are computed from data. Empirical measures have counterparts in probability theory—discussed later in Chap. 3—that are derived from more abstract objects called random variables. In inferential statistics, empirical measures are interpreted as estimates of these "theoretical" values. For the sake of brevity, we may at times omit the label "sample" or "empirical."

The terms **population mean** and **population median** are commonly used to refer to the mean/median of the whole population. However, we will not make use of those terms in this book and will only refer to the sample mean/median, which may be computed with respect to a subpopulation of any size.

> **Example.** The arithmetic mean of the values for body height collected in the CDC survey is 1.70 m. The median is also 1.70 m.

Although the arithmetic mean is a very commonly used measure of central tendency, the median has a property that makes it superior for some applications: it is a **robust** statistic, resistant to **outliers**. This means that it does not change its value much, if at all, if some values in the sample differ significantly from the rest. Outliers can be the result of data defects, so such robustness may be desired.

In order to illustrate this property, we calculate the arithmetic mean and median for the sequence of observations $x = (-1.0, -1.0, 0.0, 1.0, 2.0, 100.0)$:

$$\bar{x} = \frac{1}{6} \cdot (-1.0 - 1.0 + 0.0 + 1.0 + 2.0 + 100.0) \approx 16.8$$

$$\bar{x}_{\text{median}} = \frac{1}{2} \cdot (x_3 + x_4) = \frac{1}{2} \cdot (0.0 + 1.0) = 0.5$$

The outlier $x_6 = 100.0$ has a significant impact on the arithmetic mean but not the median.

For so-called **skewed distributions**, the proportion of values smaller or larger than the arithmetic mean can deviate substantially from 50%. However, the median splits the sequence of observations right in the middle: half the values are below the median and half are above.

> **Example.** Income statistics often lead to skewed distributions. For example, according to the data provided with the ALLBUS survey, the arithmetic mean of monthly net household income in Germany in 2018 is 3150 Euros (EUR). However, more than half (60%) of households actually have lower incomes. Therefore, the median income is the better benchmark: 2800 EUR.

There are a number of tests and measures that exist for the purpose of evaluating the skewness or asymmetry of a frequency distribution. A *rule of thumb* (cf. [8]) is the following:

> The skewness of a unimodal distribution of values $x_1, \ldots, x_N$, with mode $\bar{x}_{\text{mode}}$, median $\bar{x}_{\text{median}}$, and arithmetic mean $\bar{x}$, can be characterized as follows. The distribution is:
>
> $$\textbf{symmetric if } \bar{x}_{\text{mode}} \approx \bar{x}_{\text{median}} \approx \bar{x},$$
> $$\textbf{negatively skewed if } \bar{x}_{\text{mode}} > \bar{x}_{\text{median}} > \bar{x},$$
> $$\textbf{positively skewed if } \bar{x}_{\text{mode}} < \bar{x}_{\text{median}} < \bar{x}.$$

The income distribution derived from the ALLBUS data is positively skewed, as can be inferred from the following figure:

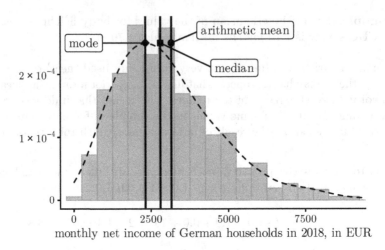

monthly net income of German households in 2018, in EUR

**Fig. 2.9.** *Histogram of a positively skewed distribution*

The mode has been determined via the maximum of a so-called kernel density estimate of the distribution (given by the dashed curve). We explain in Sect. 4.4.3 how to compute such an estimate.

### 2.3.2 Sample quantiles

Quantiles are another important class of statistical measures.

> Let $0 < \alpha < 1$, and $x = (x_1, \ldots, x_N)$ be a sample of numeric or ordinal values. A **sample quantile of order** $\alpha$ or **$\alpha$-quantile** is a value $Q_\alpha(x)$ from the sample such that below and above that value a certain proportion of observations can be found:
>
> 1. At least $\alpha \cdot N$ of the observations are less than or equal to $Q_\alpha(x)$, and
>
> 2. at least $(1 - \alpha) \cdot N$ of the observations are greater than or equal to $Q_\alpha(x)$.

The $\alpha$-quantile is only *not* uniquely determined if $\alpha \cdot N$ is an integer. Otherwise, the quantile is given by $Q_\alpha(x) = x_{\lfloor \alpha N \rfloor + 1}$, where the observations have been sorted in ascending order and $\lfloor \alpha N \rfloor$ is the largest integer less than or equal to $\alpha \cdot N$.

In practice, it is common—as with the median—to enforce uniqueness via the arithmetic mean of the candidates:

$$Q_\alpha(x) = \begin{cases} x_{\lfloor \alpha N \rfloor + 1} & \text{if } \alpha \cdot N \notin \mathbb{N} \\ \frac{1}{2} \cdot (x_{\alpha N} + x_{\alpha N + 1}) & \text{if } \alpha \cdot N \in \mathbb{N} \end{cases}$$

With this convention, the sample median is just the 0.5-quantile: $\bar{x}_{\text{median}} = Q_{0.5}(x)$.

Other quantiles that deserve a special name are the lower and upper **quartile**: $Q_{0.25}$ and $Q_{0.75}$, respectively.

**Example.** The lower quartile of the height surveyed in the CDC study is 1.63 m, and the upper quartile is 1.78 m.

### 2.3.3 Geometric and harmonic mean

The following measures of central tendency have special applications where the arithmetic mean does not produce a reasonable average value.

For a sequence of *positive* numeric observations $x = (x_1, x_2, \ldots, x_N) \in ]0, \infty[^N$, we define:

The **geometric mean**

$$\bar{x}_{\text{geom}} = (x_1 \cdot x_2 \cdots x_N)^{\frac{1}{N}} = \left( \prod_{n=1}^{N} x_n \right)^{\frac{1}{N}}$$

and the **harmonic mean**

$$\bar{x}_{\text{harm}} = \frac{N}{\frac{1}{x_1} + \frac{1}{x_2} + \cdots + \frac{1}{x_N}} = N \cdot \left( \sum_{n=1}^{N} \frac{1}{x_n} \right)^{-1}.$$

It can be proven that—for any sample $x = (x_1, \ldots, x_N)$ of positive values—the various measures of central tendency satisfy the inequalities:

$$0 < \min_n \{x_n\} \leq \bar{x}_{\text{harm}} \leq \bar{x}_{\text{geom}} \leq \bar{x}_{\text{arithm}} \leq \bar{x}_{\text{RMS}} \leq \max_n \{x_n\}$$

Here, $\bar{x}_{\text{RMS}}$ is the **root mean square**, another measure of central tendency that we haven't mentioned yet:

$$\bar{x}_{\text{RMS}} = \frac{1}{\sqrt{N}} \cdot \sqrt{\sum_{n=1}^{N} x_n^2}$$

The geometric mean is used in particular for averaging rates of exponential growth or decline, as illustrated by the following calculation. Say that some monetary investment of 1000 EUR yields variable profits. During the first year, the investment grows by $p_1 = +20.0\%$. In the second year, there is a $p_2 = -20.0\%$ loss. Finally, in the third year, there is a $p_3 = +10.0\%$ profit. The corresponding factors by which the investment grows each year are $W = (1.200, 0.800, 1.100)$. Therefore, after three years, the investor has the following amount in EUR:

$$1000 \cdot 1.200 \cdot 0.800 \cdot 1.100 \approx 1000 \cdot 1.056$$

Consequently, the total growth rate is $+5.6\%$. However, what is the average annual growth rate? That growth rate corresponds to the average factor $\bar{W}_?$ under the assumption of a fixed rate of interest with the same end result:

$$1000 \cdot \bar{W}_? \cdot \bar{W}_? \cdot \bar{W}_? = 1000 \cdot 1.200 \cdot 0.800 \cdot 1.100$$

It is not so difficult to see that this average factor is just equal to the geometric mean of the individual factors:

$$\bar{W}_? = \bar{W}_{\text{geom}} = (1.200 \cdot 0.800 \cdot 1.100)^{\frac{1}{3}} \approx 1.018$$

Therefore, the average annual growth rate is $1.8\%$. A naive use of the arithmetic mean would have led to a different and thus factually incorrect result: $\bar{W}_{\text{arithm}} = 1.033$, $\bar{p}_{\text{arithm}} = 3.3\%$.

We will encounter the harmonic mean again in the definition of the $F_1$-score, a metric for the evaluation of classification algorithms (see Sect. 6.1.3.2). Another application is in averaging speeds, as the following example will illustrate. Anna takes a public bus to work in the morning (distance $\Delta s = 5\,\text{km}$) when there is little traffic and an average speed of $v_1 = 40\,\frac{\text{km}}{\text{h}}$. The average speed on the return trip home is only $v_2 = 10\,\frac{\text{km}}{\text{h}}$ since the bus has to fight its way through traffic.

What is the average speed $\bar{v}_?$ by which Anna or the bus cover the total distance $2 \cdot \Delta s$ on the way there and back? This average speed is equal to the distance traveled divided by the total time taken $\Delta t$: $\bar{v}_? = \frac{2 \cdot \Delta s}{\Delta t}$. For the outward trip, the bus needs the time $\Delta t_1 = \frac{\Delta s}{v_1}$, and for the return trip the time $\Delta t_2 = \frac{\Delta s}{v_2}$. Thus, the average speed is given by:

$$\bar{v}_? = \frac{2 \cdot \Delta s}{\Delta t_1 + \Delta t_2} = \frac{2 \cdot \Delta s}{\frac{\Delta s}{v_1} + \frac{\Delta s}{v_2}} = \frac{2}{\frac{1}{v_1} + \frac{1}{v_2}} = \bar{v}_{\text{harm}}$$

Plugging in the actual numbers gives the result $\bar{v}_{\text{harm}} = 16\,\frac{\text{km}}{\text{h}}$. The arithmetic mean yields a higher value: $\bar{v}_{\text{arithm}} = 25\,\frac{\text{km}}{\text{h}}$. This is because Anna needs much more time for the return trip than for the outward trip. Therefore, she also sits in the slow-moving bus for a longer time, so overall she moves with a speed that is lower than the arithmetic mean would imply.

## 2.4 Measures of variation

Measures of central tendency compute a "typical" value of a distribution. However, the significance of such a calculation can vary, as the value thus obtained can be "more or less typical." For example, approximately $55\%$ of the CDC survey respondents reported being female, while $45\%$ reported being male. Thus, strictly speaking, "female" is the mode of the variable *sex*. Nevertheless, this category can hardly be described as typical because the proportion of persons of male sex is about the same.

On the other hand, the average body height of 1.70 m may well be considered a typical value because we can infer from experience or from the shape of the histogram that not many persons deviate from this value to a significant degree.

**Measures of variation** or **dispersion** indicate the extent to which values occur with equal frequency, and they also quantify the extent of dispersion around a central point.

### 2.4.1 Deviation around the mean or the median

A simple measure of variation for numeric variables is the **range**, given by the difference between the largest and smallest values in the sample. However, even single outliers can have a significant impact on this value. The **interquartile range** is more resistant to outliers than the full range. The interquartile range is defined as the difference between the upper and lower quartile: $Q_{0.75} - Q_{0.25}$.

> **Example.** The interquartile range of the body height values collected with the CDC survey is given by $178\,\text{cm} - 163\,\text{cm} = 15\,\text{cm}$.

The following measures of variation are based on the idea of calculating the average deviation from the average value.

> For a sequence of numeric obervations $x = (x_1, x_2, \ldots, x_N) \in \mathbb{R}^N$, the **sample variance** is the mean squared deviation from the arithmetic mean:
> $$s^2(x) = \frac{1}{N} \sum_{n=1}^{N} (x_n - \bar{x})^2$$
> The **sample standard deviation** is the square root of the variance:
> $$s(x) = \sqrt{s^2(x)} = \left( \frac{1}{N} \sum_{n=1}^{N} (x_n - \bar{x})^2 \right)^{\frac{1}{2}}$$

The above definition describes what should more accurately be called the *biased sample variance*. The so-called **unbiased sample variance**, also called the Bessel-corrected sample variance, is given by:

$$s^2_{\text{cor}}(x) = \frac{1}{N-1} \sum_{n=1}^{N} (x_n - \bar{x})^2 = \frac{N}{N-1} \cdot s^2(x)$$

The use of the unbiased variance is particularly recommended for small samples. For large samples, however, both formulas give approximately the same results. We will learn in Sect. 4.2.3 about the motivation and rationale behind the Bessel correction.

**Example.** The standard deviation of the body height values collected with the CDC survey is given by 11 cm.

Other measures of variation include the following.

The **mean absolute deviation around the mean** is given by:

$$\text{MAD}(x) = \frac{1}{N} \sum_{n=1}^{N} |x_n - \bar{x}|$$

The **mean absolute deviation around the median** is given by:

$$\text{MAD}_{\text{median}}(x) = \frac{1}{N} \sum_{n=1}^{N} |x_n - \bar{x}_{\text{median}}|$$

**Example.** The mean absolute deviation around the mean/median of the body height values collected with the CDC survey are both given by 9 cm.

**Example.** The following histogram shows the distribution of the length of the petals of iris plants [9]. The horizontal bars indicate the intervals $[\bar{x} - s(x), \bar{x} + s(x)]$ and $[\bar{x}_{\text{median}} \pm \text{MAD}_{\text{median}}(x)]$, respectively. Apparently, the frequency distribution is multimodal, as the histogram shows two maxima. In fact, there are even two separated groups, each of which belong to different species.

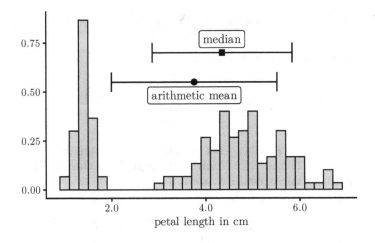

**Fig. 2.10.** *Deviation around the mean and the median*

## 2.4.2 Shannon index

For a categorical feature with $K$ possible values or categories, we can obtain the relative frequencies, $f_1, \ldots, f_K$, of the occurrence of each category by dividing the absolute frequencies, $n_1, \ldots, n_K$, by the sample size $N$. The absolute frequencies always sum to $N$, while the relative frequencies sum to one:

$$\sum_{k=1}^{K} f_k = \sum_{k=1}^{K} \frac{n_k}{N} = \frac{1}{N} \cdot \sum_{k=1}^{K} n_k = \frac{1}{N} \cdot N = 1$$

The category that occurs most frequently, i.e., the maximum value for $f_k$ or $n_k$, is the measure of central tendency that we already know as the mode. In a bar chart, the mode corresponds to the category with the longest bar. The following metric, on the other hand, is a measure of the extent to which all bars have a similar length.

Let $x$ be a sequence of observations of a categorical variable with $K$ possible values. Furthermore, let $f_1, \ldots, f_K$ be the relative frequencies that the $K$ categories occur with. The **Shannon index** is then given as follows:

$$H(x) = -\sum_{k=1}^{K} f_k \log_b(f_k)$$

Here, $b$ denotes the base of the logarithm, and we agree on the convention $0 \cdot \log_b 0 = 0$: terms with vanishing relative frequency $f_k$ are set to zero.

The **normalized Shannon index** is given as follows:

$$H_{\text{norm}}(x) = \frac{1}{\log_b(K)} \cdot H(x)$$

Common values for the base of the logarithm are $b = 2$ (binary logarithm) or $b = e = 2.718\ldots$ (natural logarithm). The Shannon index can be interpreted as an empirical estimate of the *Shannon entropy*, see Sect. 3.4.1, an abstract measure of average information content. In this context, the logarithm base determines units of measurement: the choice $b = 2$ corresponds to measuring the Shannon entropy in the unit of information commonly referred to as *bit*. For the definition of the *normalized* Shannon index, the choice of the base does not matter. In the following paragraphs, we will always use the natural logarithm.

The normalized Shannon index varies between zero and one. It vanishes if only a single category actually appears in the sample. Without loss of generality, we may assume that it is the the first one, $f_1 = 1$ and $f_k = 0$ for all $k \in \{2, \ldots, K\}$:

$$H_{\text{norm}}(x) = -\frac{1}{\ln(K)} \sum_{k=1}^{K} f_k \ln(f_k) = -\frac{1}{\ln(K)} \cdot 1 \cdot \ln(1) = 0$$

The other extreme is given by the situation when all categories occur with the same frequency. In this case, they are said to follow a **uniform distribution**. For a uniform distribution, $f_k = 1/K$ holds for all $k \in \{1, \ldots, K\}$ and the Shannon index is maximized:

$$
H_{\text{norm}}(x) = -\frac{1}{\ln(K)} \sum_{k=1}^{K} f_k \ln(f_k) = -\frac{1}{\ln(K)} \sum_{k=1}^{K} \frac{1}{K} \ln\left(\frac{1}{K}\right)
$$

$$
= -\frac{1}{\ln(K)} \cdot K \cdot \frac{1}{K} \cdot (-\ln(K)) = 1
$$

The inverse statements also apply. For example, a maximum Shannon index implies uniform distribution. The Shannon index measures the variation within a categorical frequency distribution. A low Shannon index, approaching zero, indicates that the majority of observations belong to one category (the mode). A high Shannon index, approaching its maximum, indicates that every category occurs with about the same frequency.

**Example.** The normalized Shannon index for the distribution of sex among the respondents to the CDC survey is given by:

$$
H_{\text{norm}}(\text{sex}) = -\frac{1}{\log(2)} \left( 0.55 \cdot \log(0.55) + 0.45 \cdot \log(0.45) \right) \approx 0.993
$$

The entropy of the distribution of income categories is lower, since some categories are more frequent than others:

$$
H_{\text{norm}}(\text{income category}) \approx 0.893
$$

## 2.5 Measures of association

Variables can be dependent of each other, i.e., there can be functional relationships between them. For example, let us imagine a company that produces rectangular shape workpieces and—perhaps for quality assurance—measures the size of a random sample of these workpieces. Even if the manufactured parts vary in size, the following is true for each: the area of the workpiece is the product of its side lengths. The example seems trivial because the relationship follows immediately from geometric considerations. However, this is not always the case; there may also be *hidden* functional relationships between features that are not immediately apparent. Secondly, even very strong associations do not reflect a strictly deterministic dependence between observations. For example, there is an association, demonstrated by many studies, between inhalation of tobacco smoke and the development of cancer [10]. Nevertheless, not *every* smoker develops cancer, and some—albeit few—lung cancer patients turn out to be non-smokers.

**Measures of association** are used to confirm or explore relationships between variables. To this end, we compare sequences of observations $x = (x_1, \ldots, x_N)$ and $y = (y_1, \ldots, y_N)$ of those variables. For the results to have any practical value, these observations need to represent **paired data**: every data point $(x_n, y_n)$ comes from a single statistical unit so that $x_n$ and $y_n$ are naturally matched. For example, $x_n$ and $y_n$ could be body height and body weight of the same individual.

### 2.5.1 Sample covariance and Pearson's correlation coefficient

The statistical measures defined below can be used to gauge a *linear* functional relationship between numeric variables. If such a measure comparing two variables has a high value in terms of magnitude, then the linear increase in one variable is associated with the linear increase or decrease of the other.

For two paired sequences of numeric measurements/observations, $x = (x_1, \ldots, x_N)$ and $y = (y_1, \ldots, y_N)$, their **sample covariance** is given as follows:

$$s(x, y) = \frac{1}{N} \sum_{n=1}^{N} (x_n - \bar{x}) \cdot (y_n - \bar{y})$$

The **sample Bravais–Pearson correlation coefficient** is given by:

$$r(x, y) = \frac{s(x, y)}{s(x) \cdot s(y)},$$

where we need to assume $s(x) \neq 0$ and $s(y) \neq 0$.

The sample variance can be interpreted as the covariance of a single sequence of observations with itself: namely that $s(x, x) = s^2(x)$ holds for all $x \in \mathbb{R}^N$. The Bravais–Pearson correlation coefficient is also known as **Pearson's correlation coefficient**.

We would now like to clarify how to interpret the covariance and Pearson's correlation coefficient. For this purpose, we first claim the following general properties of the arithmetic mean $\mu(\cdot)$ and the sample variance $s^2(\cdot)$, which are not difficult to prove:

$$\mu(mx + c) = m\mu(x) + c,$$
$$s^2(mx + c) = m^2 s^2(x)$$

for all $x \in \mathbb{R}^N$ and $m, c \in \mathbb{R}$. Thus, for the standard deviation, $s(mx + c) = \sqrt{s^2(mx + c)} = \sqrt{m^2 s^2(x)} = |m| \cdot s(x)$.

Let's assume now that the equation $y = mx + c$ holds: when plotting the $y$-values over the $x$-values in a scatter plot, they would lie *exactly* on a straight line with slope $m$ and intercept $c$. Thus, the variable values are associated through a perfect linear relationship. In this case, the covariance can be computed as follows:

$$s(x, y) = s(x, mx + c) = \frac{1}{N} \sum_{n=1}^{N} (x_n - \bar{x}) \cdot (mx_n + c - m\bar{x} - c)$$

$$= m \cdot \frac{1}{N} \sum_{n=1}^{N} (x_n - \bar{x}) \cdot (x_n - \bar{x}) = m \cdot s(x, x)$$

$$= m \cdot s^2(x)$$

Consequently, for the correlation coefficient (assuming $m \neq 0$):

$$r(x, y) = \frac{s(x, mx + c)}{s(x) \cdot s(mx + c)} = \frac{ms^2(x)}{|m|(s(x))^2}$$

$$= \text{sgn}(m) = \begin{cases} 1 & \text{if } m > 0 \\ -1 & \text{if } m < 0 \end{cases}$$

If the $y$ values increase linearly with $x$, then the correlation is given by $r(x, y) = +1$. But if they decrease linearly with $x$, then we have $r(x, y) = -1$. Each situation corresponds to a perfect positive or negative linear correlation of both features, respectively. We come back to the relationship between Pearson correlation and linear dependence in more detail at the end of Sect. 4.5.1.

In addition to these extreme values, the following interpretations of the effect size can be given for certain magnitudes of the Bravais–Pearson coefficient [11, p. 79]:

| $\lvert r(x, y) \rvert$ | 0.0 | 0.1 | 0.3 | 0.5 |
|---|---|---|---|---|
| effect | none | small | medium | large |

**Table 2.4.** *Size of correlation effects*

> **Example.** By analyzing $N = 164{,}798$ answers from male respondents in the CDC survey, we can determine the Pearson correlation coefficient between body mass index and the Broca index. In the scatter plot of Fig. 2.3, the data points lie on a straight line with a positive slope, to good approximation. Therefore, we can expect that there is a strong positive correlation of $r \approx 1$. In fact, $r = 0.995$ holds. Between body weight and BMI, the correlation coefficient is only $r = 0.872$—this value still indicates a strong correlation though. Body height and weight show a moderate correlation with $r = 0.387$. Body height and BMI show no significant linear correlation: $r = -0.094$.

## 2.5.2 Rank correlation coefficients

The measures of association defined in this section can be used to gauge a *monotone* functional relationship between numeric or ordinal variables. A high value in terms of magnitude indicates that the increase in one variable is associated with the increase or decrease of the other. This relationship—in contrast to the scope of the Pearson correlation—need not necessarily be linear.

Let $x = (x_1, \ldots, x_N)$ be a sequence of observations of a numeric or an ordinal variable. We can sort these observations in descending order:

$$x_{\iota(1)} \geq x_{\iota(2)} \geq \cdots \geq x_{\iota(N)},$$

where $\iota \colon \{1, \ldots, N\} \to \{1, \ldots, N\}$ is a permutation of indices. The numbers

$$\mathrm{rg}(x) = (\iota(1), \iota(2), \ldots, \iota(N))$$

are called the **ranks** of the observations. The observation with the largest value is given rank 1, and the observation with the smallest value is given rank $N$. The convention of putting the smallest value first is also common.

If no value occurs more than once in the sample, each value will have a unique rank. Otherwise, to ensure uniqueness, the first occurrence of a repeated value should receive the lower rank.s

However, this convention will lead to identical variable values being assigned a different rank, which can be undesirable. We may correct the ranking as follows: identical observations are assigned the arithmetic mean of their ranks, and we denote the result by $\overline{\mathrm{rg}}(x)$.

Another useful definition is the **percentile rank** of an observation $x_n$:

$$\%\text{-rg}(x_n) = \frac{1}{N} \cdot |\{m \in \{1, \ldots, N\} | x_m \leq x_n\}|$$

Thus, the largest value in the sample always receives the maximum percentile rank of 100%.

We consider a hypothetical example to illustrate the different definitions of a rank. Imagine five students receive the following grades at the end of the school year: $x = (B, A, B, A, C)$. We interpret these grades as a sequence of values from an ordinal variable with domain $E/F < D < C < B < A$. The corresponding ranks are given as follows, where we have re-ordered the sample by the simplest definition of the rank:

| $n$ | $x_n$ | $\mathrm{rg}(x_n)$ | $\overline{\mathrm{rg}}(x_n)$ | $\%\text{-rg}(x_n)$ |
|---|---|---|---|---|
| 2 | A | 1 | 1.5 | 100% |
| 4 | A | 2 | 1.5 | 100% |
| 1 | B | 3 | 3.5 | 60% |
| 3 | B | 4 | 3.5 | 60% |
| 5 | C | 5 | 5 | 20% |

**Table 2.5.** *Rank statistics of school grades*

Let $x = (x_1, \ldots, x_N)$ and $y = (y_1, \ldots, y_N)$ be two paired sequences of observations of numeric or ordinal values.

**Spearman's rank correlation coefficient** is defined as follows:

$$\rho(x, y) = r(\overline{\mathrm{rg}}(x), \overline{\mathrm{rg}}(y)),$$

where $r(\cdot, \cdot)$ is the Bravais–Pearson correlation coefficient.

**Kendall's rank correlation coefficient** is the following measure:

$$\tau(x, y) = \frac{1}{N(N-1)} \sum_{k,l \in \{1,\ldots,N\}} \mathrm{sgn}(x_l - x_k) \cdot \mathrm{sgn}(y_l - y_k)$$

In order to compare the above rank correlation coefficients with Pearson's and showcase some of their characteristics, we compute them for a collection of four synthetic datasets called **Anscombe's quartet** [12], shown in Fig. 2.11.

A calculation of the individual correlation coefficients for each dataset given by $x^{(i)}$ and $y^{(i)}$, $i = 1, 2, 3, 4$, yields the following results:

| $i$ | Pearson | Spearman | Kendall |
|---|---|---|---|
| 1 | 0.82 | 0.82 | 0.63 |
| 2 | 0.82 | 0.69 | 0.56 |
| 3 | 0.82 | 0.99 | 0.96 |
| 4 | 0.82 | 0.50 | 0.18 |

**Table 2.6.** *Correlation coefficients computed for Anscombe's quartet*

First, we note that for all four datasets—although they are of a very different shape, as evidenced by their scatter plots—the Bravais–Pearson correlation coefficient is identical. The rank correlation coefficients for the second dataset, on the other hand, are lower because the $y$ values do not grow or fall monotonically with the $x$ values. In addition, the rank correlation is more robust to individual outliers, as shown by the third example. However, the fourth example also reveals a difficulty: the majority of the $x$ values here are identical or nearly identical. Any small variation in these values leads to arbitrary ranks and, consequently, to equally arbitrary rank correlation coefficients.

### 2.5.3 Sample mutual information and Jaccard index

Let $x = (x_1, \ldots, x_N)$ and $y = (y_1, \ldots, y_N)$ be paired sequences of observations that take values from a list of $K$ and $L$ categories, respectively.

The association between those categorical variables is characterized by their **joint frequencies** $0 \le f_{kl} \le 1$ with $k \in \{1, \ldots, K\}$ and $l \in \{1, \ldots, L\}$. These are the relative frequencies in which the categories co-occur: $f_{kl}$ is the relative frequency with which $x_n$ is observed to be the $k$-th category and $y_n$ to be the $l$-th category, at the same time.

From the joint frequencies, we can calculate the **marginal frequencies**:

$$f_{k\bullet} = \sum_{j=1}^{L} f_{kj}, \; f_{\bullet l} = \sum_{i=1}^{K} f_{il}$$

for all $k \in \{1, \ldots, K\}$ and $l \in \{1, \ldots, L\}$. The marginal frequencies are simply the frequencies of the individual variables, represented as a sum over joint frequencies.

All frequencies add up to 100%:

$$\sum_{k=1}^{K} f_{k\bullet} = \sum_{l=1}^{L} f_{\bullet l} = \sum_{k=1}^{K} \sum_{l=1}^{L} f_{kl} = 1$$

We can summarize the joint and marginal frequencies in a **contingency table**:

$$
\begin{array}{cccc|c}
f_{11} & f_{12} & \cdots & f_{1L} & f_{1\bullet} \\
f_{21} & f_{22} & \cdots & f_{2L} & f_{2\bullet} \\
\vdots & & & \vdots & \vdots \\
f_{K1} & f_{K2} & \cdots & f_{KL} & f_{K\bullet} \\
\hline
f_{\bullet 1} & f_{\bullet 2} & \cdots & f_{\bullet L} &
\end{array}
$$

Given the above notation, we can define the following measures of association for categorical variables.

The **joint Shannon index** is given as follows (as usual we agree on $0 \cdot \ln 0 = 0$):

$$H(x, y) = -\sum_{k=1}^{K} \sum_{l=1}^{L} f_{kl} \cdot \ln(f_{kl})$$

The **sample mutual information** of the two variables/features is the following quantity:

$$\mathrm{MI}(x, y) = \sum_{k=1}^{K} \sum_{l=1}^{L} f_{kl} \cdot \ln\left(\frac{f_{kl}}{f_{k\bullet} \cdot f_{\bullet l}}\right)$$

The **normalized sample mutual information** is determined as follows:

$$\mathrm{MI}_{\mathrm{norm}}(x, y) = \frac{\mathrm{MI}(x, y)}{H(x, y)}$$

for $H(x, y) > 0$, otherwise we may assume $\mathrm{MI}_{\mathrm{norm}}(x, y) = 1$.

The joint Shannon index or mutual information of a sequence of observations $x$ with respect to itself is just its regular Shannon index, i.e., $\mathrm{MI}(x,x) = H(x,x) = H(x)$ holds. That's because in this case we have $f_{kl} = 0$ for $k \neq l$ and $f_{kl} = f_{k\bullet} = f_{\bullet l}$ for $k = l$. This fact also implies $\mathrm{MI}_{\mathrm{norm}}(x,x) = 1$.

The normalized mutual information may range between zero and one. A value of one indicates a maximum possible association between the categorical variables. A vanishing mutual information, on the other hand, corresponds to the condition that every joint frequency is the product of the corresponding marginal frequencies:

$$f_{kl} = f_{k\bullet} \cdot f_{\bullet l}$$

If we interpret the frequencies as estimates for probabilities, we will see later—in Sect. 3.1.1—that this condition is a statement about the variables' *independence*. In other words, the frequency of the $k$-th category occurring does not provide information on how often the $l$-th category occurs, and vice versa.

**Example.** The following is a contingency table based on the German 2018 ALLBUS survey. The table compares the preference for a political party with the answer to the following question: "What do you think of the following statement: the influx of refugees should be stopped."

The mutual information between the two categorical variables is given by $\mathrm{MI}_{\mathrm{norm}} = 0.035$. This may sound like a small value, so here is a comparison. The mutual information between party preference and the presence of a door phone in the respondent's home is much smaller: $\mathrm{MI}_{\mathrm{norm}} = 0.0026$. This is an expected result, as we can hardly imagine a mechanism or explanation that would imply a strong enough correlation between those two features.

All figures are in percent:

| | do not agree | tend to not agree | indiffer-ent | tend to agree | agree | $\Sigma$ |
|---|---|---|---|---|---|---|
| Union parties | 5.9 | 13.0 | 8.0 | 8.3 | 3.8 | 39.0 |
| Social Democratic Party | 5.9 | 9.3 | 4.6 | 4.7 | 2.3 | 26.9 |
| The Greens | 6.4 | 5.4 | 1.5 | 0.5 | 0.2 | 13.9 |
| The Left | 2.7 | 3.0 | 1.4 | 1.2 | 0.7 | 9.0 |
| Alternative for Germany | 0.2 | 0.3 | 0.6 | 1.1 | 3.7 | 5.8 |
| Free Democratic Party | 1.0 | 1.1 | 0.9 | 1.1 | 0.6 | 4.7 |
| others | 0.1 | 0.3 | 0.1 | 0.1 | 0.2 | 0.7 |
| $\Sigma$ | 22.1 | 32.3 | 17.1 | 17.0 | 11.5 | |

**Table 2.7.** *Contingency table of party preference vs. opinion on migration control*

Let $x = (x_1, \ldots, x_N)$ and $y = (y_1, \ldots, y_N)$ be two paired sequences of *binary* observations: $x_n, y_n \in \{0, 1\}$ for all $n \in \{1, \ldots, N\}$.

Let $f_{11}$ be the frequency of co-occurrence of a positive value, that is, the frequency of $x_n = y_n = 1$. Furthermore, let $f_{1\bullet}$ be the frequency of $x_n = 1$ and $f_{\bullet 1}$ be the frequency of $y_n = 1$.

The **Jaccard index** is then given as:

$$J(x, y) = \frac{f_{11}}{f_{\bullet 1} + f_{1\bullet} - f_{11}},$$

if not exactly $x = y = 0$, in which case the Jaccard index is not defined.

In order to showcase similarities with mutual information, we have defined the Jaccard index in terms of frequencies. It does not matter whether the relative or absolute frequencies are used; the sample size cancels out either way. However, the Jaccard index can be described more conveniently in terms of subpopulations. As noted earlier, there is a natural one-to-one correspondence between binary variables and subpopulations. Comparing with the above formula, we find that the Jaccard index of two subpopulations $S, T \subseteq \Omega$ via that correspondence is given as follows:

$$J(S, T) = \frac{|S \cap T|}{|S| + |T| - |S \cap T|} = \frac{|S \cap T|}{|S \cup T|}$$

The Jaccard index takes values between zero and one, where $J(S, T) = 0$ corresponds to subpopulations that have no statistical unit in common: $S \cap T = \emptyset$. Maximum association, $J(S, T) = 1$, is given if and only if both subpopulations are identical: $S = T$.

**Example.** In the 2013 German federal election, 16.89 million voters who had not moved away or died by 2017 cast their ballots for the alliance of center-right parties CDU and CSU, also called the Union parties. Excluding new voters and new citizens, the figure for the following election in 2017 was 14.77 million. 11.09 million voters gave their vote to the Union in both years [13]. Therefore, the Jaccard index between the subpopulations of Union voters in 2013 and those in 2017 is given by:

$$J(\text{Union 2013}, \text{Union 2017}) = \frac{11.09}{16.89 + 14.77 - 11.09} \approx 54\%$$

The table below shows the Jaccard index for selected other parties and the cohort of non-voters.

We can interpret this value as a measure that gauges voter loyalty or vote retention. If the value were 0%, this would mean that in 2017 only swing votes went to that party. In contrast, a value of 100% would mean that the pool of voters was identical in 2013 and 2017.

| political party | Jaccard index, voters 2013 vs. 2017 |
|---|---|
| Union parties | 54% |
| Social Democratic Party | 43% |
| The Greens | 34% |
| The Left | 36% |
| Alternative for Germany | 23% |
| Free Democratic Party | 22% |
| non-voters | 48% |

**Table 2.8.** *Jaccard index as a measure for vote retention*

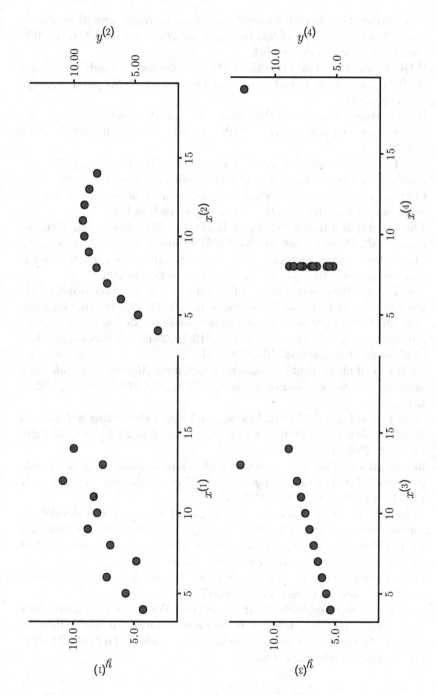

**Fig. 2.11.** *Anscombe's quartet: four different datasets with identical Pearson correlation coefficient*

# References

[1]   CDC Population Health Surveillance Branch. *Behavioral Risk Factor Surveillance System (BRFSS) Survey Data 2018*. Accessed Feb. 1, 2020. URL: https://www.cdc.gov/brfss/.

[2]   WHO Consultation on Obesity (1999: Geneva, Switzerland) and World Health Organization. *Obesity: preventing and managing the global epidemic*. Tech. rep. 2000.

[3]   Harri Siirtola. "The Cost of Pie Charts". In: *23rd International Conference Information Visualisation (IV)*. 2019, pp. 151–156. DOI: 10.1109/IV.2019.00034.

[4]   The Federal Returning Officer of Germany. *2021 Bundestag Election: final result*. Press release no. 52/21. Wiesbaden, Germany, Oct. 2021. URL: https://www.bundeswahlleiter.de/en/info/presse/mitteilungen/bundestagswahl-2021/52_21_endgueltiges-ergebnis.html.

[5]   The Federal Returning Officer of Germany. *2017 Bundestag Election: final result*. Press release no. 34/17. Wiesbaden, Germany, Oct. 2017. URL: https://www.bundeswahlleiter.de/en/info/presse/mitteilungen/bundestagswahl-2017/34_17_endgueltiges_ergebnis.html.

[6]   The Federal Returning Officer of Germany. *Official final result of the 2013 Bundestag Election*. Press release no. 34/13. Wiesbaden, Germany, Oct. 2013. URL: https://www.bundeswahlleiter.de/en/info/presse/mitteilungen/bundestagswahl-2013/2013-10-09-endgueltiges-amtliches-ergebnis-der-bundestagswahl-2013.html.

[7]   GESIS – Leibniz-Institut für Sozialwissenschaften. *Allgemeine Bevölkerungsumfrage der Sozialwissenschaften ALLBUS 2018*. 2019. DOI: 10.4232/1.13250.

[8]   Paul T. von Hippel. "Mean, Median, and Skew: Correcting a Textbook Rule". In: *Journal of Statistics Education* 13.2 (Jan. 2005). DOI: 10.1080/10691898.2005.11910556.

[9]   Ronald Aylmer Fisher. "The use of multiple measurements in taxonomic problems". In: *Annals of Eugenics* 7.2 (Sept. 1936), pp. 179–188. DOI: 10.1111/j.1469-1809.1936.tb02137.x.

[10]  Robert N. Proctor. "Tobacco and the global lung cancer epidemic". In: *Nature Reviews Cancer* 1.1 (Oct. 2001), pp. 82–86. DOI: 10.1038/35094091.

[11]  Jacob Cohen. *Statistical power analysis for the behavioral sciences*. 2nd ed. New Jersey, USA: Lawrence Earlbaum Associates, 1988.

[12]  Francis John Anscombe. "Graphs in Statistical Analysis". In: *The American Statistician* 27.1 (Feb. 1973), pp. 17–21.

[13]  infratest dimap Gesellschaft für Trend- und Wahlforschung mbH. *Bundestagswahl 2017 Deutschland Wählerwanderungen*. Accessed June 10, 2020. URL: https://wahl.tagesschau.de/wahlen/2017-09-24-BT-DE/analyse-wanderung.shtml.

# Part II

Stochastics

# 3

# Probability theory

Descriptive statistics are concerned with the investigation of a sample that is usually limited in size. A key benefit is that these investigations allow us to draw conclusions about the population as a whole. For example, based on data alone, we can conclude with some confidence that a human being cannot grow to a height of three meters.

A somewhat more practical example is the political election poll. Before the actual election, a sample of eligible voters is selected, and each person is asked which political party they plan to vote for. The hope is that the proportion of votes for a particular party in that sample is approximately the same proportion of *all* voters on election day.

Furthermore, this proportion can be interpreted as a measure for the probability of a voter—randomly selected from the population—to have a preference for the party in question. For this reason, the relative frequency can also be called the **empirical probability**.

One important goal of **stochastics** is to justify such a procedure rigorously, mathematically. The field can be divided into two subfields:

- **Probability theory**, which is the subject of this chapter, deals with the mathematical definition and investigation of the concept of probability. A central object of such an investigation are variables the values of which are not specified or known precisely but are subject to uncertainty. In other words, a probability can only be given that such a **random variable** takes values within a certain range.

- **Inferential statistics**, which will be discussed in the next chapter, builds on descriptive statistics and probability theory. Its rationale is based on the assumption that statistical observations and measures, such as frequencies, means, etc., are values or **realizations** of random variables. Conversely, the field investigates the extent to which characteristics of random variables can be estimated from sampled data. In particular, under certain simplifying

© Springer-Verlag GmbH Germany, part of Springer Nature 2023
M. Plaue, *Data Science*, https://doi.org/10.1007/978-3-662-67882-4_3

assumptions, it is possible to quantify the accuracy or error of such an estimate.

The field of (statistical) machine learning makes substantial use of stochastic considerations, such as with classification. In classification, probabilities quantify the frequency of occurrence of a class (e.g., "proportion of photographs in the training dataset showing a cat") or the confidence in a classification result ("with high probability, this photograph in the test dataset shows a cat").

## 3.1 Probability measures

In the context of probability theory, a process with (possibly) an uncertain outcome that can be repeated—in principle, at least—under identical conditions and as often as desired is called an **experiment**. Each individual repetition of an experiment is called a **trial** and yields a single outcome from a well-defined set of outcomes. A classic example of such an experiment is the flipping of a coin. At the end of each trial, either the side with heads shows, or the side with tails. Another example is that of rolling a six-sided die.

The set of all possible outcomes is also called the **sample space**. The sample space for a coin toss or a die roll can be written as $\Omega = \{\text{tails}, \text{heads}\}$ or $\Omega = \{\boxdot, \boxdot, \boxdot, \boxdot, \boxdot, \boxdot\}$, respectively.

We can interpret the relative frequency with which an outcome occurs in an experiment after many trials as the probability of that event occurring. This interpretation is called the **frequentist interpretation of probability** and is common in the natural sciences, where repeatable experiments are essential for gaining empirical knowledge. An example of a frequentist statement could be the following: "We can observe that about 50% of all atomic nuclei of the radioactive isotope iodine-131 have decayed after eight days. Therefore, the probability that a single such atomic nucleus has decayed after this time is 50%."

In comparison, the **Bayesian interpretation of probability** sees probability as a (subjective) measure for the plausibility of an event, without the event necessarily being the result of a repeatable experiment. For example, according to the Bayesian interpretation, it would make sense to say: "the probability that it will rain today is 75%." Adopting the Bayesian viewpoint is useful for reasoning and decision-making under uncertainty, with the aim of maximizing expected utility—"should I, or should I not, take an umbrella with me today?"

There are even more approaches to imbue "probability" with universal meaning [1]. Whatever the interpretation, probability can be defined rigorously as a mathematical concept. The key idea is to assign subsets of the sample space a number between zero and one: the probability that one of the outcomes in that subset is realized with each trial.

A **probability measure** is a map $\Pr(\cdot)$ that assigns a nonnegative real number to certain subsets $A \subseteq \Omega$ of a sample space $\Omega$ that has the following properties:

(1) $\Pr(\emptyset) = 0$ and $\Pr(\Omega) = 1$

(2) For every finite or countably infinite family of pairwise disjoint sets $(A_i)_{i \in I}$:

$$\Pr\left(\bigcup_{i \in I} A_i\right) = \sum_{i \in I} \Pr(A_i)$$

The above conditions form the basis for the **Kolmogorov axioms of probability theory**. A more rigorous mathematical analysis shows that it is not always possible or reasonable to assign a well-defined probability to *every* arbitrary subset of $\Omega$. Subsets of the sample space that *can* be assigned a probability are called **events**, or **measurable**. Events that contain exactly one element are called **elementary events**, or **atomic events**.

We want to construct a meaningful probability measure that describes a game of dice. When rolling a six-sided die, it should be impossible for the player to predict which side will show. Nevertheless, it is possible to estimate how likely it is that a certain side will show, or that a certain number will be rolled. Without further information, they must assume that each possible outcome is equally likely:

$$\Pr(\{\omega\}) = \frac{1}{6}$$

for all $\omega \in \Omega = \{\boxdot, \boxminus, \boxtimes, \boxminus, \boxtimes, \boxplus\}$. The probabilities of all other events are obtained by applying the rules above. For example, the probability of rolling an even number is given by:

$$\Pr(\{\omega | \omega \text{ represents an even number}\}) =$$
$$\Pr(\{\boxminus, \boxminus, \boxplus\}) =$$
$$\Pr(\{\boxminus\} \cup \{\boxminus\} \cup \{\boxplus\}) =$$
$$\Pr(\{\boxminus\}) + \Pr(\{\boxminus\}) + \Pr(\{\boxplus\}) =$$
$$\frac{1}{6} + \frac{1}{6} + \frac{1}{6} = \frac{1}{2}$$

In the argument above, we have invoked the more general **principle of indifference**. According to this principle, if no additional information other than the set of outcomes is available, then each outcome or atomic event $\{\omega\} \subset \Omega$ should be assigned equal probability, meaning that they are **uniformly distributed**.

For finite sample spaces with a uniform distribution, we have $\Pr(\{\omega\}) = \frac{1}{|\Omega|}$, which implies **Laplace's rule**:

$$\Pr(A) = \frac{|A|}{|\Omega|}$$

for all events $A \subseteq \Omega$. This formula can be used to determine the probability of rolling an even number as follows:

$$\Pr(\{\boxdot, \boxdot, \boxdot\}) = \frac{|\{\boxdot, \boxdot, \boxdot\}|}{|\{\boxdot, \boxdot, \boxdot, \boxdot, \boxdot, \boxdot\}|} = \frac{3}{6} = \frac{1}{2}$$

For another example, imagine a dart player throws at a dartboard with radius $R > 0$. A reasonable mathematical model for this situation is given by the sample space $\Omega = \{(\omega_1, \omega_2) \in \mathbb{R}^2 | (\omega_1)^2 + (\omega_2)^2 \leq R^2\}$, that is, a circular disk in the plane. Each point in the disk corresponds to a possible location where the dart can hit.

The goal of the player is to throw the dart so that it lands as close as possible to the center point $(0, 0)$. A perfect player that hits the center every single time would be described by the following probability measure, which is called a **Dirac measure**:

$$\mathrm{Pr}_0(A) = \begin{cases} 1 & \text{if } (0,0) \in A \\ 0 & \text{otherwise} \end{cases}$$

For atomic events $\{\omega\} \in \Omega$, this means in particular:

$$\mathrm{Pr}_0(\{\omega\}) = \begin{cases} 1 & \text{if } \omega = (0,0) \\ 0 & \text{otherwise} \end{cases}$$

A player who is able to hit the dartboard but otherwise has no control over the dart's trajectory corresponds to the other extreme, a uniform distribution:

$$\mathrm{Pr}_R(A) = \frac{|A|}{|\Omega|} = \frac{|A|}{\pi R^2}$$

Here we have denoted the area of a region $B \subseteq \Omega$ by $|B|$. Note that predicting whether a single point on the dartboard will be hit is no longer meaningful because $\mathrm{Pr}_R(\{\omega\}) = 0$ holds for all $\omega \in \Omega$. Consequently, this probability measure cannot be defined by determining its value for elementary events, which would be possible for a finite sample space.

In a more realistic scenario, the player is not perfect but still very good, and therefore the player hits with 90% probability the bull's eye $B(r) = \{(\omega_1, \omega_2) \in \mathbb{R}^2 | (\omega_1)^2 + (\omega_2)^2 \leq r^2\}$ with radius $r$, $0 < r < R$:

$$\mathrm{Pr}_r(A) = 0.9 \cdot \frac{|A \cap B(r)|}{|B(r)|} + 0.1 \cdot \frac{|A \cap (\Omega \setminus B(r))|}{|\Omega \setminus B(r)|}$$

**Rules for calculating with probabilities.** For a fixed probability measure $\Pr(\cdot)$, where $A, B \subseteq \Omega$ are arbitrary events:

(1) $\Pr(A \cup B) = \Pr(A) + \Pr(B) - \Pr(A \cap B)$

(2) $\Pr(A \setminus B) = \Pr(A \cup B) - \Pr(B)$, in particular $\Pr(\Omega \setminus A) = 1 - \Pr(A)$

The event $\Omega \setminus A$ is called the event **complementary** to $A$, which we will also denote as $\neg A$.

A quick example calculation for rule (1): the probability of rolling an even number or a number divisible by three with a six-sided die ($A = \{\boxdot, \boxdot, \boxdot\}$ or $B = \{\boxdot, \boxdot\}$) is given by:

$$\Pr(A \cup B) = \Pr(A) + \Pr(B) - \Pr(A \cap B)$$
$$= \frac{3}{6} + \frac{2}{6} - \Pr(\{\boxdot\}) = \frac{3}{6} + \frac{2}{6} - \frac{1}{6}$$
$$= \frac{2}{3} \approx 0.67$$

Of course, in this case we can easily calculate the probability directly:

$$\Pr(A \cup B) = \Pr(\{\boxdot, \boxdot, \boxdot, \boxdot\}) = \frac{4}{6} = \frac{2}{3}$$

Regarding (2): the probability of *not* rolling the number two is given by:

$$\Pr(\neg C) = \Pr(\Omega \setminus C) = 1 - \frac{1}{6} = \frac{5}{6} \approx 0.83,$$

where $C = \{\boxdot\}$.

**Inequalities for probabilities.** For any events $A, B \subseteq \Omega$, the following inequalities hold:

(1) $\Pr(A) \leq \Pr(B)$ if $A \subseteq B$

(2) $\Pr(A \cup B) \leq \Pr(A) + \Pr(B)$

(3) $\Pr(A \cap B) \geq \Pr(A) + \Pr(B) - 1$

The second inequality is useful only for small probabilities or frequencies (since otherwise the right-hand side can become larger than one), the third only for large ones (since otherwise the right-hand side can become smaller than zero).

**Example.** (1): The proportion of German households with no more than two cars is at least as large as the proportion of households with no more than one car.

(2): 19% of all German households are home to a dog, 23% to a cat (in 2018 [2]). These statistics imply that at most 42% of all households own either a dog, a cat, or both.

(3): In Germany, around 77% of all households own a car, and 97% own a mobile phone (in 2019 [3]). From these statistics, we can deduce that at least 74% of all German households own a car and also a mobile phone.

Since probability measures behave in many ways like other measures for the size of a set, the above inequalities and similar facts about probability can be illustrated using Venn or Euler diagrams such as the following:

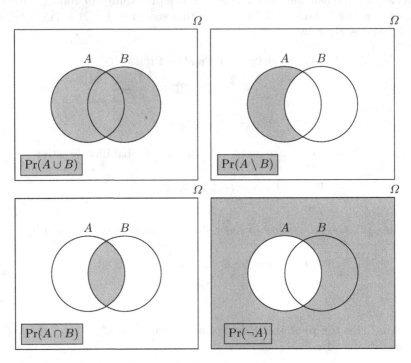

**Fig. 3.1.** *Venn diagrams of the probability of events*

### 3.1.1 Conditional probability

It is important to note that the estimated probability of an event may depend on additional information known to the person making the estimate. For example, if the game master of a game of dice rolls a single die and hides the result from the player but tells the player that the result is an even number (i.e., two, four, or six), then, from the player's perspective, the probability that the number rolled is three is no longer one-sixth but zero.

What is the probability that the player will assign to the event of rolling a two, i.e., the event $A = \{⊡\}$? The set of possible outcomes has effectively changed from $\Omega = \{⊡, \ldots, ⊞\}$ to the smaller set $B = \{⊡, ⊡, ⊞\}$. Therefore, according to Laplace's rule, the **conditional probability** is the following:

$$\Pr(A|B) = \Pr\left(\{⊡\}|\{⊡, ⊡, ⊞\}\right) = \frac{|\{⊡\}|}{|\{⊡, ⊡, ⊞\}|} = \frac{1}{3}$$

Laplace's rule applies only to uniform distributions over finite sample spaces. In general, the conditional probability is defined as follows.

Let $A, B \subseteq \Omega$ be events with $\Pr(B) > 0$. Then

$$\Pr(A|B) = \frac{\Pr(A \cap B)}{\Pr(B)}$$

is called the **probability of $A$ under the condition $B$**.

We can easily verify that this definition of conditional probability is consistent with the intuition we obtained earlier. If Laplace's rule holds, the conditional probability is indeed the proportion of atomic events in $A$ with respect to a restricted set of possible outcomes $B$:

$$\Pr(A|B) = \frac{\Pr(A \cap B)}{\Pr(B)} = \frac{|A \cap B|/|\Omega|}{|B|/|\Omega|} = \frac{|A \cap B|}{|B|}$$

To illustrate the concept of conditional probability, consider the following game. Anna, the game master, rolls a six-sided die and tells Robert whether the result is an even or an odd number. Robert must then guess whether the number is a prime number, or not. Robert wants to develop a strategy that maximizes his chances of guessing correctly. We define the following events:

$$B = \{\omega|\omega \text{ represents an even number}\} = \{⚁, ⚃, ⚅\},$$
$$A = \{\omega|\omega \text{ represents a prime number}\} = \{⚁, ⚂, ⚄\}$$

If Anna states that the number is even, then the hypothesis that it is also prime yields the following probability from Robert's perspective:

$$\Pr(A|B) = \frac{\Pr(A \cap B)}{\Pr(B)} = \frac{\Pr(\{⚁\})}{\Pr(\{⚁, ⚃, ⚅\})} = \frac{^1/_6}{^3/_6} = \frac{1}{3} \approx 0.33$$

In this case, the probability that the number is prime is only 33%. Therefore, Robert should guess that the result is *not* a prime number, which will result in a probability of winning of 67%.

If Anna announces that the number is odd, the conditional probability of the number being prime is given by:

$$\Pr(A|\neg B) = \frac{\Pr(A \cap (\Omega \setminus B))}{\Pr(\Omega \setminus B)} = \frac{\Pr(\{⚂, ⚄\})}{1 - \Pr(B)} = \frac{^2/_6}{^1/_2} = \frac{2}{3} \approx 0.67$$

In that case, Robert should guess that the result is prime.

Two events $A$ and $B$ are called **independent** if the following holds:

$$\Pr(A \cap B) = \Pr(A) \cdot \Pr(B)$$

Assuming that the probabilities of $A$ and $B$ are strictly positive, independence can also be characterized as follows:

$$\Pr(A|B) = \Pr(A) \text{ or, equivalently, } \Pr(B|A) = \Pr(B)$$

Thus, the occurrence of $B$ does not change the probability of $A$ occurring (and vice versa). Example: Anna rolls the die and tells Robert whether the number is even or odd. Robert then has to guess whether the number is divisible by three, or not:

$$B = \{\omega | \omega \text{ represents an even number}\} = \{\square, \square, \square\}$$
$$A = \{\omega | \omega \text{ represents a number divisible by three}\} = \{\square, \square\}$$

We compute the probability of $A$ conditioned on $B$:

$$\Pr(A|B) = \frac{\Pr(A \cap B)}{\Pr(B)} = \frac{\Pr(\{\square\})}{\Pr(\{\square, \square, \square\})} = \frac{1/6}{3/6} = \frac{1}{3} = \Pr(A)$$

Consequently, Robert has not gained any useful information from Anna's statement. Therefore, he should always guess that the number *is not* divisible by three because $\Pr(\neg A) = \frac{2}{3} > \Pr(A)$ holds.

The following construction can be interpreted as a numeric measure of the dependence of two events:

> The **pointwise mutual information** of two events $A$ and $B$ is given as follows:
>
> $$\mathrm{PMI}(A, B) = \log\left(\frac{\Pr(A \cap B)}{\Pr(A) \cdot \Pr(B)}\right)$$

The choice of the base of the logarithm does not play an important role, and with suitable normalization (see below) it does not play a role at all. We will use the natural logarithm. The pointwise mutual information is positive if the probability of $A$ increases when conditioned on $B$:

$$\Pr(A|B) > \Pr(A) \Leftrightarrow \ln\left(\frac{\Pr(A|B)}{\Pr(A)}\right) > 0$$

$$\Leftrightarrow \ln\left(\frac{\Pr(A \cap B)}{\Pr(A) \cdot \Pr(B)}\right) > 0$$

$$\Leftrightarrow \mathrm{PMI}(A, B) > 0$$

All probabilities must be assumed to be strictly positive. The mutual information between two events vanishes if and only if they are independent.

This measure is often normalized in a suitable manner. One possible normalization is the following:

$$\mathrm{nPMI}(A, B) = \ln\left(\frac{\Pr(A \cap B)}{\Pr(A) \cdot \Pr(B)}\right) \cdot (-\ln(\Pr(A \cap B)))^{-1}$$

$$= \frac{\ln(\Pr(A) \cdot \Pr(B))}{\ln(\Pr(A \cap B))} - 1$$

For all events $A$ and $B$, we have $-1 < \mathrm{nPMI}(A, B) \le 1$.

The limit case $\mathrm{nPMI}(A, B) \to (-1)$ represents the situation where $A$ and $B$ occur together with vanishing probability: $\Pr(A \cap B) = 0$.

The case $\mathrm{nPMI}(A, B) = 1$ corresponds to a maximum possible dependence of events: $\Pr(B) = \Pr(A) = \Pr(A \cap B)$. In this case, the events $A$ and $B$ are identical, except, possibly, for a subset with vanishing probability:

$$\Pr(A \setminus B) = \Pr(A \cup B) - \Pr(B)$$
$$= \Pr(A) + \Pr(B) - \Pr(A \cap B) - \Pr(B) = 0,$$
$$\Pr(B \setminus A) = 0$$

**Example.** Pointwise mutual information is used, for example, in **keyword extraction**, an important task in **natural language processing**. Given a collection of text documents, the association of a word (or phrase) with one of the documents can be measured as follows:

$$\mathrm{PMI(document, word)} =$$
$$\ln \left( \frac{\text{relative frequency with which the word occurs in the document}}{\text{relative frequency with which the word occurs in the whole corpus}} \right)$$

Given a document, if a word occurs more frequently in that document than in the whole corpus, this measure has a positive value. Words with high values for PMI may be interpreted as keywords, i.e., the terms that best describe the subject of the document.

An examination of the twitter feeds of the U.S. Republican Party [4] and the U.S. Democratic Party [5] (3200 tweets each, retrieved on July 03, 2022) reveals that the following top 10 words have particularly high (normalized) pointwise mutual information with respect to each feed:

| @TheDemocrats | | | | @GOP | | | |
|---|---|---|---|---|---|---|---|
| keyword | # in feed | # in corpus | nPMI | keyword | # in feed | # in corpus | nPMI |
| president | 966 | 1081 | 0.101 | joe | 756 | 759 | 0.175 |
| infrastructure | 258 | 258 | 0.098 | border | 328 | 330 | 0.149 |
| health | 214 | 218 | 0.092 | biden | 1151 | 1456 | 0.139 |
| bipartisan | 189 | 192 | 0.090 | failed | 217 | 223 | 0.135 |
| bidenharris | 192 | 198 | 0.088 | gop | 271 | 293 | 0.131 |
| act | 200 | 209 | 0.086 | policies | 152 | 159 | 0.125 |
| rescue | 118 | 118 | 0.086 | prices | 273 | 315 | 0.120 |
| plan | 312 | 340 | 0.086 | rnc | 91 | 93 | 0.119 |
| vaccinated | 123 | 124 | 0.086 | primary | 124 | 132 | 0.118 |
| climate | 85 | 85 | 0.082 | southern | 72 | 72 | 0.118 |

**Table 3.1.** *Keyword extraction via pointwise mutual information*

### 3.1.2 Bayes' theorem

In the previous section, we saw how new information can influence the subjective probability of an event. This process of gaining new information can be understood as learning under uncertainty, and it may impact the learner's decision-making. For example, if a dice player learns that an even number is very likely to be the outcome, they may choose to bet less money on a roll of three.

We can describe the process of learning under uncertainty in more detail as follows. At the outset, the probability for a certain **hypothesis** $H$ has a certain **prior probability** $\Pr(H)$. When new **evidence** in the form of an observation $E$ becomes available, it leads to a correction of the prior probability, resulting in the **posterior probability** $\Pr(H|E)$. In the example mentioned above, the hypothesis $H$ corresponds to the event "the die shows the number three," and $\Pr(E)$ corresponds to the probability of "the number rolled is even."

Bayes' theorem tells us how to compute the posterior probability.

> **Bayes' theorem.** Let $H$, $E$ be events with $\Pr(H) > 0$, $\Pr(E) > 0$. Then:
>
> $$\Pr(H|E) = \frac{\Pr(E|H)}{\Pr(E)} \cdot \Pr(H)$$

Thus, the posterior $\Pr(H|E)$ is obtained from the prior $\Pr(H)$ by multiplying with a factor that is proportional to the **likelihood** $\Pr(E|H)$, that is, the probability of observing the evidence under the hypothesis.

The proof of the theorem is not difficult and follows immediately from the definition of conditional probability:

$$\Pr(H|E) = \frac{\Pr(H \cap E)}{\Pr(E)} = \frac{\Pr(E|H) \cdot \Pr(H)}{\Pr(E)}$$

An alternative version of Bayes' theorem useful for many calculations is the following:

$$\Pr(H|E) = \frac{\Pr(E|H)}{\Pr(E|H) \cdot \Pr(H) + \Pr(E|\neg H) \cdot \Pr(\neg H)} \cdot \Pr(H)$$

> **Example. Sensitivity** and **specificity** are common measures to evaluate the quality of diagnostic tests. These tests may detect a particular disease or an infection with a particular pathogen. Sensitivity is the proportion of sick or infected patients who indeed test positive. Specificity is the proportion of healthy patients for whom the test is negative, as expected. These proportions can be interpreted as empirical probabilities. Thus:

$$\text{sensitivity} = \Pr(E|H)$$
$$\text{specificity} = \Pr(\neg E|\neg H) = 1 - \Pr(E|\neg H)$$

Here we abbreviated the hypothesis "patient is sick/infected" as $H$, and the evidence "test turns out positive" as $E$.

Suppose a patient tests positive for a particular pathogen. What is the probability $\Pr(H|E)$ that the patient is actually infected with the pathogen in question? This probability depends on the prevalence of the pathogen among the population being tested. Prevalence refers to the spread of the pathogen among the test subjects and determines the prior probability.

Let us compute an example. Rapid antigen tests for detecting an infection with the pathogen SARS-CoV-2, the causative agent of the disease COVID-19, have a typical sensitivity of around 50% and a specificity of 99% [6]. Among patients randomly screened at a general practitioner, we might want to assume a prevalence of around $\Pr(H) = 3\%$. The above numbers imply the following posterior probability for an actual infection with SARS-CoV-2 given a positive test:

$$\Pr(H|E) = \frac{\Pr(E|H)}{\Pr(E|H) \cdot \Pr(H) + \Pr(E|\neg H) \cdot \Pr(\neg H)} \cdot \Pr(H)$$
$$= \frac{0.5}{0.5 \cdot 0.03 + (1 - 0.99) \cdot (1 - 0.03)} \cdot 0.03$$
$$\approx 61\%$$

Conversely, we can ask for the probability for a patient to test negative and actually not be infected:

$$\Pr(\neg H|\neg E) = \frac{\Pr(\neg E|\neg H)}{\Pr(\neg E|\neg H) \cdot \Pr(\neg H) + \Pr(\neg E|H) \cdot \Pr(H)} \cdot \Pr(\neg H)$$
$$= \frac{0.99}{0.99 \cdot (1 - 0.03) + (1 - 0.5) \cdot 0.03} \cdot (1 - 0.03)$$
$$\approx 98\%$$

The prevalence of a disease or infection can significantly impact the probability that a patient is actually infected based on a positive test result. For example, if a patient exhibits typical symptoms of COVID-19 during a period of high transmission, we might assume a prior probability of 30%. This assumption would lead to a probability of 96% that a positive rapid test indicates the patient is infected and a probability of 82% that a negative test correctly indicates the patient is not infected.

Bayes' theorem forms the foundation for many machine learning methods. In this context, the theorem is interpreted to mean that the initial belief of a learning algorithm about the probability of a hypothesis $H$ is given by $\Pr(H)$. This belief is then updated based on newly collected data $E$, to become $\Pr(H|E)$ after the learning process. If new data is repeatedly presented to the algorithm, further adjustments are made.

We want to sketch the function of a simple email spam filter in order to illustrate how a machine learning algorithm based on Bayesian inference may work. This algorithm could be fed the following data:

- In general, 45% of all email messages can be considered as spam. Thus, the prior probability for the hypothesis "this message is spam" is given by $\Pr(H) = 0.45$.

- Among all these spam messages, 5% have the word "Viagra" in the subject line, so the likelihood for this piece of evidence is $\Pr(E|H) = 0.05$.

- Among all the remaining messages, only a vanishingly small percentage have the word "Viagra" in the subject line: $\Pr(E|\neg H) = 0.001$.

Overall, this results in the following posterior probability:

$$\Pr(H|E) = \frac{\Pr(E|H)}{\Pr(E|H) \cdot \Pr(H) + \Pr(E|\neg H) \cdot \Pr(\neg H)} \cdot \Pr(H)$$
$$= \frac{0.05}{0.05 \cdot 0.45 + 0.001 \cdot (1 - 0.45)} \cdot 0.45$$
$$\approx 98\%$$

This means that if the word "Viagra" is included in the email subject, the algorithm is convinced that this message is most likely spam.

Now let the evidence $E$ be given by the fact that the email message was sent from an address that is included in the recipient's address book. Only a vanishingly small proportion of spam messages originate from known senders: $\Pr(E|H) = 0.001$. In contrast, a significantly higher proportion of non-spam messages originate from known senders: $\Pr(E|\neg H) = 0.2$. From these statistics we get:

$$\Pr(H|E) = \frac{\Pr(E|H)}{\Pr(E|H) \cdot \Pr(H) + \Pr(E|\neg H) \cdot \Pr(\neg H)} \cdot \Pr(H)$$
$$= \frac{0.001}{0.001 \cdot 0.45 + 0.2 \cdot (1 - 0.45)} \cdot 0.45$$
$$\approx 0.4\%$$

Consequently, the algorithm concludes that an email from a known sender is *not* spam with a very high probability.

A note on hypotheses with a prior probability of one or zero: If $\Pr(H) = 1$ holds, then $\Pr(H|E) = 1$ always holds regardless of the evidence $E$. This means that a hypothesis assumed to be true a priori is incontrovertible. Similarly, if $\Pr(H) = 0$ is assumed, the Bayesian method cannot learn from the evidence.

Evidence that is observed with a probability of one is also problematic: If $\Pr(E) = 1$, it always follows that $\Pr(H|E) = \Pr(H)$, meaning that the evidence provides no additional information. Therefore, it is generally best to avoid extreme probabilities in model building, a principle known as **Cromwell's rule** [7, Sect. 6.7].

## 3.2 Random variables

When rolling a die, it has a specific position and orientation in space at any given time during the experiment. However, the process of rolling a die is complex and the randomness comes from our lack of knowledge of the exact trajectory.

From this perspective, the sample space $\Omega$ consists of many more states than just the six sides of the die that can be rolled. It would be impractical and unnecessarily complicated to determine a probability measure for all possible trajectories and initial conditions. Ultimately, we are only interested in the final number. We can model this situation formally by introducing a map, called a *random variable*, $X$, which assigns one of the six possible outcomes to each history of the die roll: $X: \Omega \to \{1, 2, \ldots, 6\}$.

In inferential statistics, the distribution of values of the random variables studied is essential, and precise knowledge of the underlying sample space is not necessary.

### 3.2.1 Discrete and continuous random variables

A random variable is a quantity that takes on values in a given, limited range only with a certain probability. Formally, a random variable can be defined as follows.

> A **(real-valued) random variable** $X$ is a function on the sample space $X: \Omega \to \mathbb{R}$ such that for any value $u \in \mathbb{R}$ the set $X^{-1}(]-\infty, u])$ is measurable.

When a random variable assumes a particular value $u = X(\omega)$, that value is called a **realization** or an **observed value** of that random variable. The set of all possible realizations is given by the range or image, $\text{range}(X) = X(\Omega)$. Inferential statistics is based on the assumption that the observations in a sample are realizations of random variables.

The quantity $\Pr(X^{-1}(]-\infty, u]))$ is the probability that the random variable $X$ takes on a value less than or equal to $u$. Thus, we may write more succinctly:

$$\Pr(X \le u) := \Pr(X^{-1}(]-\infty, u]))$$

Similarly, $\Pr(X = u)$ stands for $\Pr(X^{-1}(\{u\}))$. We can introduce such abbreviations for other sets and types of intervals. For example, we can write

$\Pr(X \in [a, b])$ or $\Pr(a \le X \le b)$ for an interval $[a, b] \subset \mathbb{R}$ instead of the clunkier $\Pr(X^{-1}([a, b]))$.

> The **cumulative distribution function** of a random variable $X$ is defined as follows:
> $$F_X \colon \mathbb{R} \to [0, 1], \; F_X(u) = \Pr(X \le u)$$

The cumulative distribution function thus indicates the probability with which a random variable takes on values up to a given threshold. Furthermore, if the cumulative distribution function is known, it is also possible to calculate the probability with which the random variable takes on values in a given interval: $\Pr(a < X \le b) = F_X(b) - F_X(a)$ for all $a, b \in \mathbb{R}$ with $a \le b$.

It can be shown that cumulative distribution functions always satisfy the following properties.

> **Properties of the cumulative distribution function.** Let $X$ be a random variable and $F_X(\cdot)$ be its cumulative distribution function. Then, the following statements always hold:
>
> 1. $F_X$ is monotonically increasing.
> 2. $F_X$ is right-continuous, i.e.: $\lim_{\xi \searrow u} F_X(\xi) = u$ for all $u \in \mathbb{R}$.
> 3. We have $\lim_{\xi \to -\infty} F_X(\xi) = 0$ and $\lim_{\xi \to \infty} F_X(\xi) = 1$.

Statistical variables can be characterized as qualitative/categorical or quantitative/numeric. Similarly, random variables can be distinguished according to whether they take on a discrete number of values (e.g., from the set $\{0, 1, 2, 3, \dots\}$) or an entire continuum of values (e.g., from the interval $[0, 1]$).

> A random variable is called **discrete** if its range is given by a finite or a countably infinite set of values.

> A random variable is called **continuous** if its cumulative distribution function is continuous.

> A random variable is **absolutely continuous** if its cumulative distribution function is continuous and, moreover, continuously differentiable with the possible exception of at most finitely many points.

A few notes on the mathematical foundations: The conditions on absolute continuity mentioned above can be relaxed, but the characterization provided is usually sufficient for data analysis. A more in-depth mathematical analysis also shows that every "sufficiently regular" random variable can be represented as the sum of a discrete and an absolutely continuous random variable (**Lebesgue decomposition**, cf. [8, Proposition 4.5.1]). In that sense, with the exception of "pathological cases," the types of random variables mentioned above cover all possibilities that are relevant in practice.

In what follows, instead of saying "absolutely continuous," we will simply say "continuous," since we will only be dealing with this type of continuous random variable. For a discrete random variable $X$, we define its **support** as follows:

$$\text{supp}(X) = \{u \in \mathbb{R} | \Pr(X = u) > 0\}$$

The cumulative distribution function of a discrete random variable must be a piecewise constant function, with points of discontinuity given by the elements of the random variable's support.

We are now able to sketch the shape of typical cumulative distribution functions of discrete (on the left) or continuous random variables (on the right):

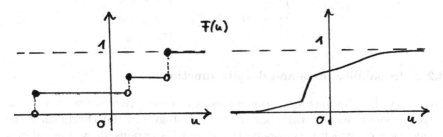

Wins and losses in games of chance are examples of quantities that can be modeled by random variables. Let's consider an example. Robert wins 5.00 of some currency (say, Euro) when rolling the number 6. Otherwise, he loses 1.00 EUR. The payoff is a random variable $X$ with

$$X(\omega) = \begin{cases} -1.00 & \text{if } \omega \in \{\boxdot, \boxdot, \boxdot, \boxdot, \boxtimes\} \\ 5.00 & \text{if } \omega = \boxplus \end{cases}$$

This random variable yields only the two possible values $t_1 = -1.00$ or $t_2 = 5.00$. Therefore, it is a discrete random variable. The associated cumulative distribution function is the following piecewise constant function:

$$F_X(u) = \begin{cases} 0 & \text{if } u < -1.00 \\ \frac{5}{6} & \text{if } -1.00 \le u < 5.00 \\ 1 & \text{if } 5.00 \le u \end{cases}$$

We return to the example of the dart player. As a random variable, we consider the distance of the thrown dart to the center of the dartboard: $X \colon \Omega \to \mathbb{R}$, $X(\omega_1, \omega_2) = \sqrt{(\omega_1)^2 + (\omega_2)^2}$. The cumulative distribution function of a random variable also depends on the underlying probability measure. For the perfect dart player (corresponding to the Dirac measure), $X$ is discrete with the following cumulative distribution function:

$$F_X^{(0)}(u) = \begin{cases} 0 & \text{if } u < 0 \\ 1 & \text{if } 0 \le u \end{cases}$$

If any two areas of equal size are hit with equal probability, the random variable is continuous and distributed as follows:

$$F_X^{(R)}(u) = \begin{cases} 0 & \text{if } u < 0 \\ \left(\frac{u}{R}\right)^2 & \text{if } 0 \le u < R \\ 1 & \text{if } R \le u \end{cases}$$

For the last probability measure given in the example (the bull's eye $u \le r$ is hit in 90% of all cases), we also have a continuous distribution:

$$F_X^{(r)}(u) = \begin{cases} 0 & \text{if } u < 0 \\ 0.9 \cdot \left(\frac{u}{r}\right)^2 & \text{if } 0 \le u < r \\ 0.9 + 0.1 \cdot \frac{u^2 - r^2}{R^2 - r^2} & \text{if } r \le u < R \\ 1 & \text{if } R \le u \end{cases}$$

### 3.2.2 Probability mass and density functions

Frequency distributions are a central focus of descriptive statistics and can be represented visually using bar charts or histograms. Bar charts show the frequency with which a category occurs, and histograms show how often a numeric variable takes on values in a certain interval. In probability theory, probability mass functions show the probability of a discrete random variable taking on certain values, while the probability of a continuous random variable taking on values within certain intervals is given by the area under its probability density function.

#### 3.2.2.1 Probability mass functions of discrete random variables

A discrete random variable is characterized by the probabilities with which it yields each value in its support.

> The **probability mass function** of a discrete random variable $X$ is given as follows:
> $$p_X\colon \operatorname{supp}(X) \to [0,1], \; p_X(u) = \Pr(X = u)$$

The cumulative distribution function can be reconstructed from the probability mass function as follows:

$$F_X(u) = \sum_{\substack{\kappa \le u, \\ \kappa \in \operatorname{supp}(X)}} p_X(\kappa)$$

Conversely, consider any map $p\colon T \to [0,1]$ with at most countably infinite domain $T = \{t_1, t_2 \ldots\} \subset \mathbb{R}$ that satisfies the condition:

$$\sum_{\kappa \in T} p(\kappa) = 1$$

We want to call any such function a probability mass function as well.

As an example, consider again the piecewise constant cumulative distribution function:

$$F(u) = \begin{cases} 0 & \text{if } u < -1.00 \\ \frac{5}{6} & \text{if } -1.00 \le u < 5.00 \\ 1 & \text{if } 5.00 \le u \end{cases}$$

The graph of this function is a step function that looks like this:

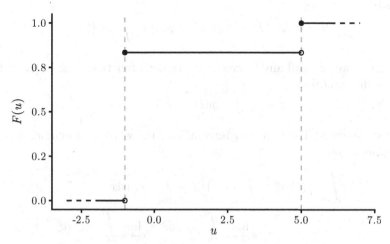

**Fig. 3.2.** *Cumulative distribution function of a discrete random variable*

The location of each step is determined by the points of discontinuity $t_1 = -1.00$ and $t_2 = 5.00$. The associated probability mass function simply reflects the height of each step:

$$p \colon \{-1.00, 5.00\} \to [0, 1], \; p(u) = \begin{cases} \frac{5}{6} & \text{if } u = -1.00 \\ \frac{1}{6} & \text{if } u = 5.00 \end{cases}$$

### 3.2.2.2 Probability density functions of continuous random variables

The cumulative distribution function of a discrete random variable can be represented as a sum of the probability mass function. For continuous random variables, a similar construction is possible using an integral instead of a sum.

A **probability density function** $p_X \colon \mathbb{R} \to [0, \infty[$ of a continuous random variable $X$ satisfies the following condition:

$$F_X(u) = \int_{-\infty}^{u} p_X(\xi) \, d\xi$$

Instead of referring to it as a probability density function, we can also use the shorter terms **probability density**, **density function**, or simply just "density" to refer to this concept. For continuous random variables $X$—that is, "continuous" in the less rigorous, or stronger, sense that we use the term—the cumulative distribution function $F_X$ is piecewise continuously differentiable. Hence, a density always exists. Furthermore, at each point $u \in \mathbb{R}$ where $F_X$ is differentiable, this density is uniquely determined by the derivative of the cumulative distribution function: $p_X(u) = \frac{d}{du} F_X(u)$.

Once we know its density function, we can quite conveniently calculate the probability with which a continuous random variable takes on values in a given interval:

$$\Pr(a \le X \le b) = F_X(b) - F_X(a) = \int_a^b p_X(\xi)\, d\xi$$

Conversely, we can call any piecewise continuous function $p\colon \mathbb{R} \to [0, \infty[$ that satisfies the condition

$$\int_{-\infty}^{\infty} p(\xi)\, d\xi = 1$$

a density function. The improper integral over the whole real number line $\mathbb{R}$ is defined as follows:

$$\int_{-\infty}^{\infty} p(\xi)\, d\xi = \int_{-\infty}^{0} p(\xi)\, d\xi + \int_0^{\infty} p(\xi)\, dx$$

$$= \lim_{u \to -\infty} \int_u^0 p(\xi)\, d\xi + \lim_{u \to \infty} \int_0^u p(\xi)\, d\xi$$

Consequently, both partial integrals must exist. Unlike a probability mass function, it is quite possible for a density function to have $p(u) > 1$ at some point. A typical probability mass function (on the left) and density function can be sketched as follows:

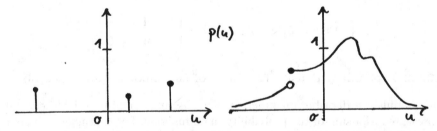

If a continuous/discrete random variable $X$ is distributed according to a probability density/mass function $p(\cdot)$, we will write $X \sim p(\cdot)$.

An important example of a probability density function is the following.

The **standard normal distribution** is given by this density function:

$$p\colon \mathbb{R} \to [0, \infty[,\ p(u) = \frac{1}{\sqrt{2\pi}} e^{-\frac{1}{2}u^2}$$

The following figure shows the graph of that function and the corresponding cumulative distribution function $F(\,\cdot\,)$:

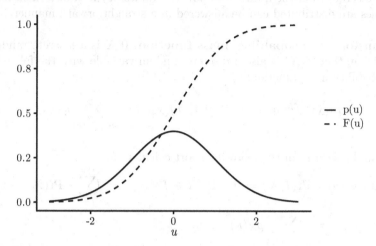

**Fig. 3.3.** *Standard normal probability density and cumulative distribution function*

For this function to actually represent a density, its integral must be equal to one. This fact is implied by the value of the **Gaussian integral**—see, for example, [9], or Sect. B.3.4 in the appendix of this book:

$$\int_{-\infty}^{\infty} e^{-\xi^2}\, d\xi = \sqrt{\pi}$$

The probability that a standard normally distributed random variable $X$ yields a value, for example, between $-1.96$ and $+1.96$ is given by the following integral:

$$\Pr(-1.96 \le X \le 1.96) = \int_{-1.96}^{1.96} \frac{1}{\sqrt{2\pi}} e^{-\frac{1}{2}\xi^2}\, d\xi$$

In contrast to the integral over the entire number line, this value cannot be calculated analytically and given in a closed form. However, the value can be approximated by means of numerical integration: $\Pr(-1.96 \le X \le 1.96) \approx 0.95$.

### 3.2.3 Transformations of random variables

We want to investigate functions of a random variable $X \colon \Omega \to \mathbb{R}$. More precisely, suppose we have a sufficiently regular (e.g., piecewise continuous) function defined within the range of $X$, i.e.: $f \colon I \to \mathbb{R}$ with $\mathrm{range}(X) \subseteq I$. Then, we can consider the **transformed random variable** $f(X)$:

$$f(X) \colon \Omega \to \mathbb{R},\ f(X)(\omega) := (f \circ X)(\omega)$$

The values of the transformed random variable $Y = f(X)$ follow a different distribution than the values of the original variable:

$$F_Y(u) = \Pr(Y \le u) = \Pr(f(X) \le u) = \Pr(X \in f^{-1}(]-\infty, u]))$$

### 3.2.3.1 Transformations of discrete random variables

In the discrete case, the question of how the values of the transformed random variables are distributed can be answered in a straightforward manner.

> **Transformed probability mass function.** If $X$ is a discrete random variable, then $f(X)$ is also a discrete random variable with the following probability mass function:
>
> $$p_{f(X)}\colon f(\operatorname{supp}(X)) \to [0,1], \; p_{f(X)}(v) = \sum_{\kappa \in f^{-1}(v)} p_X(\kappa)$$

This can be seen from the following short calculation:

$$p_{f(X)}(v) = \Pr(f(X) = v) = \Pr(X \in f^{-1}(v)) = \sum_{\kappa \in f^{-1}(v)} \Pr(X = \kappa)$$

$$= \sum_{\kappa \in f^{-1}(v)} p_X(\kappa)$$

for all $v \in \operatorname{supp}(f(X)) = f(\operatorname{supp}(X))$.

We illustrate with an example. Let $X$ be a random variable that is distributed as follows:

$$\Pr(X = u) = \begin{cases} 0.1 & \text{if } u = -3 \\ 0.2 & \text{if } u = 0 \\ 0.7 & \text{if } u = 3 \\ 0 & \text{otherwise} \end{cases}$$

Now let $f\colon \mathbb{R} \to \mathbb{R}$, $f(u) = u^2$ be the standard quadratic function. The values of the random variable $f(X) = X^2$ are then distributed as follows:

$$\Pr(X^2 = v) = \begin{cases} \Pr(X = 0) & \text{if } v = 0 \\ \Pr(X = -3) + \Pr(X = 3) & \text{if } v = 9 \\ 0 & \text{otherwise} \end{cases}$$

$$= \begin{cases} 0.2 & \text{if } v = 0 \\ 0.8 & \text{if } v = 9 \\ 0 & \text{otherwise} \end{cases}$$

### 3.2.3.2 Transformations of continuous random variables

We will now discuss the calculation of the density function for transformations of a continuous random variable. It is worth noting that the transformed random variable does not necessarily have to be continuous again. For example, the zero function, $f(u) = 0$ for all $u$, results in the discrete random variable $f(X) = 0$.

However, if $f\colon \mathbb{R} \to \mathbb{R}$ represents a continuously differentiable, monotonically increasing function with positive derivative, then:

$$\Pr(f(X) \le u) = \Pr(X \le f^{-1}(u))$$
$$= \int_{-\infty}^{f^{-1}(u)} p_X(\xi) \, d\xi$$

for all $u \in \mathbb{R}$. We continue the calculation with the substitution $\xi(t) = f^{-1}(t)$:

$$\Pr(f(X) \le u) = \int_{-\infty}^{u} p_X(f^{-1}(t)) \cdot \frac{\mathrm{d}}{\mathrm{d}t}(f^{-1}(t)) \, \mathrm{d}t = \int_{-\infty}^{u} \frac{p_X(f^{-1}(t))}{f'(f^{-1}(t))} \, \mathrm{d}t$$

If $f$ is monotonically *decreasing*, then we have the same calculation with the signs reversed. Thus, if the transformation is monotonically increasing or decreasing, we have the following handy formula(s) for the density function of $f(X)$:

$$p_{f(X)} \colon \mathbb{R} \to [0, \infty[\, , \ p_{f(X)}(v) = p_X(f^{-1}(v)) \cdot \left| \frac{\mathrm{d}}{\mathrm{d}v} f^{-1}(v) \right| = \frac{p_X(f^{-1}(v))}{|f'(f^{-1}(v))|}$$

Linear functions provide simple but important examples for such transformations. According to the above formula, the density function of $Y = m \cdot X + c$ with constants $m \in \mathbb{R}, c \in \mathbb{R}, m \neq 0$ is given by:

$$p_Y(v) = \frac{1}{|m|} \cdot p_X \left( \frac{v - c}{m} \right)$$

If $f$ is not monotonic, the transformation formula needs to be applied to intervals where it is. These considerations lead to the following formula.

---

**Transformed probability density function.** Let $X$ be a continuous random variable and $f \colon I \to \mathbb{R}$ with range$(X) \subseteq I$ be a continuously differentiable function with nonvanishing derivative.

The probability density function of the transformed random variable $f(X)$ is then given as follows:

$$p_{f(X)} \colon \mathbb{R} \to [0, \infty[\, , \ p_{f(X)}(v) = \sum_{k=1}^{K} |g_k'(v)| \cdot p_X(g_k(v))$$

where $g_1(\cdot), \ldots, g_K(\cdot)$ are all the solutions of the equation $f(g(v)) = v$.

---

As an example, we consider once more the standard quadratic function $f(u) = u^2$. First, we have to exclude the point $u = 0$ because $f'(0) = 0$. However, since this point is merely an isolated point, we can simply specify a suitable value there later. We want to transform a random variable $X$ that is normally distributed:

$$p_X \colon \mathbb{R} \to [0, \infty[\, , \ p_X(u) = \frac{1}{\sqrt{2\pi}} \cdot e^{-\frac{1}{2}u^2}$$

The equation $f(g(v)) = v \Leftrightarrow (g(v))^2 = v$ has two solutions if $v > 0$, namely $g_1(v) = -\sqrt{v}$ and $g_2(v) = \sqrt{v}$. Consequently, the density function of $X^2$ is given as follows:

$$p_{X^2}(v) = |g_1'(v)| \cdot p_X(g_1(v)) + |g_2'(v)| \cdot p_X(g_2(v))$$

$$= \frac{1}{2\sqrt{v}} \cdot (p_X(-\sqrt{v}) + p_X(\sqrt{v})) = \frac{1}{\sqrt{v}} \cdot p_X(\sqrt{v}) =$$

$$= \frac{1}{\sqrt{2\pi v}} \cdot e^{-\frac{1}{2}v}$$

There is no real solution to $(g(v))^2 = v$ for $v < 0$, in which case we get an empty sum that is simply zero. Putting it all together yields:

$$p_{X^2} \colon \mathbb{R} \to [0, \infty[, \ p_{X^2}(v) = \begin{cases} 0 & \text{if } v \leq 0 \\ \frac{1}{\sqrt{2\pi v}} \cdot e^{-\frac{1}{2}v} & \text{if } v > 0 \end{cases}$$

This distribution is called the **chi-squared distribution with one degree of freedom**. We explain in Sect. 3.5.1 what a chi-squared distribution with an arbitrary number of degrees of freedom looks like.

## 3.3 Joint distribution of random variables

Random variables can often have functional dependencies. One important task in inferential statistics is modeling and describing these relationships based on empirical data. This is the subject of regression analysis, which will be discussed further in Sect. 4.5.

### 3.3.1 Joint probability mass and density functions

For two random variables $X, Y$ (and mutatis mutandis for more than two random variables or other types of intervals), we write:

$$\Pr(X \leq u, Y \leq v) := \Pr(X^{-1}\left(]-\infty, u]\right) \cap Y^{-1}\left(]-\infty, v]\right))$$

This is the probability with which the random variables do not exceed certain threshold values at the same time.

Given the random variables $X_1, \ldots, X_D$, their **joint cumulative distribution function** is the following:

$$F_{X_1, \ldots, X_D} \colon \mathbb{R}^D \to \mathbb{R}, \ F_{X_1, \ldots, X_D}(u_1, \ldots, u_D) = \Pr(X_1 \leq u_1, \ldots, X_D \leq u_D)$$

If $X_1, \ldots, X_D$ are all discrete random variables, their **joint probability mass function** is given by:

$$p_{X_1, \ldots, X_D}(u_1, \ldots, u_D) = \Pr(X_1 = u_1, \ldots, X_D = u_D)$$

for all $u_1 \in \text{supp}(X_1), \ldots, u_D \in \text{supp}(X_D)$.

If $X_1, \ldots, X_D$ are all continuous random variables, the multiple integral over a **joint probability density function** yields the joint cumulative distribution function:

$$F_{X_1,\ldots,X_D}(u_1,\ldots,u_D) = \int_{-\infty}^{u_1} \cdots \int_{-\infty}^{u_D} p_{X_1,\ldots,X_D}(\xi_1,\ldots,\xi_D)\,\mathrm{d}\xi_1 \cdots \mathrm{d}\xi_D$$

for all $u_1,\ldots,u_D \in \mathbb{R}$. This fact implies that at every point where those partial derivatives exist:

$$p_X(u_1,\ldots,u_D) = \frac{\partial^D F_X}{\partial u_1 \cdots \partial u_D}(u_1,\ldots,u_D)$$

For discrete random variables, the individual probability mass functions can be reconstructed from the joint probability mass function by summing over the remaining variables. For example, for two variables $X, Y$:

$$p_X(u) = \sum_{\kappa \in \mathrm{supp}(Y)} p_{X,Y}(u,\kappa)$$

$$p_Y(v) = \sum_{\kappa \in \mathrm{supp}(X)} p_{X,Y}(\kappa,v)$$

Similarly, the densities of continuous random variables can be "integrated out" from their joint density function, yielding a **marginal probability density** for each:

$$p_X(u) = \int_{-\infty}^{\infty} p_{X,Y}(u,\xi)\,\mathrm{d}\xi$$

$$p_Y(v) = \int_{-\infty}^{\infty} p_{X,Y}(\xi,v)\,\mathrm{d}\xi$$

Finally, we can also consider the joint distribution of a discrete and a continuous random variable.

The **mixed joint probability density function** for a continuous random variable $X$ and a discrete random variable $Y$ is a function $p_{X,Y}: \mathbb{R} \times \mathrm{supp}(Y) \to [0,\infty[$ with the following property:

$$F_{X,Y}(u,v) = \Pr(X \le u, Y \le v) = \sum_{\substack{\kappa \le v, \\ \kappa \in \mathrm{supp}(Y)}} \int_{-\infty}^{u} p_{X,Y}(\xi,\kappa)\,\mathrm{d}\xi$$

for all $u \in \mathbb{R}$, $v \in \mathrm{supp}(Y)$.

In this case, the marginal distributions can be determined by summation or integration with respect to the other variable:

$$p_X(u) = \sum_{\kappa \in \mathrm{supp}(Y)} p_{X,Y}(u, \kappa)$$

$$p_Y(v) = \int_{-\infty}^{\infty} p_{X,Y}(\xi, v) \, \mathrm{d}\xi$$

This concept can be generalized, mutatis mutandis, to any (finite) number of discrete and continuous random variables.

### 3.3.2 Conditional probability mass and density functions

The concept of conditional probability can also be applied to probability mass and density functions. A conditional probability mass or density function describes the dependence of a random variable $Y$ on another random variable $X$. Estimating joint conditional distributions from data is a goal of regression analysis, as discussed in Section 4.5.

For two discrete random variables $X, Y$ and a realization $u \in \mathrm{supp}(X)$ of $X$, the **probability mass function of $Y$ under the condition $X = u$** is given as follows:

$$p_{Y|X} \colon \mathrm{supp}(Y) \to [0, 1], \; v \mapsto p_{Y|X}(v|u) = \frac{p_{X,Y}(u, v)}{p_X(u)}$$

The conditional probability mass function simply describes the probability of $Y$ taking on the values in its support under the condition that $X$ obtains a certain value $u$:

$$p_{Y|X}(v|u) = \frac{p_{X,Y}(u, v)}{p_X(u)} = \frac{\Pr(X = u, Y = v)}{\Pr(X = u)} = \Pr(Y = v | X = u)$$

For continuous random variables, we have the difficulty that $\Pr(X = u) = 0$ always holds, so the conditional probability is not defined. Nevertheless, we can apply a similar rationale.

For two continuous random variables $X, Y$ and a realization $u \in \mathbb{R}$ of $X$ with $p_X(u) > 0$, the **probability density function of $Y$ under the condition $X = u$** is given as follows:

$$p_{Y|X} \colon \mathbb{R} \to [0, \infty[, \; v \mapsto p_{Y|X}(v|u) = \frac{p_{X,Y}(u, v)}{p_X(u)}$$

The conditional density function indeed satisfies the defining properties of a density function: $p_{Y|X}(v|u) \geq 0$ holds for all $v \in \mathbb{R}$, and also:

$$\int_{-\infty}^{\infty} p_{Y|X}(\xi|u) \, \mathrm{d}\xi = \frac{1}{p_X(u)} \cdot \int_{-\infty}^{\infty} p_{X,Y}(u, \xi) \, \mathrm{d}\xi = \frac{1}{p_X(u)} \cdot p_X(u) = 1$$

Similarly, we can construct conditional probability mass and density functions for other combinations of discrete and continuous random variables. If $Y$ is a discrete random variable and $X$ a continuous random variable, the probability mass function of $Y$ under the condition $X = u$ is the following:

$$p_{Y|X}(\cdot \,|u)\colon \operatorname{supp}(Y) \to [0, \infty[, \; p_{Y|X}(v|u) = \frac{p_{X,Y}(u,v)}{p_X(u)}$$

As before, we cannot directly interpret this conditional mass function as a conditional probability. We will nevertheless notate it as such. That is, $\Pr(Y = v|X = u) := p_{Y|X}(v|u)$ because in the limit of small intervals $X \in [u, u + h]$ we have:

$$
\begin{aligned}
\lim_{h \searrow 0} \Pr(Y = v| u \le X \le u + h) &= \lim_{h \searrow 0} \frac{\Pr(Y = v, u \le X \le u + h)}{\Pr(u \le X \le u + h)} \\
&= \lim_{h \searrow 0} \frac{F_{X,Y}(u + h, v) - F_{X,Y}(u, v)}{F_X(u + h) - F_X(u)} \\
&= \lim_{h \searrow 0} \frac{h^{-1} \cdot (F_{X,Y}(u + h, v) - F_{X,Y}(u, v))}{h^{-1} \cdot (F_X(u + h) - F_X(u))} \\
&= \frac{p_{X,Y}(u, v)}{p_X(u)}
\end{aligned}
$$

whenever $p_X(\cdot)$ and $p_{X,Y}(\cdot, v)$ are continuous at $u$.

### 3.3.3 Independent random variables

We consider two random variables $X$ and $Y$ to be independent if for all intervals $[a, b]$ and $[c, d]$ the events $X^{-1}([a, b])$ and $Y^{-1}([c, d])$ are independent:

$$\Pr(a \le X \le b, c \le Y \le d) = \Pr(a \le X \le b) \cdot \Pr(c \le Y \le d)$$

Any "reasonable" sets $A, B$ can be written as a countable union of intervals. Accordingly, for independent random variables, we may assume more generally:

$$\Pr(X \in A, Y \in B) = \Pr(X \in A) \cdot \Pr(Y \in B)$$

Since the distribution of values of a random variable is determined by its cumulative distribution function, the condition of independence can be characterized as follows.

The random variables $X_1, \ldots, X_D$ are called **(mutually) independent** if for all $u_1, \ldots, u_D \in \mathbb{R}$:

$$F_{X_1, \ldots, X_D}(u_1, \ldots, u_D) = F_{X_1}(u_1) \cdot F_{X_2}(u_2) \cdots F_{X_D}(u_D) = \prod_{d=1}^{D} F_{X_d}(u_d)$$

Independence of random variables can also be inferred from their probability mass and density functions:

**Independence of discrete or continuous random variables.** The discrete random variables $X_1, \ldots, X_D$ are independent if and only if

$$p_{X_1, \ldots, X_D}(u_1, \ldots, u_D) = \prod_{d=1}^{D} p_{X_d}(u_d)$$

holds for all $u_1 \in \operatorname{supp}(X_1), \ldots, u_D \in \operatorname{supp}(X_D)$.

The continuous random variables $X_1, \ldots, X_D$ are mutually independent if and only if

$$p_{X_1, \ldots, X_D}(u_1, \ldots, u_D) = \prod_{d=1}^{D} p_{X_d}(u_d)$$

holds for all points of continuity $u_1, \ldots, u_D \in \mathbb{R}$.

Another proposition is the following:

**Transformations of independent random variables.** If two random variables $X$ and $Y$ are independent, then the transformed variables $f(X)$ and $g(Y)$ are also independent.

This result follows from the following calculation:

$$
\begin{aligned}
F_{f(X),f(Y)}(u, v) &= \Pr(X \in f^{-1}(]-\infty, u]), Y \in g^{-1}(]-\infty, v])) \\
&= \Pr(X \in f^{-1}(]-\infty, u])) \cdot \Pr(Y \in g^{-1}(]-\infty, v])) \\
&= \Pr(f(X) \in ]-\infty, u]) \cdot \Pr(g(Y) \in ]-\infty, v]) \\
&= F_{f(X)}(u) \cdot F_{g(Y)}(v)
\end{aligned}
$$

## 3.4 Characteristic measures of random variables

Discrete random variables correspond to categorical variables in descriptive statistics, while continuous random variables correspond to numeric variables. If we draw a sufficiently large sample from a distribution of a discrete random variable, the resulting bar chart should approximate the underlying mass function. Similarly, if we draw a sample from the distribution of a continuous random variable, the histogram of the observations should approximate the density function.

In this way, probability mass and density functions are analogous to frequency distributions in descriptive statistics. Probability theory also has counterparts to statistical measures of central tendency, variation, and association.

### 3.4.1 Median, expected value, and variance

In this section, we define measures of central tendency and variation for random variables.

Let $X$ be a random variable with cumulative distribution function $F_X \colon \mathbb{R} \to [0,1]$. The **quantile function** of $X$ is given as follows:

$$Q[X] \colon \mathbb{R} \to [0,1], \; Q[X](r) = \inf\{u \in \mathbb{R} | F_X(u) \geq r\}$$

Let $0 < \alpha < 1$. An $\boldsymbol{\alpha}$**-quantile** of $X$ is any number $q_\alpha$ with

$$\Pr(X \leq q_\alpha) \geq \alpha \text{ and } \Pr(X \geq q_\alpha) \geq 1 - \alpha$$

In particular, $Q[X](\alpha)$ is an $\alpha$-quantile. A **median** of $X$ is an $\alpha$-quantile with $\alpha = 1/2$.

Similar to descriptive statistics, these definitions imply the subtle technicality that a random variable can have multiple medians or $\alpha$-quantiles, even for a fixed value of $\alpha$. However, we can enforce uniqueness by casually speaking of *the* quantile $q_\alpha[X] := Q[X](\alpha)$ and *the* median $m[X] := Q[X](1/2)$.

For a continuous random variable, the $\alpha$-quantile can be expressed most conveniently via its probability density function $p_X(\cdot)$:

$$\int_{-\infty}^{q_\alpha[X]} p_X(\xi)\,\mathrm{d}\xi = \alpha$$

Especially for the median:

$$\int_{-\infty}^{m[X]} p_X(\xi)\,\mathrm{d}\xi = \frac{1}{2}$$

Let $X$ be a discrete or continuous random variable with probability mass function $p_X \colon \mathrm{supp}(X) \to [0,1]$ or density function $p_X \colon \mathbb{R} \to [0,\infty[$, respectively.

The **expected value**, or **expectation**, $E[X]$ is defined as follows:

$$E[X] = \sum_{\kappa \in \mathrm{supp}(X)} \kappa \cdot p_X(\kappa)$$

or

$$E[X] = \int_{-\infty}^{\infty} \xi \cdot p_X(\xi)\,\mathrm{d}\xi$$

The expected value and the median of random variables are closely related to the arithmetic mean and the sample median of descriptive statistics: they represent a "typical" observed value. Consequently, the expected value can also be called the random variable's **mean** or **average**.

. Let us compute an example for the expected value of a random variable. For a game of dice with winnings of $t_1 = 5\,\mathrm{EUR}$ when a six is rolled, and $t_2 = 1\,\mathrm{EUR}$ loss otherwise, the expected winnings are:

$$E[X] = t_1 \cdot \Pr(X = t_1) + t_2 \cdot \Pr(X = t_2) = (-1) \cdot \frac{5}{6} + 5 \cdot \frac{1}{6} = 0$$

On average, the player would neither lose money nor win money in the long run. However, it should be noted that if he plays only once, his risk of losing is much greater than his probability of winning: 5/6 versus 1/6, i.e., five times higher. This circumstance is compensated by the much higher payout when winning.

Unlike the arithmetic mean, the expected value need not always exist or be finite. For example, the **St. Petersburg paradox** leads to a random variable for which the expected value is not finite. In this scenario, game master Anna flips a coin multiple times until heads appears. If heads appears after $k$ coin tosses, Robert receives $2^{k-1}$ euros. For example, if heads appears on the first toss, he receives 1.00 EUR, while if heads appears on the second toss he receives 2.00 EUR. The potential payout doubles with each toss of tails. If heads and tails are equally probable and independent events with each flip, the winnings can be modeled by a discrete random variable $X$ with the following probability mass function:

$$p_X\left(2^{k-1}\right) = \left(\frac{1}{2}\right)^k, \ k \in \{1, 2, \dots\}$$

The expected profit from participating in the game is therefore:

$$E[X] = \sum_{k=1}^{\infty} 2^{k-1} \cdot \left(\frac{1}{2}\right)^k = \sum_{k=1}^{\infty} \frac{1}{2} = \infty$$

The paradox lies in the observation that if Robert were to base his decision-making solely on his expected profits then he would accept any entry fee for participating in the game, no matter how high.

An example of a continuous random variable with no well-defined expected value is given by the probability density function

$$p(u) = \frac{1}{\pi} \cdot \frac{1}{u^2 + 1}$$

since the improper integral

$$\int_{-\infty}^{\infty} \frac{1}{\pi} \cdot \frac{\xi}{\xi^2 + 1} \, d\xi$$

cannot be assigned a finite value.

After transformation, the values of a random variable usually follow a different distribution. There is a simple rule for calculating the expected value of the transformed variable.

**Expectation of transformed random variables.** Let $X$ be a discrete or continuous random variable, and $f: I \to \mathbb{R}$ with range$(X) \subseteq I$ a (sufficiently regular, nonconstant) function.

The expected value of the transformed random variable $f(X)$ can be calculated as follows:

$$E[f(X)] = \sum_{\kappa \in \text{supp}(X)} f(\kappa) \cdot p_X(\kappa) \text{ if } X \text{ is discrete, or}$$

$$E[f(X)] = \int_{-\infty}^{\infty} f(\xi) \cdot p_X(\xi) \, \mathrm{d}\xi \text{ if } X \text{ is continuous.}$$

The summand/integrand can be assumed to vanish where $f(\cdot)$ is undefined (the probability mass/density function vanishes at those points so they do not contribute).

Sometimes, these formulas are used to *define* the expected value. However, they should rather be viewed as theorems that follow from the transformation formulas for probability mass and density functions. We show this by the example of a continuous random variable $X$ and a monotonically increasing transformation $f: \mathbb{R} \to \mathbb{R}$ with positive derivative:

$$E[f(X)] = \int_{-\infty}^{\infty} \xi \cdot p_{f(X)}(\xi) \, \mathrm{d}\xi = \int_{-\infty}^{\infty} \xi \cdot p_X(f^{-1}(\xi)) \cdot \left( \frac{\mathrm{d}}{\mathrm{d}\xi} f^{-1}(\xi) \right) \, \mathrm{d}\xi$$

$$= \int_{-\infty}^{\infty} f(t) \cdot p_X(t) \, \mathrm{d}t$$

In the above calculation, the substitution $\xi(t) = f(t)$ has been applied.

The formulas for calculating expected values also apply to functions of random variables over the joint distribution:

**Expectation of jointly transformed random variables.** Let $X$ and $Y$ be two discrete or two continuous random variables. Let $f: I \times J \to \mathbb{R}$ be a (sufficiently regular, nonconstant) function with $\text{range}(X) \subseteq I$, $\text{range}(Y) \subseteq J$.

The expected value of the random variable $f(X, Y)$ can be calculated as follows:

$$E[f(X, Y)] = \sum_{\kappa_1 \in \text{supp}(X)} \sum_{\kappa_2 \in \text{supp}(Y)} f(\kappa_1, \kappa_2) \cdot p_{X,Y}(\kappa_1, \kappa_2)$$

if $X$ and $Y$ are discrete, or

$$E[f(X, Y)] = \int_{-\infty}^{\infty} \int_{-\infty}^{\infty} f(\xi_1, \xi_2) \cdot p_{X,Y}(\xi_1, \xi_2) \, \mathrm{d}\xi_1 \mathrm{d}\xi_2$$

if $X$ and $Y$ are continuous.

The following important properties of expected values can be derived from the formulas which have been presented so far.

**Linearity and monotonicity of the expected value.** The expected value is a linear map:

$$E[a \cdot X + b \cdot Y + c] = a \cdot E[X] + b \cdot E[Y] + c$$

for all random variables $X, Y$ and numbers $a, b, c \in \mathbb{R}$.

In addition, the following monotonicity property holds. If $Y$ is almost surely at least as large as $X$, i.e., if $\Pr(Y - X \geq 0) = 1$ holds, then the analogous inequality holds for the expected values:

$$E[Y] \geq E[X]$$

The condition $\Pr(Y - X \geq 0) = 1$ is satisfied, for example, if $Y$ is always at least as large as $X$, i.e., $Y(\omega) \geq X(\omega)$ holds for all $\omega \in \Omega$. In particular, the expected value of a nonnegative random variable is also nonnegative: $Y \geq 0 \Rightarrow E[Y] \geq 0$.

The following measures are analogues to the corresponding empirical measures of variation.

The **variance** of a random variable $X$ is given by the expected squared deviation from the mean:

$$\sigma^2[X] = E[(X - E[X])^2]$$

The **standard deviation** of $X$ is given by $\sigma[X] = \sqrt{\sigma^2[X]}$.

An alternative notation is $\mathrm{var}[X] = \sigma^2[X]$. Unlike the expected value, the variance is not a linear function, but the following rule applies. For all constants $m, c \in \mathbb{R}$:

$$\sigma^2[m \cdot X + c] = m^2 \cdot \sigma^2[X]$$

We determine the expectation and variance of a standard normally distributed random variable $X$ as an example. The expected value is zero since it is an integral over an odd function with $f(-\xi) = -f(\xi)$:

$$E[X] = \int_{-\infty}^{\infty} \xi \cdot p_X(\xi)\, d\xi = \int_{-\infty}^{\infty} \frac{\xi}{\sqrt{2\pi}} \cdot e^{-\frac{\xi^2}{2}}\, d\xi = 0$$

We use integration by parts to compute the variance:

$$\mathrm{var}[X] = E[(X - E[X])^2] = E[X^2]$$

$$= \int_{-\infty}^{\infty} \xi^2 \cdot p_X(\xi)\, d\xi = \frac{1}{\sqrt{2\pi}} \int_{-\infty}^{\infty} \xi \cdot \left( \xi \cdot e^{-\frac{\xi^2}{2}} \right) d\xi$$

$$= -\frac{\xi}{\sqrt{2\pi}} \cdot e^{-\frac{\xi^2}{2}} \bigg|_{\xi=-\infty}^{\infty} + \frac{1}{\sqrt{2\pi}} \int_{-\infty}^{\infty} e^{-\frac{\xi^2}{2}}\, d\xi$$

$$= 1$$

Furthermore, we want to consider linear functions of the standard normally distributed random variable: $Y = \sigma \cdot X + \mu$ where $\mu, \sigma \in \mathbb{R}$, $\sigma > 0$. According to the transformation formulas for the probability density function, for all $u \in \mathbb{R}$:

$$p_Y(u) = \frac{1}{\sqrt{2\pi}\sigma} e^{-\frac{(u-\mu)^2}{2\sigma^2}}$$

To determine the expected value and variance of $Y$, we can use the behavior of the expected value and variance under transformations:

$$E[Y] = E[\sigma \cdot X + \mu] = \sigma \cdot E[X] + \mu = \mu,$$
$$\text{var}[Y] = \text{var}[\sigma \cdot X + \mu] = \sigma^2 \cdot \text{var}[X] = \sigma^2$$

We call a random variable distributed like $Y$ **normally distributed with mean $\mu$ and variance $\sigma^2$**. For the probability density function, we use the following notation:

$$Y \sim p_Y(\cdot) = \mathcal{N}(\cdot \mid \mu, \sigma^2)$$

For $\mu = 0$ and different values of $\sigma$, the normal distribution is shown in Fig. 4.2 on the top left. With smaller values for $\sigma$, the function graph, also called a Gaussian **bell curve**, becomes narrower: values sampled from a normally distributed random variable with smaller variance will (usually) show a smaller dispersion. From this perspective, the standard normal distribution is the normal distribution with zero mean and unit variance: $X \sim \mathcal{N}(\cdot \mid 0, 1)$.

The following general inequalities can be useful to produce bounds on the expectation or variance of a random variable if the proper value may be difficult to compute. Despite the expectation being a linear operation, in general, transformations and taking the mean do not commute: $E[f(X)] \neq f(E[X])$. However, the following is true.

**Jensen's inequality.** Let $X$ be a random variable and $f \colon I \to \mathbb{R}$ a convex function where $I \subseteq \mathbb{R}$ is an interval with $\text{range}(X) \subseteq I$. Given these conditions, the following inequality holds:

$$E[f(X)] \geq f(E[X])$$

For concave transformations, the opposite inequality holds. Examples: $E[X^2] \geq (E[X])^2$ or $E[\sqrt{X}] \leq \sqrt{E[X]}$.

For a continuous random variable $X$ and a differentiable convex function $f \colon I \to \mathbb{R}$, we may sketch a proof. For convenience, we write $\mu = E[X]$. A differentiable function is convex if and only if the graph of the function lies above any tangent. In particular, for all $\xi \in \mathbb{R}$:

$$f(\xi) \geq f(\mu) + f'(\mu) \cdot (\xi - \mu)$$

We can multiply this inequality by the probability density of $X$ because it is nonnegative, and we can integrate on both sides because of the monotonicity of the integral:

$$\int_{-\infty}^{\infty} f(\xi) \cdot p_X(\xi) \, d\xi \geq f(\mu) \cdot \underbrace{\int_{-\infty}^{\infty} p_X(\xi) \, d\xi}_{=1} + f'(\mu) \cdot \underbrace{\int_{-\infty}^{\infty} (\xi - \mu) \cdot p_X(\xi) \, d\xi}_{=0}$$

which simplifies to $E[f(X)] \geq f(\mu)$, as desired.

We can also show that bounded random variables have bounded variance.

> **Popoviciu's inequality.** Suppose that $X$ is a random variable such that $\Pr(a \leq X \leq b) = 1$. Then, the following inequality holds:
>
> $$\sigma^2[X] \leq \frac{1}{4} \cdot (b - a)^2$$

To show the result, we examine the rescaled variable:

$$Y = \frac{X - a}{b - a}$$

For convenience, we write $\mu = E[Y]$. We note that $Y$ (almost surely) takes values within the unit interval, $[0, 1]$. Thus, the quantity $E[(1 - Y) \cdot Y] = E[Y] - E[Y^2]$ is nonnegative, which means that $E[Y^2] \leq E[Y] = \mu$. We can use this fact to put a bound on the variance:

$$\sigma^2[Y] = E[(Y - \mu)^2] = E[Y^2] - 2\mu \cdot E[Y] + E[Y]^2$$
$$= E[Y^2] - \mu^2 \leq \mu - \mu^2 = \mu \cdot (1 - \mu) \leq \frac{1}{4}$$

The last inequality follows from the fact that $E[Y] = \mu$ is also bounded by zero and one. The general inequality follows from scaling $Y$ back to $X$.

Finally, we can also define the following measures that correspond to the Shannon index.

> The **Shannon entropy** of a discrete random variable $X$ is given by:
>
> $$H[X] = - \sum_{\kappa \in \mathrm{supp}(X)} p_X(\kappa) \cdot \ln(p_X(\kappa))$$

> The **differential entropy** of a continuous random variable $X$:
>
> $$H[X] = - \int_{-\infty}^{\infty} p_X(\xi) \cdot \ln(p_X(\xi)) \, d\xi$$

As usual, we agree on $0 \cdot \ln 0 = 0$ to hold.

## 3.4.2 Covariance and correlation

The following measures correspond to the measures of association of the same name in descriptive statistics.

The **covariance** of two random variables $X, Y$ is given as follows:

$$\sigma[X,Y] = E\left[(X - E[X]) \cdot (Y - E[Y])\right]$$

The **correlation** between the two variables:

$$\rho[X,Y] = \frac{\sigma[X,Y]}{\sigma[X] \cdot \sigma[Y]},$$

if $\sigma[X] > 0$ and $\sigma[Y] > 0$.

The random variables $X$ and $Y$ are called **uncorrelated** if $\sigma[X,Y] = 0$ holds.

An alternative notation is $\text{cov}[X,Y] = \sigma[X,Y]$. The covariance of a random variable with itself is its variance: $\sigma[X,X] = \sigma^2[X]$.

An alternative calculation rule that is often used can be derived from the linearity of the expected value:

$$\begin{aligned}
\sigma[X,Y] &= E[X \cdot Y] - E[E[X] \cdot Y] - E[E[Y] \cdot X] + E[X] \cdot E[Y] \\
&= E[X \cdot Y] - E[X] \cdot E[Y]
\end{aligned}$$

In particular, the following applies to the variance:

$$\sigma^2[X] = E[X^2] - (E[X])^2$$

As an example, we consider two continuous random variables $X$ and $Y$ with the following joint density function:

$$p_{X,Y}(u,v) = \frac{1}{2\pi\sqrt{1-\rho^2}} \exp\left(-\frac{u^2 - 2\rho \cdot u \cdot v + v^2}{2(1-\rho^2)}\right)$$

where $\rho \in \mathbb{R}$, $0 \le \rho < 1$ is an additional parameter. This is a member of the family of multivariate normal distributions, which we will discuss in Sect. 5.4.2. We want to calculate the covariance of $X$ and $Y$, $\sigma[X,Y] = E[X \cdot Y] - E[X] \cdot E[Y]$. In order to do so, we will use the formula for computing the general Gaussian integral:

$$\int_{-\infty}^{\infty} e^{-a\xi^2 + b\xi + c}\, d\xi = \sqrt{\frac{\pi}{a}} \cdot e^{\frac{b^2}{4a} + c}$$

for all $a, b, c \in \mathbb{R}$ with $a > 0$. The marginal density of $Y$ is given as follows:

$$\begin{aligned}
p_Y(v) &= \int_{-\infty}^{\infty} p_{X,Y}(\xi, v)\, d\xi \\
&= \frac{1}{2\pi\sqrt{1-\rho^2}} \int_{-\infty}^{\infty} \exp\left(-\frac{\xi^2}{2(1-\rho^2)} + \frac{\rho v}{(1-\rho^2)} \cdot \xi - \frac{v^2}{2(1-\rho^2)}\right) d\xi \\
&= \frac{1}{\sqrt{2\pi}} e^{-\frac{v^2}{2}} = \mathcal{N}(v|0,1)
\end{aligned}$$

This is the density of a standard normal distribution. For reasons of symmetry, $X$ follows a standard normal distribution as well. In particular, we have $E[X] = E[Y] = 0$.

Now, we just need to calculate the expected value of the product.

$$\sigma[X,Y] = E[X \cdot Y] - E[X] \cdot E[Y] = E[X \cdot Y]$$

$$= \int_{-\infty}^{\infty} \int_{-\infty}^{\infty} \xi_1 \cdot \xi_2 \cdot p_{X,Y}(\xi_1, \xi_2)\, d\xi_1 d\xi_2$$

$$= \int_{-\infty}^{\infty} \int_{-\infty}^{\infty} \frac{\xi_1 \cdot \xi_2}{2\pi\sqrt{1-\rho^2}} \cdot \exp\left(-\frac{(\xi_1)^2 - 2\rho \cdot \xi_1 \cdot \xi_2 + (\xi_2)^2}{2(1-\rho^2)}\right)\, d\xi_1 d\xi_2$$

The following linear transformation with determinant $\det(D\xi(s,t)) = \sqrt{1-\rho^2}$ leads to a decoupling of the variables:

$$\xi(s,t) = \begin{pmatrix} \xi_1(s,t) \\ \xi_2(s,t) \end{pmatrix} = \frac{1}{\sqrt{2}}\begin{pmatrix} \sqrt{1+\rho} & -\sqrt{1-\rho} \\ \sqrt{1+\rho} & \sqrt{1-\rho} \end{pmatrix} \cdot \begin{pmatrix} s \\ t \end{pmatrix}$$

The integral is then transformed via a few algebraic transformations to the following:

$$\sigma[X,Y] = \int_{-\infty}^{\infty} \int_{-\infty}^{\infty} \xi_1(s,t) \cdot \xi_2(s,t) \cdot p_{X,Y}(\psi(s,t)) \cdot \det(D\xi(s,t))\, dsdt$$

$$= \frac{1}{4\pi} \cdot \int_{-\infty}^{\infty} \int_{-\infty}^{\infty} \left((1+\rho) \cdot s^2 - (1-\rho)t^2\right) \cdot e^{-\frac{1}{2}(s^2+t^2)} \cdot dsdt$$

This result can now be integrated step by step—first with respect to one variable, then the other—until, finally:

$$\sigma[X,Y] = \rho$$

To understand the interpretation of covariance, consider a hypothetical company that inspects square-shaped workpieces with an average side length of $\bar{x} = 1.00\,\text{m}$. After the production process, quality control measures the side length of the workpieces again with high precision and finds that they vary with a standard deviation of $s(x) = 0.05\,\text{m}$.

We assume that the side length can be modeled as a random variable $X$, with expected value $E[X] = \bar{x}$, and variance $\sigma^2[X] = s^2(x)$. We will learn later in Section 4.2 under which circumstances this assumption may be justified.

The average area of a workpiece is not as one might expect, $(E[X])^2 = 1.00\,\text{m}^2$ but is actually a little higher even if the side lengths are symmetrically distributed around the mean:

$$E[X^2] = \sigma^2[X] + (E[X])^2 = 1.0025\,\text{m}^2$$

However, this is only true if the manufacturing process led to variations in the side length, but both side lengths were always exactly identical. A more realistic model might assume that the side lengths are realizations of different random variables, $X$ and $Y$, that have the same expected value but are uncorrelated, i.e., $E[X \cdot Y] = E[X] \cdot E[Y]$ holds. In this case, the average area of the workpieces is in fact given by $E[X \cdot Y] = 1.00\,\mathrm{m}^2$.

Furthermore, the following relationship between independence and correlation of random variables is important.

> **Correlation of independent random variables.** Independent random variables are always pairwise uncorrelated.

For two continuous random variables $X, Y$ this can be shown as follows:

$$
\begin{aligned}
E[X \cdot Y] &= \int_{-\infty}^{\infty} \int_{-\infty}^{\infty} \xi_1 \cdot \xi_2 \cdot p_{X,Y}(\xi_1, \xi_2)\, \mathrm{d}\xi_1 \mathrm{d}\xi_2 \\
&= \int_{-\infty}^{\infty} \int_{-\infty}^{\infty} \xi_1 \cdot \xi_2 \cdot p_X(\xi_1) \cdot p_Y(\xi_2)\, \mathrm{d}\xi_1 \mathrm{d}\xi_2 \\
&= \left( \int_{-\infty}^{\infty} \xi_1 \cdot p_X(\xi_1)\, \mathrm{d}\xi_1 \right) \cdot \left( \int_{-\infty}^{\infty} \xi_2 \cdot p_Y(\xi_2)\, \mathrm{d}\xi_2 \right) \\
&= E[X] \cdot E[Y]
\end{aligned}
$$

The converse is *not* true in general: uncorrelated random variables need not be independent. For a counterexample, consider a continuous variable $X$ with the following density function, a uniform distribution on the interval $[-2, 2]$:

$$
p_X(u) = \begin{cases} \frac{1}{4} & \text{if } -2 \le u \le 2 \\ 0 & \text{otherwise} \end{cases}
$$

The random variables $X$ and $Y = X^2$ are not independent, because on the one hand the following holds:

$$
\begin{aligned}
F_{X,Y}(1,1) &= \Pr(X \le 1, X^2 \le 1) = \Pr(X \le 1, -1 \le X \le 1) \\
&= \Pr(-1 \le X \le 1) = \frac{1}{2}
\end{aligned}
$$

On the other hand:

$$
\begin{aligned}
F_X(1) \cdot F_Y(1) &= \Pr(X \le 1) \cdot \Pr(-1 \le X \le 1) = \frac{3}{4} \cdot \frac{1}{2} \\
&\neq F_{X,Y}(1,1)
\end{aligned}
$$

Nevertheless, the variables are uncorrelated because their covariance vanishes:

$$
\begin{aligned}
\sigma[X, Y] &= E[X \cdot Y] - E[X] \cdot E[Y] = E[X^3] - E[X] \cdot E[X^2] \\
&= \frac{1}{4} \int_{-2}^{2} \xi^3\, \mathrm{d}\xi - \left( \frac{1}{4} \int_{-2}^{2} \xi\, \mathrm{d}\xi \right) \cdot \left( \frac{1}{4} \int_{-2}^{2} \xi^2\, \mathrm{d}\xi \right) \\
&= 0
\end{aligned}
$$

### 3.4.3 Chebyshev's inequality

Our work so far is rewarded by the following result, which has far-reaching consequences.

> **Chebyshev's inequality.** For any random variable $X$ for which expected value $E[X]$ and variance $\sigma^2[X]$ exist and are finite, the following holds for any $r \in \mathbb{R}$, $r > 0$:
>
> $$\Pr(|X - E[X]| \geq r) \leq \frac{\sigma^2[X]}{r^2}$$

In other words, the smaller the variance, the less likely it is that the random variable will take on values that are far from the mean. This result aligns with our intuitive understanding of a measure of dispersion.

An alternative form of the inequality can be obtained by substituting $r = z\sigma[X]$:

$$\Pr(|X - E[X]| \geq z \cdot \sigma[X]) \leq \frac{1}{z^2}$$

for all $z \in \mathbb{R}$, $z > 1$.

For example, the probability that a random variable takes on a value that deviates from the mean by more than six standard deviations is less than 3%. In other words, the probability of finding an observed value of the random variable within six standard deviations of its expected value is very high, at least 97%. This result holds for distributions of any shape as long as the expected value and variance are finite values.

Chebyshev's inequality can be proved as follows. First, for the sake of clarity of the formulas, we set $\mu = E[X]$ and $\sigma = \sigma[X]$.

Furthermore, we introduce the concept of an **indicator function**. For each event $A \subseteq \Omega$, we can define it as follows:

$$I_A \colon \Omega \to \mathbb{R}, \ I_A(\omega) = \begin{cases} 1 & \text{if } \omega \in A \\ 0 & \text{otherwise} \end{cases}$$

The indicator function is a discrete random variable with the following cumulative distribution function:

$$F_{I_A}(u) = \begin{cases} 0 & \text{if } u < 1 \\ \Pr(A) & \text{if } 1 \leq u \end{cases}$$

We can express the probability of any event as the expectation of its indicator function: $E[I_A] = \Pr(A)$.

Let us apply this fact to the event $A := \{\omega \in \Omega \,|\, |X(\omega) - \mu| \geq r\}$ and set $Y := I_A$, which implies:

$$E[Y] = \Pr(A) = \Pr(|X(\omega) - \mu| \geq r)$$

On the other hand, for all $\omega \in \Omega$, $|X(\omega) - \mu|^2 \geq r^2 \cdot Y(\omega)$ holds. We can use the monotonicity and the linearity of the expected value to conclude:

$$\sigma^2 = E[|X - \mu|^2] \geq E[r^2 Y] = r^2 E[Y] = r^2 \cdot \Pr(|X(\omega) - \mu| \geq r)$$

Dividing both sides of the inequality by $r^2$ gives the desired result.

Another useful form of Chebyshev's inequality is given by:

$$\Pr(|X - \mu| < \varepsilon) \geq 1 - \frac{\sigma^2}{\varepsilon^2}$$

for all $\varepsilon \in \mathbb{R}$, $\varepsilon > 0$.

## 3.5 Sums and products of random variables

For discrete random variables $X, Y$, we can compute the probability mass function of their sum $Z = X + Y$.

> **Probability mass function of the sum of discrete random variables.**
> For discrete random variables $X, Y$, the following holds:
>
> $$p_{X+Y}(u) = \sum_{\kappa \in \mathrm{supp}(X)} p_{X,Y}(\kappa, u - \kappa)$$
>
> for all $u \in \mathrm{supp}(X + Y)$.

Moreover, if $X$ and $Y$ are independent random variables, we have:

$$p_{X+Y}(u) = \sum_{\kappa \in \mathrm{supp}(X)} p_X(\kappa) \cdot p_Y(u - \kappa)$$

We now want to calculate the probability density function of the sum $Z = X + Y$ of two continuous random variables $X$ and $Y$. First, we express the distribution of $Z$ in terms of the joint distribution of $X$ and $Y$ as follows:

$$F_Z(u) = \Pr(X + Y \leq u) = \iint\limits_{B(u)} p_{X,Y}(\xi_1, \xi_2)\, d\xi_1 d\xi_2$$

Here, $B(u)$ is the domain of integration $\{(\xi_1, \xi_2) \in \mathbb{R}^2 | \xi_1 + \xi_2 \leq u\}$. Written as a multiple integral, substitution of $\xi_2(t) = t - \xi_1$ and swapping the order of integration yields:

$$F_Z(u) = \int_{-\infty}^{\infty} \left( \int_{-\infty}^{u - \xi_1} p_{X,Y}(\xi_1, \xi_2)\, d\xi_2 \right) d\xi_1$$

$$= \int_{-\infty}^{\infty} \left( \int_{-\infty}^{u} p_{X,Y}(\xi_1, t - \xi_1)\, dt \right) d\xi_1$$

$$= \int_{-\infty}^{u} \left( \int_{-\infty}^{\infty} p_{X,Y}(\xi_1, t - \xi_1) \, d\xi_1 \right) dt$$

Finally, we observe that the density function is the derivative of the cumulative distribution function $p_Z(u) = \frac{d}{du} F_Z(u)$, that is, the inner integrand.

**Density function of the sum of continuous random variables.** For continuous random variables $X, Y$, the density function of their sum is given as follows:

$$p_{X+Y} \colon \mathbb{R} \to [0, \infty[, \; p_{X+Y}(u) = \int_{-\infty}^{\infty} p_{X,Y}(\xi, u - \xi) \, d\xi$$

An interesting special case occurs when $X$ and $Y$ are independent random variables:

$$p_{X+Y}(u) = \int_{-\infty}^{\infty} p_X(\xi) \cdot p_Y(u - \xi) \, d\xi$$

Generally, the operation

$$(f * g)(u) := \int_{-\infty}^{\infty} f(\xi) \cdot g(u - \xi) \, d\xi$$

for integrable functions $f, g \colon \mathbb{R} \to \mathbb{R}$ is called a **convolution**. Thus, using this terminology, the density function of the sum of two independent continuous random variables is given by the convolution of their density functions: $p_{X+Y} = p_X * p_Y$.

As an example, let's calculate the density function of the sum of two independent random variables $X$ and $Y$, each normally distributed with zero mean. We denote the variance with $\sigma^2$ and $\tau^2$, respectively:

$$p_X(u) = \mathcal{N}(u|0, \sigma^2) = \frac{1}{\sqrt{2\pi}\sigma} \cdot e^{-\frac{u^2}{2\sigma^2}}, \; p_Y(u) = \mathcal{N}(u|0, \tau^2) = \frac{1}{\sqrt{2\pi}\tau} \cdot e^{-\frac{u^2}{2\tau^2}}$$

Computing the convolution of these two functions leads to a general Gaussian integral:

$$p_{X+Y}(u) = \frac{1}{2\pi\sigma\tau} \int_{-\infty}^{\infty} e^{-\frac{\xi^2}{2\sigma^2}} \cdot e^{-\frac{(u-\xi)^2}{2\tau^2}} \, d\xi$$

$$= \frac{1}{2\pi\sigma\tau} \int_{-\infty}^{\infty} \exp\left( -\frac{\sigma^2 + \tau^2}{2\sigma^2\tau^2} \xi^2 + \frac{u}{\tau^2} \xi - \frac{u^2}{2\tau^2} \right) d\xi$$

$$= \frac{1}{\sqrt{2\pi} \cdot \sqrt{\sigma^2 + \tau^2}} \cdot e^{-\frac{u^2}{2(\sigma^2 + \tau^2)}} = \mathcal{N}(u|0, \sigma^2 + \tau^2)$$

Thus, the sum of two normally distributed random variables with zero mean is again normally distributed with zero mean. The new variance is given by the sum of variances. In general, the expected values also simply add up:

$$p_{X_1} = \mathcal{N}(\cdot \,|\mu_1, \sigma_1^2),\ p_{X_2} = \mathcal{N}(\cdot \,|\mu_2, \sigma_2^2) \Rightarrow$$
$$p_{X_1+X_2} = \mathcal{N}(\cdot \,|\mu_1 + \mu_2, \sigma_1^2 + \sigma_2^2)$$

For another application of the convolution formula, we want to calculate the density function of the sum of two independent random variables $X$ and $Y$ that both follow a chi-squared distribution with one degree of freedom:

$$\chi_1^2 \colon \mathbb{R} \to [0, \infty[,\ \chi_1^2(v) = \begin{cases} 0 & \text{if } v \leq 0 \\ \frac{1}{\sqrt{2\pi v}} \cdot e^{-\frac{1}{2}v} & \text{if } v > 0 \end{cases}$$

The convolution of this function with itself is calculated as follows:

$$p_{X+Y}(u) = \int_{-\infty}^{\infty} \chi_1^2(\xi) \cdot \chi_1^2(u - \xi)\, \mathrm{d}\xi$$

We have $p_{X+Y}(u) = 0$ for $u \leq 0$, so we can assume $u > 0$ in the following. The integrand vanishes if $\xi \leq 0$ or $u \leq \xi$. Consequently, with the substitution $\xi(t) = \frac{u}{2} \cdot (t + 1)$, for $u > 0$:

$$p_{X+Y}(u) = \frac{e^{-\frac{1}{2}u}}{2\pi} \int_0^u \frac{1}{\sqrt{\xi \cdot (u - \xi)}}\, \mathrm{d}\xi = \frac{e^{-\frac{1}{2}u}}{2\pi} \int_{-1}^1 \frac{1}{\sqrt{1 - t^2}}\, \mathrm{d}t$$

$$= \frac{e^{-\frac{1}{2}u}}{2\pi} \cdot \arcsin t \, \Big|_{t=-1}^{1} = \frac{1}{2} \cdot e^{-\frac{1}{2}u}$$

We recall that the chi-squared distribution is identical to the distribution of the square of a standard normally distributed variable. Thus, the above distribution corresponds to the sum of two such squares: this is the chi-squared distribution with *two* degrees of freedom.

For products of independent continuous random variables, we have the following formula.

**Density function of the product of continuous random variables.**
For two continuous random variables $X$ and $Y$, the following formula holds:

$$p_{X \cdot Y}(u) = \int_{-\infty}^{\infty} \frac{1}{|\xi|} \cdot p_{X,Y}\left(\xi, \frac{u}{\xi}\right) \mathrm{d}\xi$$

The proof works similarly to the sum of random variables. First, the following applies to the cumulative distribution function:

$$F_{X \cdot Y}(u) = \Pr(X \cdot Y \leq u) = \iint_{B(u)} p_{X,Y}(\xi_1, \xi_2)\, \mathrm{d}\xi_1 \mathrm{d}\xi_2$$

with $B(u) = \{(\xi_1, \xi_2) \in \mathbb{R}^2 | \xi_1 \cdot \xi_2 \leq u\}$. We decompose the domain of integration into two sub-domains:

$$B_+(u) = \left\{ (\xi_1, \xi_2) \in \mathbb{R}^2 \,\middle|\, \xi_2 \leq \frac{u}{\xi_1}, \, \xi_1 > 0 \right\},$$

$$B_-(u) = \left\{ (\xi_1, \xi_2) \in \mathbb{R}^2 \,\middle|\, \xi_2 \geq \frac{u}{\xi_1}, \, \xi_1 < 0 \right\}$$

Furthermore, with the substitution $\xi_2(t) = \frac{t}{\xi_1}$:

$$\iint\limits_{B_+(u)} p_{X,Y}(\xi_1, \xi_2) \, d\xi_1 d\xi_2 = \int_0^\infty \left( \int_{-\infty}^{u/\xi_1} p_{X,Y}(\xi_1, \xi_2) \, d\xi_2 \right) d\xi_1$$

$$= \int_{-\infty}^u \left( \int_0^\infty \frac{1}{\xi_1} \cdot p_{X,Y}\left(\xi_1, \frac{t}{\xi_1}\right) d\xi_1 \right) dt$$

In a similar way, we achieve the result:

$$\iint\limits_{B_-(u)} p_{X,Y}(\xi_1, \xi_2) \, d\xi_1 d\xi_2 = \int_u^\infty \left( \int_{-\infty}^0 \frac{1}{\xi_1} \cdot p_{X,Y}\left(\xi_1, \frac{t}{\xi_1}\right) d\xi_1 \right) dt$$

Therefore, we can conclude the desired result:

$$p_{X \cdot Y}(u) = \frac{d}{du} F_{X \cdot Y}(u)$$

$$= \int_0^\infty \frac{1}{\xi} \cdot p_{X,Y}\left(\xi, \frac{u}{\xi}\right) d\xi - \int_{-\infty}^0 \frac{1}{\xi} \cdot p_{X,Y}\left(\xi, \frac{u}{\xi}\right) d\xi$$

$$= \int_{-\infty}^\infty \frac{1}{|\xi|} \cdot p_{X,Y}\left(\xi, \frac{u}{\xi}\right) d\xi$$

As an application of the product formula, we consider a standard normally distributed random variable $Z$ as well as a random variable $Y$ that follows a chi-squared distribution with two degrees of freedom, that is:

$$p_Y(u) = \begin{cases} 0 & \text{if } u \leq 0 \\ 1/2 \cdot e^{-u/2} & \text{if } u > 0 \end{cases}$$

Let's assume that $Z$ and $Y$ are independent. We want to determine the distribution of the following random variable:

$$T = \frac{\sqrt{2} \cdot Z}{\sqrt{Y}} = \sqrt{2} \cdot Z \cdot Y^{-\frac{1}{2}}$$

First, with $f(u) = u^{-1/2}$:

$$p_{Y^{-1/2}}(v) = p_{f(Y)}(v) = \frac{p_Y(f^{-1}(v))}{|f'(f^{-1}(v))|}$$

$$= \begin{cases} 0 & \text{if } v \leq 0 \\ \frac{1}{v^3} \cdot e^{-\frac{1}{2v^2}} & \text{if } v > 0 \end{cases}$$

From the product formula, assuming independence, and substituting $\xi(t) = \sqrt{1 + u^2} \cdot t^{-1}$:

$$p_{Z \cdot Y^{-1/2}}(u) = \int_{-\infty}^{\infty} \frac{1}{|\xi|} \cdot p_Y(\xi) \cdot p_Z\left(\frac{u}{\xi}\right) d\xi = \frac{1}{\sqrt{2\pi}} \int_0^{\infty} \frac{1}{\xi^4} \cdot e^{-\frac{1+u^2}{\xi^2}} d\xi$$

$$= \frac{1}{\sqrt{2\pi}} \cdot (1 + u^2)^{-\frac{3}{2}} \cdot \int_0^{\infty} t^2 \cdot e^{-\frac{1}{2}t^2} dt = \frac{1}{2} \cdot (1 + u^2)^{-\frac{3}{2}}$$

And finally via the transformation formula for linear functions:

$$p_{\sqrt{2} \cdot Z \cdot Y^{-1/2}}(u) = \frac{1}{\sqrt{2}} \cdot p_{Z \cdot Y^{-1/2}}\left(\frac{u}{\sqrt{2}}\right) = \frac{1}{2\sqrt{2}} \cdot \left(1 + \frac{u^2}{2}\right)^{-\frac{3}{2}}$$

This distribution is called Student's $t$-distribution[1] with two degrees of freedom.

### 3.5.1 Chi-squared and Student's $t$-distribution

In the examples of the previous sections, we calculated the density functions of $Y_1 = (X_1)^2$ and $Y_2 = (X_1)^2 + (X_2)^2$ for standard normally distributed and independent random variables $X_1, X_2$. Generalizing these results yields the following family of distributions.

The sum $Y_N = (X_1)^2 + \cdots + (X_N)^2$ of squares of independent standard normally distributed random variables $X_1, \ldots, X_N$ follows a **chi-squared distribution with $N$ degrees of freedom**:

$$p_{Y_N}(\cdot) = \chi_N^2(\cdot) \colon \mathbb{R} \to [0, \infty[,$$

$$\chi_N^2(u) = \begin{cases} 0 & \text{if } u \leq 0 \\ 2^{-\frac{N}{2}} \cdot \left(\Gamma\left(\frac{N}{2}\right)\right)^{-1} \cdot u^{\frac{N}{2}-1} \cdot e^{-\frac{u}{2}} & \text{if } u > 0 \end{cases}$$

Here, we have introduced the **gamma function**:

$$\Gamma \colon \, ]0, \infty[ \, \to \mathbb{R}, \, \Gamma(z) = \int_0^{\infty} \xi^{z-1} \cdot e^{-\xi} d\xi$$

For each nonnegative integer $n \in \mathbb{N}$, we have

$$\Gamma(n + 1) = n!$$

where $n! = n \cdot (n - 1) \cdots 2 \cdot 1$ is the **factorial** of $n \geq 1$, and by definition $0! = 1$. The gamma function thus extends the domain of the factorial beyond integer values. Specifically, for half-integer values, we have:

$$\Gamma\left(n + \frac{1}{2}\right) = \frac{(2n)! \cdot \sqrt{\pi}}{n! \cdot 4^n}$$

---

[1] actually William Sealy Gosset (*1876 – †1937), *Student* is a pseudonym

At the top of Fig. 3.4, there is a function graph of the chi-squared distribution for a few selected values for the number of degrees of freedom.

For $N = 1$ and $N = 2$, we have already calculated the chi-squared distribution. By inserting these values for $N$, we can convince ourselves that the above formula is correct for these cases. The other cases can be proven by induction. The nonconstant part is important; the constant factor in front is implied by the normalization condition $\int_{-\infty}^{\infty} \chi_N^2(\xi)\,d\xi = 1$. Assuming that the formula is correct for $\chi_N^2(u)$ with $u > 0$, the substitution $\xi(t) = \frac{u}{2} \cdot (t+1)$ implies the formula for $N + 1$ in a way that is very similar to the already calculated base case $\chi_2^2 = \chi_1^2 * \chi_1^2$:

$$(\chi_{N+1}^2)(u) = (\chi_N^2 * \chi_1^2)(u) \propto \int_0^u \xi^{\frac{N}{2}-1} \cdot e^{-\frac{\xi}{2}} \cdot (u - \xi)^{-\frac{1}{2}} \cdot e^{-\frac{u-\xi}{2}}\,d\xi$$

$$= \left(\frac{u}{2}\right)^{\frac{N+1}{2}-1} \cdot e^{-\frac{u}{2}} \cdot \int_{-1}^1 (1+t)^{\frac{N}{2}-1} \cdot (1-t)^{-\frac{1}{2}}\,dt$$

$$\propto \left(\frac{u}{2}\right)^{\frac{N+1}{2}-1} \cdot e^{-\frac{u}{2}}$$

For two degrees of freedom, we have already introduced the Student's $t$-distribution. When there is an arbitrary number of degrees of freedom, then the Student's $t$-distribution is given as follows.

Let $Z$ be a standard normally distributed random variable, and $Y_N$ a chi-squared distributed random variable with $N$ degrees of freedom. Suppose that $Z$ and $Y_N$ are independent, and consider the following random variable:

$$T_N = \frac{\sqrt{N} \cdot Z}{\sqrt{Y_N}}$$

Then, $T_N$ follows the **Student's $t$-distribution with $N$ degrees of freedom**:

$$p_{T_N}(\cdot) = t_N(\cdot): \mathbb{R} \to [0, \infty[,\ t_N(u) = \frac{\Gamma\left(\frac{N+1}{2}\right)}{\sqrt{N\pi} \cdot \Gamma\left(\frac{N}{2}\right)} \cdot \left(1 + \frac{u^2}{N}\right)^{-\frac{N+1}{2}}$$

We will do without a more detailed derivation. The distribution is of special importance in statistical test theory; more about its application in Sect. 4.3.3.

At the bottom, Fig. 3.4 shows a Student's $t$-distribution for $N = 5$ degrees of freedom. The $t$-distribution is very similar to the standard normal distribution—in fact, in the limit of a large number of degrees of freedom, for all $u \in \mathbb{R}$:

$$\lim_{N \to \infty} t_N(u) = \frac{1}{\sqrt{2\pi}} e^{-\frac{1}{2}u^2}$$

For sufficiently large values of $N$, the two distributions are hardly distinguishable in practice; a common rule of thumb is $N > 30$.

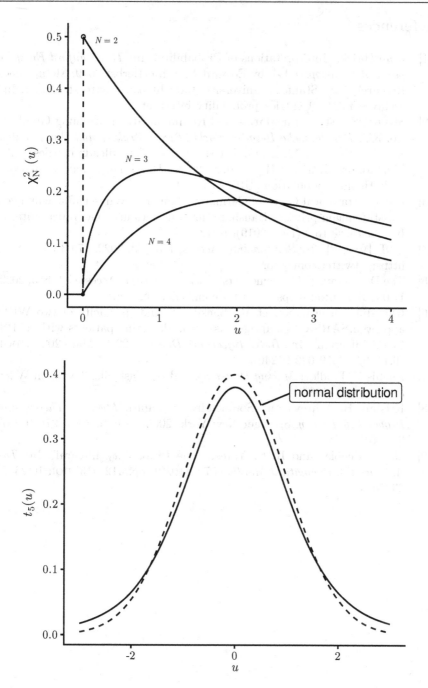

**Fig. 3.4.** *Chi-squared distribution (top) and Student's t-distribution (bottom)*

# References

[1] Alan Hájek. "Interpretations of Probability". In: *The Stanford Encyclopedia of Philosophy*. Ed. by Edward N. Zalta. Herbst 2019. Metaphysics Research Lab, Stanford University. URL: https://plato.stanford.edu/archives/fall2019/entries/probability-interpret/.

[2] SKOPOS Institut für Markt- und Kommunikationsforschung GmbH & Co. KG. *Der deutsche Heimtiermarkt 2018 – Struktur und Umsatzdaten*. Accessed July 10, 2020. URL: https://www.zzf.de/fileadmin/files/ZZF/Marktdaten/ZZF_IVH_Folder_2018_Deutscher_Heimtiermarkt_und_Heimtierpopulation_UPDATE.pdf.

[3] Federal Statistical Office of Germany. "Laufende Wirtschaftsrechnungen Ausstattungen privater Haushalte mit ausgewählten Gebrauchsgütern". In: *Fachserie* 15.2 (Dec. 2019), p. 12.

[4] GOP [@GOP]. *Twitter timeline*. Accessed July 3, 2022, 17:03 CET. URL: https://twitter.com/gop.

[5] The Democrats [@TheDemocrats]. *Twitter timeline*. Accessed July 3, 2022, 17:03 CET. URL: https://twitter.com/TheDemocrats.

[6] Chandima Jeewandara et al. "Sensitivity and specificity of two WHO approved SARS-CoV2 antigen assays in detecting patients with SARS-CoV2 infection". In: *BMC Infectious Diseases* 22.1 (Mar. 2022). DOI: 10.1186/s12879-022-07240-6.

[7] Dennis V. Lindley. *Making Decisions*. 2nd ed. Nashville, TN: John Wiley & Sons, Oct. 1985.

[8] Krishna B. Athreya and Soumendra N. Lahiri. *Measure Theory and Probability Theory*. Springer New York, 2006. DOI: 10.1007/978-0-387-35434-7.

[9] C. P. Nicholas and R. C. Yates. "The Probability Integral". In: *The American Mathematical Monthly* 57.6 (1950), pp. 412–413. DOI: 10.2307/2307644.

# 4

# Inferential statistics

**Inferential statistics** is based on the assumption that data are realizations of random variables. For example, we may assume that the outcome of tossing a fair coin is determined by a discrete random variable $X_1$ that takes the values zero (corresponding to tails) or one (heads) with equal probability: $\Pr(X_1 = 0) = \Pr(X_1 = 1) = 1/2$. If we flip the coin again, we may assume that the outcome can be described by a random variable $X_2$ that is independent of $X_1$ but follows the same distribution: $\Pr(X_2 = 0) = \Pr(X_2 = 1) = 1/2$.

If we perform the experiment repeatedly, the sequence of observations of heads or tails becomes the subject of investigation of descriptive statistics: a sample of a binary variable. Given the nature of the **data generating process**, we expect that both possible values occur with approximately the same frequency: $f_1 \approx 1 - f_0 \approx 0.5$.

Conversely, the data allow us to draw conclusions about the underlying true distribution. For example, if we observe very different frequencies for heads and tails, such as a 70% frequency for heads compared to 30% for tails, then—if the sample size is sufficiently large—we would be convinced that we need to correct our original assumption of equal probability. In other words, we may need to abandon our assumption that the coin is fair.

In the above example, the relative frequency serves as an estimator of the probability. Other important estimators include the arithmetic mean as an estimator of the expected value and the sample variance as an estimator of the "true" variance.

Intuitively, our confidence in such estimates grows with the sample size. For example, if the imbalance described above were found among only ten coin tosses—that is, seven tosses of heads and three of tails—we may not yet be convinced that we have a biased trick coin at hand. There is still the realistic possibility that the so-called **null hypothesis** of a fair coin is true. Colloquially speaking, the outcome of the experiment could also have been the result of "pure chance." On the other hand, seventy occurrences of heads among one hundred

© Springer-Verlag GmbH Germany, part of Springer Nature 2023
M. Plaue, *Data Science*, https://doi.org/10.1007/978-3-662-67882-4_4

coin tosses would be much stronger evidence for the alternative hypothesis that the coin is biased!

Inferential statistics allows us to mathematically justify this intuition of a *law of large numbers*. From these insights, statistical procedures can be derived that not only estimate measures such as probability, expected value, and variance but also gauge our confidence in the correctness of the estimate.

In addition, in this chapter, we will cover the topic of statistical modeling: a process for estimating the parameters of families of probability mass and density functions from a sample in a way that the distribution represents the best fit to the observed data.

## 4.1 Statistical models

For the purpose of statistical inference, we are particularly interested in families of distributions. That is, the probability mass or density function $p$ not only depends on the argument $u$ but also on a number of parameters $\theta_1, \ldots, \theta_K$. We express this dependency via the notation $p(u|\theta_1, \ldots, \theta_K)$. We also call such a family a **(parametric) statistical model**. A statistical model is usually required to satisfy the condition that the model parameters uniquely determine each probability mass or density function within the family: it is impossible for two different parameter assignments to lead to the same distribution.

### 4.1.1 Models of discrete random variables

The following statistical models represent families of probability mass functions over the natural numbers. For each parameter assignment, it is a nonnegative function $p \colon \mathbb{N} \to \mathbb{R}$, $k \mapsto p(k)$, the values of which sum to one. Bar graphs for a selection of parameters are shown in Fig. 4.1.

> Given the number of possible outcomes $L \in \mathbb{N}$, $L > 0$, the **discrete uniform distribution** is characterized by the following probability mass function:
>
> $$\mathcal{U}(k|L) = \begin{cases} \frac{1}{L} & \text{if } k \in \{1, \ldots, L\} \\ 0 & \text{otherwise} \end{cases}$$

More generally, uniformly distributed random variables are characterized by the fact that every possible outcome with nonzero probability is equally probable. For a uniformly distributed discrete random variable $X$ with $\text{supp}(X) = \{t_1, \ldots, t_L\}$, the expectation can be calculated as follows:

$$E[X] = \sum_{\kappa \in \text{supp}(X)} \kappa \cdot \frac{1}{L} = \frac{1}{L} \sum_{k=1}^{L} t_k$$

Thus, for a uniformly distributed variable, the expected value is formally given by the arithmetic mean of all the values in its support.

The variance can also be determined using the formula known from descriptive statistics:

$$\sigma^2[X] = E[(X - E[X])]^2 = \frac{1}{L}\sum_{l=1}^{L}\left(t_l - \frac{1}{L}\sum_{k=1}^{L}t_k\right)^2$$

The principle of indifference mentioned at the beginning of Chap. 3 states that—in the absence of further information—all possible outcomes should be assumed equally likely; i.e., the outcomes follow a uniform distribution.

Among all discrete distributions over some fixed, finite support, the uniform distribution is the unique distribution that has maximum Shannon entropy.

The following distribution describes a type of experiment called a **Bernoulli trial**, which only has two possible outcomes: success (denoted by $k = 1$) or failure (denoted by $k = 0$).

For the probability of success $p \in [0, 1]$, the **Bernoulli distribution** is given by the following probability mass function:

$$\mathcal{B}(k|p, 1) = \begin{cases} 1 - p & \text{if } k = 0 \\ p & \text{if } k = 1 \\ 0 & \text{otherwise} \end{cases}$$

An example of a Bernoulli trial is a coin toss, where either heads or tails can appear. Typically, both outcomes are equally likely, meaning $p$ is approximately $\frac{1}{2}$. However, it is possible to have a biased coin where heads is more likely, or less likely, to occur than tails, resulting in a value of $p$ that is different from $\frac{1}{2}$.

Connecting this topic back to descriptive statistics, the observations of a Bernoulli experiment produce the values of a binary variable.

The expectation and variance of a Bernoulli-distributed random variable $X$ can be calculated as follows:

$$E[X] = 0 \cdot (1 - p) + 1 \cdot p = p,$$
$$\sigma^2[X] = (0 - p)^2 \cdot (1 - p) + (1 - p)^2 \cdot p = p(1 - p)$$

The following distribution describes the probability of finding heads exactly $k$ times among a series of $N$ coin tosses. It will play a significant role in Sect. 4.2.1 when we establish the connection between the frequency of an observation and the probability of its occurrence.

For $N$ trials with probability of success $p \in [0, 1]$, the **binomial distribution** is given by the following probability mass function:

$$\mathcal{B}(k|p, N) = \begin{cases} \binom{N}{k} p^k (1-p)^{N-k} & \text{if } k \in \{0, \dots, N\} \\ 0 & \text{otherwise} \end{cases}$$

For the case that $p = 0$ or $p = 1$, the convention $0^0 = 1$ shall apply. The term $\binom{N}{k}$ is not a vector but denotes the **binomial coefficient**. When repeating a Bernoulli trial $N$ times, the binomial coefficient $\binom{N}{k}$ denotes the number of possible outcomes that produce exactly $k$ successful trials:

$$\binom{N}{k} = \frac{N!}{k!(N-k)!}$$

In the above formula, the exclamation marks denote the factorial: for every natural number $n \geq 1$, $n! = n \cdot (n-1) \cdots 2 \cdot 1$. By definition, $0! = 1$.

The Bernoulli distribution is identical to the binomial distribution for the case $N = 1$.

The expectation and variance of a binomially distributed random variable $X \sim \mathcal{B}(\cdot | p, N)$ are given as follows:

$$E[X] = Np, \ \sigma^2[X] = Np(1-p)$$

The **geometric distribution** with probability of success $p \in \mathbb{R}, 0 < p \leq 1$, is given by the following probability mass function, $k \in \mathbb{N}$:

$$\mathrm{Geom}(k|p) = p(1-p)^k$$

The probability that in a sequence of coin flips the first occurrence of heads happens after exactly $k$ times of tossing tails is given by $\mathrm{Geom}(k|p)$.

The expected value and the variance of a geometrically distributed random variable $X \sim \mathrm{Geom}(\cdot | p)$ can be computed as follows:

$$E[X] = \frac{1-p}{p}, \ \sigma^2[X] = \frac{1-p}{p^2}$$

For $\lambda \in \mathbb{R}, \lambda > 0$, the **Poisson distribution** is given by the following probability mass function, $k \in \mathbb{N}$:

$$\mathrm{Pois}(k|\lambda) = \frac{\lambda^k}{k!} e^{-\lambda}$$

This formula is for the expected value and variance of a Poisson-distributed random variable $X \sim \mathrm{Pois}(\cdot | \lambda)$:

$$E[X] = \sigma^2[X] = \lambda$$

The Poisson distribution results from the binomial distribution in the limit of a large number of Bernoulli trials with probability of success $p = \frac{\lambda}{N}$. We can show this by the following calculation:

$$
\begin{aligned}
\lim_{N\to\infty} \mathcal{B}\left(k \,\Big|\, \frac{\lambda}{N}, N\right) &= \lim_{N\to\infty} \left[ \frac{N!}{k!(N-k)!} \cdot \left(\frac{\lambda}{N}\right)^k \cdot \left(1 - \frac{\lambda}{N}\right)^{N-k} \right] \\
&= \frac{\lambda^k}{k!} \cdot \lim_{N\to\infty} \left[ \frac{N(N-1)\cdots(N-k+1)}{N^k} \cdot \left(1 - \frac{\lambda}{N}\right)^{N-k} \right] \\
&= \frac{\lambda^k}{k!} \cdot \lim_{N\to\infty} \left[ \left(1 - \frac{\lambda}{N}\right)^{-k} \left(1 - \frac{\lambda}{N}\right)^{N} \right] \\
&= \frac{\lambda^k}{k!} \cdot \lim_{N\to\infty} \left(1 - \frac{\lambda}{N}\right)^{N} = \frac{\lambda^k}{k!} e^{-\lambda} = \mathrm{Pois}(k|\lambda)
\end{aligned}
$$

In the last step, we have used the identity $e^x = \lim_{N\to\infty} \left(1 + \frac{x}{N}\right)^N$. The rate parameter $\lambda$ can then be interpreted as the average frequency of a successful trial.

### 4.1.2 Models of continuous random variables

Models for continuous random variables are listed in this section. For each assignment of parameters, the model represents a nonnegative function $p \colon \mathbb{R} \to \mathbb{R}$, $u \mapsto p(u)$, the integral of which is equal to one. Fig. 4.2 shows graphs of those density functions for selected parameters.

For $a, b \in \mathbb{R}$, $a < b$, the **continuous uniform distribution** is given by the following probability density function, $u \in \mathbb{R}$:

$$
\mathcal{U}(u|a,b) = \begin{cases} \frac{1}{b-a} & \text{if } u \in [a,b] \\ 0 & \text{otherwise} \end{cases}
$$

The expected value and variance of a uniformly distributed continuous random variable $X \sim \mathcal{U}(\,\cdot\,|a,b)$ are given as follows:

$$E[X] = \frac{1}{2}(b+a), \quad \sigma^2[X] = \frac{1}{12}(b-a)^2$$

The **normal distribution** with location parameter $\mu \in \mathbb{R}$ and scale parameter $\sigma \in \mathbb{R}$, $\sigma > 0$, is given by the following probability density function, $u \in \mathbb{R}$:

$$\mathcal{N}(u|\mu,\sigma^2) = \frac{1}{\sigma\sqrt{2\pi}} e^{-\frac{1}{2}\left(\frac{u-\mu}{\sigma}\right)^2}$$

The expected value and the variance of a normally distributed random variable $X \sim \mathcal{N}(\cdot \,|\, \mu, \sigma^2)$ are determined by the location parameter and the scale parameter, respectively:

$$E[X] = \mu, \ \sigma^2[X] = \sigma^2$$

The normal distribution with an expected value $\mu = 0$ and variance $\sigma^2 = 1$ is called the **standard normal distribution**.

Linear combinations of normally distributed random variables are normally distributed: If $X_1, \ldots, X_N$ are normally distributed, so is $Y = \sum_{n=1}^{N} a_n X_n$ with arbitrary numbers $a_1, \ldots, a_N \in \mathbb{R}$ (unless all those coefficients are zero).

If, additionally, the random variables $X_1, \ldots, X_N$ are independent, then:

$$Y = \sum_{n=1}^{N} a_n X_n \sim \mathcal{N}(\cdot \,|\, \mu_Y, \sigma_Y^2) \text{ with } \mu_Y = \sum_{n=1}^{N} a_n \mu_n \text{ and } \sigma_Y^2 = \sum_{n=1}^{N} (a_n \sigma_n)^2,$$

where $\mu_1, \ldots, \mu_N$ are the expected values and $\sigma_1^2, \ldots, \sigma_N^2$ are the variances of $X_1, \ldots, X_N$.

If $X_1, \ldots, X_N$ are independent variables that follow the same normal distribution with an identical mean $\mu$ and variance $\sigma^2$, then this formula implies:

$$\sum_{n=1}^{N} X_n = X_1 + \cdots + X_N \sim \mathcal{N}(\cdot \,|\, N \cdot \mu, N \cdot \sigma^2)$$

Later on, the following random variable will be of special interest:

$$Z = \sqrt{N} \cdot \frac{\frac{1}{N} \sum_{n=1}^{N} X_n - \mu}{\sigma} = \frac{-N\mu + \sum_{n=1}^{N} X_n}{\sqrt{N}\sigma}$$

Under the conditions imposed on $X_1, \ldots, X_N$, the random variable $Z$ is always a standard normally distributed variable:

$$p_Z(u) = \sqrt{N}\sigma \cdot \mathcal{N}\left(\sqrt{N}\sigma \cdot u + N \cdot \mu \,\Big|\, N \cdot \mu, N \cdot \sigma^2\right)$$

$$= \sqrt{N}\sigma \cdot \frac{1}{\sqrt{2\pi}\sqrt{N}\sigma} \exp\left(-\frac{1}{2}\left(\frac{\sqrt{N}\sigma u + N\mu - N\mu}{\sqrt{N}\sigma}\right)^2\right)$$

$$= \frac{1}{\sqrt{2\pi}} e^{-\frac{u^2}{2}} = \mathcal{N}(u\,|\,0, 1)$$

The central limit theorem, discussed in Sect. 4.2.4, makes the surprising statement that under fairly general circumstances, and for a large $N$, the variable $Z$ is approximately normally distributed even when $X_1, \ldots, X_N$ are not. In other words, even if we know very little about the distribution of the variables $X_1, \ldots, X_N$, we know very well how the derived statistic $Z$ is distributed. This result explains the significance of the normal distribution in inferential statistics.

Another useful way to understand the special role of the normal distribution in statistics is to note that, among all distributions of continuous random variables with a fixed mean and variance, it is the distribution with maximum[1] differential entropy. The principle of indifference states that when there is a complete lack of information then a uniform distribution of all possible observations should be assumed. The **principle of maximum entropy** is a generalization of the principle of indifference: it advises choosing the distribution with maximum entropy among those that are consistent with the data. Therefore, the normal distribution is obtained from this principle when only the mean and variance of a continuous variable are known.

> The **Cauchy–Lorentz distribution** with location parameter $x_0 \in \mathbb{R}$ and scale parameter $\gamma \in \mathbb{R}$, $\gamma > 0$, is given by the following probability density function, $u \in \mathbb{R}$:
>
> $$\mathcal{L}(u|x_0, \gamma) = \frac{1}{\pi\gamma} \cdot \frac{1}{1 + \left(\frac{u - x_0}{\gamma}\right)^2}$$

The Cauchy–Lorentz distribution (also known as the Cauchy distribution) is an example for a class of continuous distributions for which neither the expected value nor the variance exist. For all parameter assignments, the normalization condition that characterizes a probability density is satisfied:

$$
\int_{-\infty}^{\infty} \mathcal{L}(\xi|x_0, \gamma) \, d\xi = \int_{-\infty}^{\infty} \frac{1}{\pi\gamma} \cdot \frac{1}{1 + \left(\frac{\xi - x_0}{\gamma}\right)^2} \, d\xi = \frac{1}{\pi} \cdot \int_{-\infty}^{\infty} \frac{1}{1 + t^2} \, dt
$$

$$
= \frac{1}{\pi} \cdot \left( \int_{-\infty}^{0} \frac{1}{1 + t^2} \, dt + \int_{0}^{\infty} \frac{1}{1 + t^2} \, dt \right)
$$

$$
= \frac{1}{\pi} \cdot \left( \lim_{t \to -\infty} (\arctan(0) - \arctan(t)) + \right.
$$

$$
\left. \lim_{t \to \infty} (\arctan(t) - \arctan(0)) \right)
$$

$$
= \frac{1}{\pi} \cdot \left( 0 - \left(-\frac{\pi}{2}\right) + \frac{\pi}{2} - 0 \right) = 1
$$

However, the corresponding improper integrals for the expected value and variance do *not* converge. The median and mode of the Cauchy distribution do exist though. Both are given by the location parameter $x_0$.

Although the graph of the Cauchy density looks similar to that of the normal distribution (a bell curve, see Fig. 4.2 on the left), unlike the latter, it is a **fat-tailed probability density**: values far from the median $x_0$ (i.e., **outliers**) of a Cauchy-distributed random variable are more likely to occur.

---

[1] This property of the normal distribution can be proved using the calculus of variations: maximize the functional $H[p] = -\int_{-\infty}^{\infty} p(\xi) \cdot \ln(p(\xi)) \, d\xi$, subject to the constraints $\int_{-\infty}^{\infty} p(\xi) \, d\xi = 1$, $\int_{-\infty}^{\infty} \xi \cdot p(\xi) \, d\xi = \mu$ and $\int_{-\infty}^{\infty} (\xi - \mu)^2 \cdot p(\xi) \, d\xi = \sigma^2$.

For the parameters $x_{\min}, \alpha \in \,]0, \infty[$, the **Pareto distribution** is given by the following probability density function:

$$\text{Par}(u|x_{\min}, \alpha) = \begin{cases} \alpha x_{\min}^{\alpha} \cdot u^{-(\alpha+1)} & \text{if } u \geq x_{\min} \\ 0 & \text{otherwise} \end{cases}$$

A Pareto-distributed random variable $X \sim \text{Par}(\,\cdot\,|x_{\min}, \alpha)$ has the following expected value:

$$E[X] = \begin{cases} \infty & \text{if } 0 < \alpha \leq 1 \\ \frac{\alpha x_{\min}}{\alpha - 1} & \text{if } 1 < \alpha \end{cases}$$

The variance is given as follows:

$$\sigma^2[X] = \begin{cases} \infty & \text{if } 0 < \alpha \leq 2 \\ \frac{\alpha (x_{\min})^2}{(\alpha-1)(\alpha-2)} & \text{if } 2 < \alpha \end{cases}$$

## 4.2 Laws of large numbers

Our intuition tells us that the relative frequency of a value occurring in a large enough sample should be a good estimate for the probability of that occurrence. For example, when we roll a six-sided die, we can imagine that each outcome is a realization of a discrete random variable $X$ with $\text{supp}(X) = \{1, \ldots, 6\}$. If we continue to roll the die, we will usually find that each number appears roughly one-sixth of the time. Therefore, we conclude that $X$ follows a uniform distribution.

The same holds true for other characteristics of random variables and their distributions, such as the mean or variance. Our goal is to estimate these characteristics from observations of the data. We expect that these estimates become more accurate as we have more observations or a larger sample size. In this section, we will provide mathematical arguments to support this intuition, known as the **laws of large numbers**.

### 4.2.1 Bernoulli's law of large numbers

The relative frequency with which an event occurs upon repeated observation is also known as the empirical probability of that event. In the following example, we want to show **Bernoulli's law of large numbers**: the true probability can be well estimated through the empirical probability. More importantly, we can also determine the accuracy of this estimate as a function of the sample size.

To formalize this concept, we consider a discrete random variable $X_1$ that takes the value one with probability $p$ (where $0 \leq p \leq 1$) and zero with probability $1 - p$. In other words, the variable follows a Bernoulli distribution. For example,

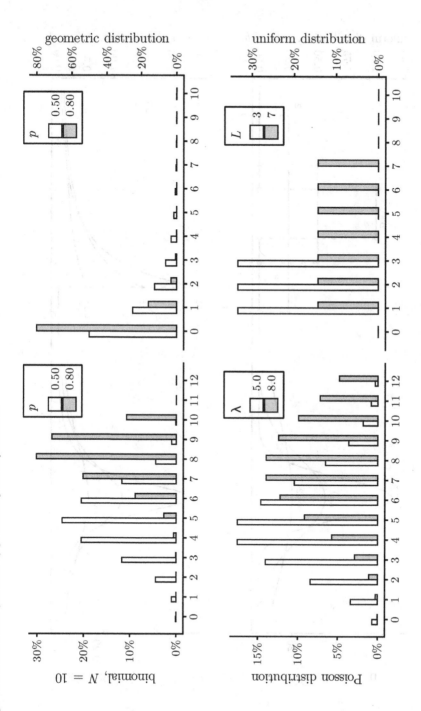

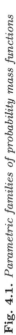

**Fig. 4.1.** *Parametric families of probability mass functions*

**Fig. 4.2.** *Parametric families of probability density functions*

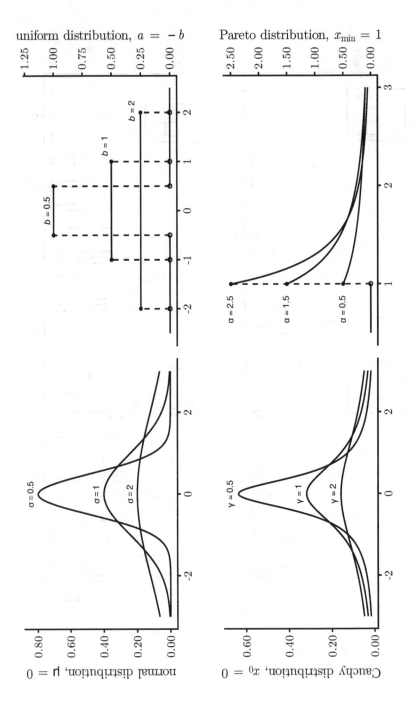

we can imagine tossing a coin that may not be fair, i.e., $p \neq 1/2$. The events $(X_1)^{-1}(1)$ and $(X_1)^{-1}(0)$ correspond to the possible outcomes of heads and tails, respectively, which we also call success or failure.

Furthermore, we can assume that if we toss the same coin again under the same conditions then heads and tails will occur with the same probabilities as before, and these events are independent of the outcome of the first coin toss. If we consider the second coin flip as a second random variable $X_2$, then the following formulas can be used to determine the joint distribution of $X_1$ and $X_2$:

$$\Pr(X_1 = 1) = \Pr(X_2 = 1) = p,$$
$$\Pr(X_1 = 1, X_2 = 1) = \Pr(X_1 = 1) \cdot \Pr(X_2 = 1) = p^2$$
$$\Pr(X_1 = 1, X_2 = 0) = p \cdot (1 - p)$$
$$\Pr(X_1 = 0, X_2 = 1) = (1 - p) \cdot p$$
$$\Pr(X_1 = 0, X_2 = 0) = (1 - p) \cdot (1 - p)$$

After a total of $N$ coin tosses, described by independent and Bernoulli-distributed random variables $X_1, \ldots, X_N$, the absolute frequency of occurrence of heads is given by the random variable $n(N) := \sum_{l=1}^{N} X_l$. We want to determine the distribution of $n(N)$. Among all sequences of coin flips with $k$ heads and $(N - k)$ tails, the probability that a specific sequence occurs is given by:

$$\Pr(X_1 = 1, \ldots, X_k = 1, X_{k+1} = 0, \ldots X_N = 0) = p^k (1 - p)^{N-k}$$

There are a total of $\binom{N}{k} = \frac{N!}{k!(N-k)!}$ possible combinations with exactly $k$ occurrences of heads. Therefore, the probability that $k$ out of $N$ Bernoulli trials are successful, regardless of the order of the outcomes, is given by:

$$\Pr(n(N) = k) = \binom{N}{k} \cdot p^k (1 - p)^{N-k}$$

Therefore, $n(N)$ is a binomially distributed random variable. The expected value of the relative frequency $f(N) = N^{-1} \cdot n(N)$ of the occurrence of heads among $N$ coin flips is thus determined as follows:

$$E[f(N)] = E\left[\frac{n(N)}{N}\right] = \frac{1}{N} \cdot E[n(N)] = \frac{1}{N} \cdot Np = p$$

This means that the expected value for the relative frequency $f(N)$ of a successful trial does indeed correspond to the probability of success. Accordingly, we say that $f(N)$ is an **unbiased estimator** of the parameter $p$.

To determine the accuracy of the empirical probability, we need to compute the variance:

$$\sigma^2[f(N)] = \frac{1}{N^2} \cdot \sigma^2[n(N)] = \frac{1}{N^2} \cdot Np(1 - p) = \frac{p(1 - p)}{N}$$

We can now apply Chebyshev's inequality:

$$\Pr(|f(N) - E[f(N)]| \geq r) \leq \frac{\sigma^2[f(N)]}{r^2},$$

consequently

$$\Pr(|f(N) - p| \geq r) \leq \frac{p(1-p)}{Nr^2} \leq \frac{1}{4Nr^2}.$$

We can illustrate the practical consequences of this inequality through an example. Suppose we want to estimate the probability $p$ for the occurrence of heads such that the result is at most $r = 0.05$ away from the true probability. Since we are observing a random process, we can never be completely sure that this error bound will be met. For instance, we might be very unlucky and have heads fall a hundred times in a row, even though the coin is fair. However, we can estimate the probability of being this unlucky. Additionally, we can require that the probability that the estimate exceeds the given error bound be at most, say, $\alpha = 1\%$. From these parameters, we can compute a minimum sample size: $N \geq \frac{1}{4\alpha r^2}$. Plugging in the example values gives: $N \geq 10{,}000$.

To summarize, if the coin is flipped 10,000 times, the observed relative frequency and the actual probability of an occurrence of heads will differ by at most $\pm 0.05$ with a probability of at least 99%. Thus, with this sample size, we can be very confident that the estimate is sufficiently accurate.

Moreover, Chebyshev's inequality implies that for all $\varepsilon > 0$:

$$\lim_{N \to \infty} \Pr(|f(N) - p| < \varepsilon) = 1$$

That is, for any given error bound, an estimate that violates that bound becomes increasingly unlikely with a larger sample size. In the upper limit of an arbitrarily large sample, a precise estimate becomes almost certain, as the relative frequency $f(N)$ is a **consistent estimator** of the parameter $p$.

In summary, the following result holds:

**Bernoulli's law of large numbers.** The relative frequency of success of a series of independent Bernoulli trials is an unbiased and consistent estimator for the probability of success.

**Example.** A total of $N = 436{,}323$ individuals that participated in the CDC survey [1] specified their sex. Of these individuals, 238,911 reported being female. This corresponds to a relative frequency of $f = \frac{238{,}911}{436{,}323} = 54.8\%$.

It would only be possible to determine the actual proportion of female adults among U.S. citizens with complete accuracy by surveying approximately all 300 million people living in the U.S.

From a statistical estimation perspective, the probability that our figure of $f = 54.8\%$ differs from the actual proportion by more than $\Delta f = 1.0\%$ is at most $\alpha = \frac{1}{4N(\Delta f)^2} = 0.57\%$. It is worth noting that the probability of the estimate being off by more than $\Delta f = 0.1\%$ could be as high as $\alpha = 57\%$. Therefore, it is not appropriate to present the result with three significant digits. Instead, it should be written as $f = 55\%$.

Other independent statistics lead to different results for the proportion of persons of female sex in the U.S. population with variations that cannot be explained by the estimation error alone. For example, a study by the United Nations concludes that the proportion of female adults in the U.S. is given by $f = 50.5\%$ (as of 2018 [2]). While estimates of statistical errors are useful, it is important to keep in mind that systematic factors, such as study design or data quality, can also be significant.

### 4.2.2 Chebyshev's law of large numbers

Just as the relative frequency (under certain conditions, such as independence) is a good estimator for the probability of an event or outcome of a binary variable, we might expect the arithmetic mean to be a good estimator for the expected value of the random variable that produces the numeric data we observe.

It is important to note that this estimate itself is again a random variable. If we roll a die 50 times and record the average number, and then we repeat the experiment, we will likely get slightly different values. If we repeat the experiment many times, the arithmetic means we recorded will follow some distribution. For large samples, however, we expect them to show only a small dispersion and to be centered around the true expected value. This is the key message of **Chebyshev's law of large numbers**, which we will detail below.

Suppose that we observe numeric values $x_1, \ldots, x_N$ in a sample that are the realizations of random variables $X_n$, $n = 1, \ldots, N$. We assume that these random variables are mutually independent and follow the same distribution, a property commonly known as **independent and identically distributed**, or **i.i.d.** for short. This assumption is reasonable when the observations are the result of independently set up and identically prepared trials or when they represent a random selection from a population. However, it is important to note that this assumption may not always be justified.

In addition, we assume that the expected value $E[X_n]$ and variance $\sigma^2[X_n]$ of each variable $X_n$ exist and are finite. Since the variables have the same distribution, these values are the same for each, and we can use the notation $\mu = E[X_n]$ and $\sigma^2 = \sigma^2[X_n]$.

Next, we consider the following random variable that produces the arithmetic mean:

$$\bar{X} = \bar{X}(N) = \frac{1}{N} \sum_{n=1}^{N} X_n$$

We want to determine the expected value and variance of this variable in order to confirm that it is an unbiased and consistent estimator of the expectation $\mu$.

By virtue of the linearity of the expected value, the arithmetic mean is an unbiased estimator of $\mu$:

$$E[\bar{X}(N)] = E\left[\frac{1}{N}\sum_{n=1}^{N} X_n\right] = \frac{1}{N}\sum_{n=1}^{N} E[X_n] = \mu$$

Furthermore, since the $X_n$ are assumed to be mutually independent, they are pairwise uncorrelated:

$$E[X_k \cdot X_l] = E[X_k] \cdot E[X_l]$$

for all $k, l = 1, \ldots, N$ with $k \neq l$.

It is not difficult to check that for pairwise uncorrelated random variables, the variance can be represented as follows since all cross terms vanish:

$$\sigma^2(X_1 + \cdots + X_N) = \sigma^2(X_1) + \cdots + \sigma^2(X_N)$$

Therefore, the variance of the arithmetic mean estimator is given as follows:

$$\sigma^2[\bar{X}(N)] = \sigma^2\left[\frac{1}{N}\sum_{n=1}^{N} X_n\right] = \frac{1}{N^2}\sum_{n=1}^{N} \sigma^2[X_n] = \frac{\sigma^2}{N}$$

Once more, we can apply Chebyshev's inequality:

$$\Pr(|\bar{X}(N) - E[\bar{X}(N)]| < \varepsilon) \geq 1 - \frac{\sigma[\bar{X}(N)]^2}{\varepsilon^2} \Rightarrow$$

$$\Pr(|\bar{X}(N) - \mu| < \varepsilon) \geq 1 - \frac{\sigma^2}{\varepsilon^2 N} \Rightarrow$$

$$\lim_{N\to\infty} \Pr(|\bar{X}(N) - \mu| < \varepsilon) = 1$$

This shows that the arithmetic mean is a consistent estimator of the expected value. For large samples, the expected value $\mu$ of the underlying distribution and the arithmetic mean of the sample are unlikely to differ significantly.

Let us summarize:

**Chebyshev's law of large numbers.** Let $X_1, \ldots, X_N$ be independent and identically distributed random variables with finite expected value $\mu$ and finite variance.

Under these assumptions, the arithmetic mean $\bar{X} = \frac{1}{N}(X_1 + \cdots + X_N)$ is an unbiased and consistent estimator of the expected value $\mu$.

### 4.2.3 Variance estimation and Bessel correction

We now examine the relationship between the empirical variance and the variance of the true underlying distribution. The sample variance is estimated from a set of independent and identically distributed random variables $X_1, \ldots, X_N$ as follows:

$$S^2(N) = \frac{1}{N} \sum_{n=1}^{N} (X_n - \bar{X}(N))^2$$

Its expected value can be calculated as follows:

$$E[S^2(N)] = \frac{1}{N} \sum_{n=1}^{N} E[(X_n - \bar{X})^2]$$

$$= \frac{1}{N} \sum_{n=1}^{N} (E[X_n^2] - 2 \cdot E[X_n \cdot \bar{X}]) + E[\bar{X}^2])$$

$$= \frac{1}{N} \sum_{n=1}^{N} (\mu^2 + \sigma^2) - \frac{2}{N} \sum_{n=1}^{N} E\left[\frac{X_n}{N} \sum_{k=1}^{N} X_k\right] + E\left[\left(\frac{1}{N} \sum_{k=1}^{N} X_k\right)^2\right]$$

$$= \mu^2 + \sigma^2 - \frac{2}{N^2} \sum_{n=1}^{N} (\mu^2 + \sigma^2 + (N-1)\mu^2) +$$

$$+ \frac{1}{N^2} \left(N(\mu^2 + \sigma^2) + (N^2 - N)\mu^2\right)$$

$$= \left(1 - \frac{1}{N}\right) \cdot \sigma^2$$

Here, we have used the following relation which follows from the $X_n$ being pairwise uncorrelated:

$$E[X_k \cdot X_n] = \begin{cases} E[X_n^2] = \mu^2 + \sigma^2 & \text{if } k = n \\ E[X_n] \cdot E[X_k] = \mu^2 & \text{if } k \neq n \end{cases}$$

Thus, the expected value of the sample variance is *not* exactly equal to the variance: $S^2(N)$ is a *biased* estimator. Instead of $S^2(N)$, we can use an unbiased estimator for the variance, which is given as follows:

$$S_{\text{cor}}^2(N) = \frac{1}{N-1} \sum_{n=1}^{N} (X_n - \bar{X}(N))^2 = \frac{N}{N-1} S^2(N)$$

The factor $N/(N-1)$ is also called **Bessel correction**. The Bessel-corrected variance is an unbiased estimator: $E[S_{\text{cor}}^2(N)] = \sigma^2$.

However, both estimators yield similar values for large samples and are thus both **asymptotically unbiased**:

$$E[S_{\text{cor}}^2(N)] = \lim_{N \to \infty} E[S^2(N)] = \sigma^2$$

Just like the arithmetic mean and the empirical probability, both estimators are consistent:

$$\lim_{N\to\infty} \Pr(|S^2(N) - \sigma^2| < \varepsilon) = \lim_{N\to\infty} \Pr(|S_{\text{cor}}^2(N) - \sigma^2| < \varepsilon) = 1$$

We recall that the standard deviation is given by the square root of the variance. Estimation of the standard deviation is consistent as well. This is not a trivial result but a consequence of the so-called **continuous mapping theorem**, which essentially states that continuous functions preserve limits even for random variables [3, Theorem 3.2.10]. In general, the estimator for the standard deviation is biased.

Let us summarize the results.

> **Variance estimation.** Let $X_1, \ldots, X_N$ be independent and identically distributed random variables with a finite expected value and finite variance $\sigma^2$.
>
> The sample variance $S^2 = \frac{1}{N}\sum_{n=1}^{N}(X_n - \bar{X})^2$ is an asymptotically unbiased and consistent estimator. The Bessel-corrected sample variance $S_{\text{cor}}^2 = \frac{1}{N-1}\sum_{n=1}^{N}(X_n - \bar{X})^2$ is an unbiased and consistent estimator of the variance $\sigma^2$.
>
> The sample standard deviation, $\sqrt{S^2}$, is a consistent estimator of the standard deviation $\sigma$.

### 4.2.4 Lindeberg–Lévy central limit theorem

Chebyshev's law of large numbers states that, under fairly general conditions, the arithmetic mean of a large sample is very likely to be found close to the true mean of the underlying distribution. Perhaps surprisingly, we can be quite specific on how the averages of large samples distribute around the true expectation:

> **Lindeberg–Lévy central limit theorem.** Let $X_1, \ldots, X_N$ be independent and identically distributed random variables with finite expected value $\mu$ and finite variance $\sigma^2 > 0$.
>
> Then, for large samples, the mean $\bar{X}(N) = \frac{1}{N}(X_1 + \cdots + X_N)$ is approximately normally distributed:
>
> $$\lim_{N\to\infty} \Pr\left(a \le \sqrt{N} \cdot \frac{\bar{X}(N) - \mu}{\sigma} \le b\right) = \int_a^b \mathcal{N}(\xi|0, 1)\, d\xi$$

In general, a sequence of random variables $Z_1, Z_2, \ldots$ is said to **converge in distribution** against the random variable $Z$ if their cumulative distribution functions converge pointwise against the distribution of $Z$: $\lim_{N\to\infty} F_{Z_N}(u) =$

$F_Z(u)$ for all $u \in \mathbb{R}$—in abbreviated notation: $Z_N \xrightarrow{D} Z$. Mathematically, the central limit theorem states that the sequence of random variables

$$Z_N := \sqrt{N}\sigma^{-1}(\bar{X}(N) - \mu)$$

always converges in distribution to a standard normally distributed random variable $Z$:

$$Z_N \xrightarrow{D} Z \text{ with } Z \sim \mathcal{N}(\,\cdot\,|0, 1)$$

Statistically, the central limit theorem states that if we repeatedly collect a sufficiently large sample from the same population, the mean values of those samples will be normally distributed. This theorem has practical significance because it allows us to make precise statements about the accuracy of statistical estimations. A common misconception is that this theorem is the reason why many empirical distributions can supposedly be approximated by a normal distribution in practice. However, this is not the case.

Although the proof of the theorem requires advanced analytical tools that we will not discuss in this book (see, e.g., [3, Theorem 3.4.1]), we can understand its practical implications through a numerical example. Imagine that we repeatedly draw samples of size $N$ from some probability density $p(\,\cdot\,)$. We can represent these samples as follows:

$$x^{(1)} = \left(x_1^{(1)}, \ldots, x_N^{(1)}\right),$$

$$x^{(2)} = \left(x_1^{(2)}, \ldots, x_N^{(2)}\right), \ldots$$

Even if we know nothing about the distribution $p(\,\cdot\,)$, except that it has a finite expected value and variance, we can be confident that the associated arithmetic means $\bar{x}^{(1)}, \bar{x}^{(2)}, \ldots$ follow a normal distribution.

In the upper left of Figure 4.3, a density function $p(\,\cdot\,)$ is shown with expected value $\mu_0 \approx 0.64$ and standard deviation $\sigma_0 \approx 1.80$. To emphasize that the theorem holds for any shape of the underlying distribution, notice how the density function does not resemble a bell curve.

The other figures show histograms of a large number of means $\bar{x}^{(1)}, \bar{x}^{(2)}, \ldots$ computed from samples of size $N$ drawn from the distribution $p(\,\cdot\,)$ through numeric simulation. The limiting distribution, a normal distribution with location parameter $\mu = \mu_0$ and dispersion parameter $\sigma = \frac{\sigma_0}{\sqrt{N}}$, is also shown for comparison.

For the small sample size $N = 1$, we cannot expect the limit theorem to apply, and we simply reproduce the underlying distribution $p(\,\cdot\,)$. However, even for the relatively small sample size $N = 5$, the distribution of the arithmetic means $\bar{x}^{(k)}$ shows the typical bell shape of the normal distribution. For $N = 20$, this bell curve is much narrower, once more illustrating Chebyshev's law of large numbers—for large samples, it is very likely that the arithmetic mean is close to the true mean.

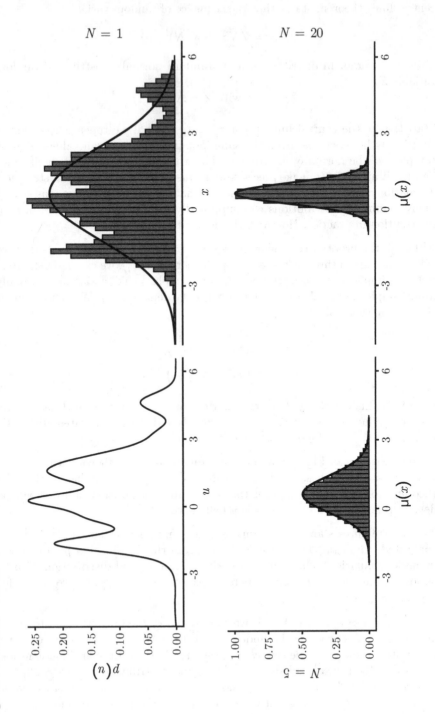

**Fig. 4.3.** *Lindeberg–Lévy central limit theorem: distribution of arithmetic means with growing sample size*

## 4.3 Interval estimation and hypothesis testing

In the derivation of the laws of large numbers, we saw that if we view the collection of a sample as the outcome of a trial with an uncertain result, we can make statements about the distribution of sample statistics, like the arithmetic mean, with repeated collection. This allows us to explore:

- Given a maximum probability of error, what is the minimum accuracy of our estimation of a distribution's parameter?

- What statistical effects, such as deviations of an estimated parameter from some default value, are unlikely to be the result of pure chance and can be considered statistically significant?

### 4.3.1 Interval estimation

The central limit theorem is important because it allows us to specify a **confidence interval** rather than just a **point estimate**. This means that instead of a single estimated value, we specify an interval that we can be reasonably sure contains the true parameter.

For example, suppose we want to ensure that our estimate of the mean is correct with 95% confidence for sufficiently large samples. We can achieve this by setting the confidence interval $[\bar{x}]\gamma = [\bar{x}_{\min}, \bar{x}_{\max}]$ with a confidence level of $\gamma = 0.95$:

$$\lim_{N \to \infty} \Pr(\bar{X}(N) \in [\bar{x}_{\min}, \bar{x}_{\max}]) = \gamma$$

We make the following ansatz with the number $z \in \mathbb{R}$, $z > 0$, which is yet to be determined:

$$\gamma = \lim_{N \to \infty} \Pr\left(\bar{X}(N) \in \left[\mu - z \cdot \frac{\sigma}{\sqrt{N}}, \mu + z \cdot \frac{\sigma}{\sqrt{N}}\right]\right)$$

The central limit theorem allows us to conclude:

$$\gamma = \lim_{N \to \infty} \Pr\left(-z \leq \sqrt{N} \cdot \frac{\bar{X}(N) - \mu}{\sigma} \leq z\right)$$
$$= \int_{-z}^{z} \mathcal{N}(\xi|0, 1)\, d\xi$$

Thus, $z = z(\gamma)$ is determined by the integral limits that produce an area of $\gamma$ under the standard normal bell curve. Consequently,

$$z(\gamma) = \Phi^{-1}\left(\frac{1 + \gamma}{2}\right),$$

where $\Phi(\cdot)$ is the standard normal cumulative distribution function:

$$\Phi(u) = \int_{-\infty}^{u} \mathcal{N}(\xi|0, 1)\, d\xi = \frac{1}{\sqrt{2\pi}} \int_{-\infty}^{u} e^{-\frac{\xi^2}{2}}\, d\xi$$

for all $u \in \mathbb{R}$. Values of $z(\gamma)$ can be determined numerically, and some frequently used ones are tabulated below. The table also lists the one-sided critical value $z^*(\gamma) = \Phi^{-1}(\gamma)$, the application of which we will explain later:

| $\gamma$ | 90.0% | 95.0% | 99.0% | 99.9% |
|---|---|---|---|---|
| $z(\gamma)$ | 1.64 | 1.96 | 2.58 | 3.29 |
| $z^*(\gamma)$ | 1.28 | 1.64 | 2.33 | 3.09 |

**Table 4.1.** *Critical values as a function of confidence level*

The **interval estimate of the mean** at confidence level $\gamma$ based on a sufficiently large sample $x = (x_1, \ldots, x_N)$ is given as follows:

$$[\bar{x}]_\gamma = \left[ \bar{x} - z(\gamma) \cdot \frac{s(x)}{\sqrt{N}}, \bar{x} + z(\gamma) \cdot \frac{s(x)}{\sqrt{N}} \right]$$

In order to arrive at the above formula, we have replaced the mean $\mu$ and the standard deviation $\sigma$ with their empirical estimates. This is a reasonable approximation for sufficiently large samples and can be justified by **Slutsky's theorem**: If $A_N \xrightarrow{D} a$ and $B_N \xrightarrow{D} b$ with *constants* $a, b \in \mathbb{R}$, then $A_N \cdot Z_N + B_N \xrightarrow{D} aZ + b$ holds if $Z_N \xrightarrow{D} Z$. Thus, for sufficiently large samples, we can replace constant summands and factors, such as $\mu$ and $\sigma$, with their consistent estimators. Common rules of thumb for the sample to be sufficiently large in order to apply these approximations are $N > 30$ or $N > 50$.

Instead of the confidence level $\gamma$, the **significance level**, or **probability of error**, $\alpha = 1 - \gamma$ can be specified.

**Example.** The 99.9% confidence interval for the average body height of male respondents in the CDC survey is given as follows:

$$[\mu(\text{body height})]_{0.999} = [177.98 \,\text{cm}; 178.10 \,\text{cm}]$$

This high statistical accuracy is due to the large sample size $N$ with more than 190,000 male persons who were interviewed.

We want to demonstrate how interval estimation works for a smaller sample size. To this end, we randomly sample a set of $N = 50$ body height values one hundred times and compute the corresponding 95% confidence interval. The result can be seen in Fig. 4.4 above. Notice that most of the confidence intervals, shown as vertical error bars, also contain the true value of 178 cm, shown as a horizontal dashed line. However, five of the intervals do not contain it—this is expected by construction and is consistent with the specified error probability of $\alpha = 5\%$. There is always the possibility that the interval estimate will miss the true mean of the population, especially at low confidence levels.

The sample of male respondents' height examined in the example above has an empirical standard deviation of $s(x) = 7.85$ cm. By Chebyshev's inequality, we can expect that at least $\delta = 95\%$ of the data points lie within the following interval around the expected value:

$$\left[\bar{x} - \frac{1}{\sqrt{1-\delta}} \cdot s(x), \bar{x} + \frac{1}{\sqrt{1-\delta}} \cdot s(x)\right] = \left[178\,\text{cm} \pm \frac{1}{\sqrt{0.05}} \cdot 7.85\,\text{cm}\right]$$

$$= [143\,\text{cm}, 213\,\text{cm}]$$

A more accurate estimate can be obtained using quantiles:

$$[Q_{0.025}(x), Q_{0.975}(x)] = [163\,\text{cm}, 193\,\text{cm}]$$

Intervals of this type are called **prediction intervals**. It is important to distinguish prediction intervals from confidence intervals. We expect most values produced by some distribution to remain within the prediction interval. The confidence interval, on the other hand, represents a range that is very likely to contain some key characteristic of the distribution, such as the expected value or a model parameter.

**Example.** Based on data from the ALLBUS survey 2018 [4], we can compute the 95% confidence interval for the average monthly net household income in Germany:
$$[\mu(\text{income})]_{0.95} = [3066\,\text{EUR}, 3236\,\text{EUR}]$$
In this case, the statistical inaccuracies are no longer negligible. Notice that the sample size is relatively small ($N = 2530$ individuals), and the standard deviation with which the income disperses around the mean is relatively high: 2179 EUR.

### 4.3.2 Z-test

One important application of the central limit theorem is in hypothesis testing. Suppose that we have reason to believe that the expected value $E[X]$ of a random variable $X$ is *not* equal to zero. In that case, we want to disprove the **null hypothesis** $E[X] = 0$ in favor of the **alternative hypothesis** $E[X] \neq 0$. At a confidence level of $\gamma$, we can reject the null hypothesis $E[X] = 0$ if the arithmetic mean $\bar{x}$ of the sequence of observed values is not contained in the following interval:

$$\left[-z(\gamma) \cdot \frac{s(x)}{\sqrt{N}}, z(\gamma) \cdot \frac{s(x)}{\sqrt{N}}\right]$$

As usual, we assume that the sample $x = (x_1, \ldots, x_N)$ is sufficiently large and represents a sequence of realizations of independent random variables with the same distribution as the model variable $X$. If the null hypothesis states that the expected value is some specific value $\mu$ that is not necessarily zero, then the interval becomes:

$$\left[ \mu - z(\gamma) \cdot \frac{s(x)}{\sqrt{N}}, \mu + z(\gamma) \cdot \frac{s(x)}{\sqrt{N}} \right]$$

Let us consider the example of a possibly biased coin. We record the result of each coin flip, which yields a sequence of binary values, where $x_n = 1$ represents heads and $x_n = 0$ represents tails. The arithmetic mean of that sequence is equal to the relative frequency $f$ of heads:

$$\bar{x} = \frac{1}{N} \sum_{n=1}^{N} x_n = f$$

The probability of success $p$ is equal to the expected value of the underlying Bernoulli-distributed random variable: $E[X] = p \approx f$.

Suppose we have reason to believe that the coin is not fair. The null hypothesis claims that the coin is fair, i.e., $E[X] = 1/2$. In a first experiment, we observe that after ten tosses, the coin lands with heads on top seven times. At a confidence level of $\gamma = 0.95$, the confidence interval for this experiment is the following:

$$\left[ 0.5 - 1.96 \cdot \frac{0.42}{\sqrt{10}}, 0.5 + 1.96 \cdot \frac{0.42}{\sqrt{10}} \right] = [0.24, 0.76]$$

The observed proportion of 0.7 is still within this interval. Therefore, the null hypothesis of a fair coin cannot be rejected at the given confidence level. The sample size is relatively small, so it is actually recommended to use the **Student's $t$-test**, which we discuss in the next section. A $t$-test would correct the score $z(0.95) \approx 1.96$ to $z_9(0.95) \approx 2.26$, and thus result in an even wider confidence interval.

If, on the other hand, we observed seventy out of one hundred coin tosses with the outcome of heads, the following confidence interval would be the result:

$$\left[ 0.5 - 1.96 \cdot \frac{0.46}{\sqrt{100}}, 0.5 + 1.96 \cdot \frac{0.46}{\sqrt{100}} \right] = [0.41, 0.59]$$

In this case, the observed proportion of 0.7 is *not* contained in the confidence interval. Thus, the null hypothesis should be rejected, and we may conclude—at the given confidence level—that the coin is biased.

The procedure outlined above is the so-called **one-sample $Z$-test**, where we compare the mean of a single population to a fixed value. Another important test is concerned with whether the difference of the means of two populations vanishes, or whether one mean is larger or smaller than the other. Fortunately, the difference of two normally distributed and independent random variables $X \sim \mathcal{N}(\cdot | \mu_X, \sigma_X^2)$ and $Y \sim \mathcal{N}(\cdot | \mu_Y, \sigma_Y^2)$ is also normally distributed:

$$Y - X \sim \mathcal{N}(\cdot | \mu_Y - \mu_X, \sigma_X^2 + \sigma_Y^2)$$

The following test criteria can be formulated as a result.

**Two-sample $Z$-test.** Let $x = (x_1, \ldots, x_{N_x})$ and $y = (y_1, \ldots, y_{N_y})$ be sequences of observations that we may assume to be realizations of identically distributed random variables $X_1, \ldots, X_{N_x}$ and $Y_1, \ldots, Y_{N_y}$, respectively. We further assume that $X_1, \ldots, X_{N_x}, Y_1, \ldots, Y_{N_y}$ are mutually independent in addition to having a finite mean and variance.

**Two-sided $Z$-test.** At a confidence level of $0 < \gamma < 1$, we reject the null hypothesis $E[X] = E[Y]$ if the following holds:

$$|\bar{y} - \bar{x}| > z(\gamma) \cdot \sqrt{\frac{s^2(x)}{N_x} + \frac{s^2(y)}{N_y}}$$

where $z(\gamma) = \Phi^{-1}\left(\frac{1+\gamma}{2}\right)$ is the **critical value**.

**One-sided $Z$-test.** At a confidence level of $0 < \gamma < 1$, we reject the null hypothesis $E[X] \leq E[Y]$ if the following holds:

$$\bar{y} - \bar{x} > z^*(\gamma) \cdot \sqrt{\frac{s^2(x)}{N_x} + \frac{s^2(y)}{N_y}}$$

where $z^*(\gamma) = \Phi^{-1}(\gamma)$. For example, if $\gamma = 0.95$, then $z^*(\gamma) = 1.645$.

---

**Example.** According to the ALLBUS survey, the average monthly net income of one-person households in Germany is 1668 EUR. However, the difference between the income of female and male respondents appears significant: 1535 EUR versus 1788 EUR.

We may therefore conjecture that women in Germany have lower income compared to men, and we want to test if the data are consistent with this hypothesis. A total of $N_y = 269$ men and $N_x = 267$ women provided information on their income. The respective standard deviations are $s(y) = 1083$ EUR and $s(x) = 827$ EUR. At a confidence level of 95%, the one-sided $Z$-test requires rejecting the null hypothesis "women in Germany earn at least as much as men" if the observed income difference is greater than the following value:

$$1.645 \cdot \sqrt{\frac{s^2(x)}{N_x} + \frac{s^2(y)}{N_y}} = 1.645 \cdot \sqrt{\frac{(827\,\text{EUR})^2}{169} + \frac{(1083\,\text{EUR})^2}{267}}$$

$$\approx 137\,\text{EUR}$$

The value is below the observed income difference of 253 EUR. The data thus provides sufficient reason to reject the null hypothesis in favor of the alternative hypothesis "women in Germany earn less than men." The gender pay gap is in fact a global and well-documented phenomenon [5].

### 4.3.3 Student's *t*-test

So far, we have always assumed that the sample size is "sufficiently large" in order to apply limit theorems appropriately. We now present a method that may also apply to small samples.

Let us recall the central limit theorem, which determines the limiting distribution of the test score that is the deviation from the mean in multiples of the standard deviation:

$$\lim_{N \to \infty} \Pr\left(-z \le \sqrt{N} \cdot \frac{\bar{X}(N) - \mu}{\sigma} \le z\right) = \int_{-z}^{z} \mathcal{N}(\xi|0,1)\,d\xi$$

The standard deviation $\sigma$ is assumed to be known. In practical applications, the standard deviation is usually *not* known and must be estimated from the data. Fortunately, the above formula remains true if we replace the true standard deviation by its consistent estimator:

$$\lim_{N \to \infty} \Pr\left(-z \le \sqrt{N} \cdot \frac{\bar{X}(N) - \mu}{\sqrt{S_{\text{cor}}^2(N)}} \le z\right) = \int_{-z}^{z} \mathcal{N}(\xi|0,1)\,d\xi$$

The above formula holds exactly only in the limit, and we may assume that it holds well enough for large samples. Without further assumptions, it appears difficult to make more accurate statements about the distribution of the test score that could be applied to small samples. It turns out that if we assume the random variables $\bar{X}(N)$ and $S_{\text{cor}}^2(N)$ to be independent, then we know the distribution of both the data and the test score exactly for any sample size $N$.

> **Conditions on a *t*-distributed test score.** Let $X_1, \ldots, X_N$, $N > 1$, be independent and identically distributed random variables with finite expected value $\mu$ and finite variance $\sigma^2$.
>
> If the estimators for the expected value $\bar{X}(N)$ and the variance $S_{\text{cor}}^2(N)$ are independent random variables, then the following holds:
>
> 1. $X_1, \ldots, X_N$ are normally distributed,
>
> 2. the test variable
> $$T(N) = \sqrt{N} \cdot \frac{\bar{X}(N) - \mu}{\sqrt{S_{\text{cor}}^2(N)}}$$
> follows a *t*-distribution with $N - 1$ degrees of freedom.

We briefly outline a proof. **Geary's theorem** and its extension by Lukacs [6, 7] state that the usual unbiased estimators for expectation and variance are independent random variables if and only if $X_1, \ldots, X_N$ follow a normal distribution.

The random variable that determines the distribution of the test score can be written as follows:

$$T(N) = \sqrt{N-1} \cdot \frac{Z}{\sqrt{Y_{N-1}}}$$

with

$$Z = \sqrt{N} \cdot \frac{\bar{X}(N) - \mu}{\sigma} \quad \text{and} \quad Y_{N-1} = \frac{(N-1) \cdot S^2_{\text{cor}}(N)}{\sigma^2}.$$

Since the random variables that produce the observed data follow exactly a normal distribution, $Z$ is standard normally distributed, a fact which we already established in Sect. 4.1.2.

We define the following new random variable:

$$\tilde{X}_n := \frac{1}{\sqrt{n \cdot (n+1)}\sigma} \cdot \left( -n \cdot X_{n+1} + \sum_{k=1}^{n} X_k \right)$$

for all $n \in \{1, \ldots, N-1\}$. It is not difficult to check that $\tilde{X}_1, \ldots, \tilde{X}_N$ are standard normally distributed. Moreover, their independence can be proved, and the following holds:

$$Y_{N-1} = \sum_{n=1}^{N-1} \left( \tilde{X}_n \right)^2$$

Therefore, $Y_{N-1}$ is distributed according to a chi-squared distribution with $N-1$ degrees of freedom, and thus $T(N)$ follows a $t$-distribution also with $N-1$ degrees of freedom.

Let us use $\Phi_{N-1}(\cdot)$ to denote the corresponding distribution function:

$$\Phi_{N-1}(u) = \int_{-\infty}^{u} t_{N-1}(\xi)\, d\xi$$

for $u \in \mathbb{R}$. The confidence intervals that we had used for interval estimates and hypothesis tests in previous sections can be easily modified by replacing the normal distribution with a $t$-distribution. At a confidence level of $\gamma$ and for samples of size $N$, the critical values are given as follows:

$$z_{N-1}(\gamma) = \Phi_{N-1}^{-1}\left( \frac{1+\gamma}{2} \right) \quad \text{and} \quad z^*_{N-1}(\gamma) = \Phi_{N-1}^{-1}(\gamma)$$

For sufficiently large samples, these critical values lead to the same confidence intervals as those used in the $Z$-test since the $t$-distribution is identical to the normal distribution in the limit of a large number of degrees of freedom:

$$\lim_{N \to \infty} z_{N-1}(\gamma) = z(\gamma), \quad \lim_{N \to \infty} z^*_{N-1}(\gamma) = z^*(\gamma)$$

At a 95% confidence level, the following critical values arise as a function of sample size:

| $N$ | 5 | 15 | 30 | 50 | 100 | 1000 | $\infty$ |
|---|---|---|---|---|---|---|---|
| $z_{N-1}(0.95)$ | 2.78 | 2.14 | 2.05 | 2.01 | 1.98 | 1.96 | 1.96 |
| $z^*_{N-1}(0.95)$ | 2.13 | 1.76 | 1.70 | 1.68 | 1.66 | 1.65 | 1.64 |

**Table 4.2.** *Critical values as a function of sample size*

**Example.** We can assume that the height of male participants in the CDC survey is approximately normally distributed, see Fig. 4.6. The large sample size of more than 190,000 individuals allows for a very accurate estimate of the average height: 178 cm. We want to demonstrate how this value can be estimated with a $t$-test on the basis of much smaller samples. To do this, we randomly select a set of five values $x_1, \ldots, x_5$ for height and compute the associated 95% confidence interval:

$$\left[ \bar{x} \pm z_4(0.95) \cdot \frac{s(x)}{\sqrt{5}} \right] \approx [\bar{x} \pm 1.24 \cdot s(x)]$$

We do this one hundred times, each time producing a different sample of size $N = 5$. Most of the confidence intervals also contain the true value of 178 cm, see Fig. 4.4 below. We can also clearly see that the standard deviation estimated from a very small sample is itself subject to large statistical variation. Four of the hundred interval estimates do *not* contain the true value, reflecting the chosen error probability of $\alpha = 5\%$.

### 4.3.4 Effect size

Especially for very large samples, it can be quite common to detect statistically significant differences in the mean or other types of effects that are considered significant according to statistical tests. However, those effects might still be small in magnitude.

**Example.** The CDC dataset allows for comparisons between different U.S. states. For example, we can compare the average height of male respondents in Rhode Island $\bar{x}$ with those in New York $\bar{y}$. We obtain the following test score for the two-sided $Z$-test at a confidence level of 95%:

$$1.96 \cdot \sqrt{\frac{s^2(x)}{N_x} + \frac{s^2(y)}{N_y}} = 1.96 \cdot \sqrt{\frac{(7.68 \,\text{cm})^2}{2391} + \frac{(8.23 \,\text{cm})^2}{15843}}$$

$$\approx 0.33 \,\text{cm}$$

This value is below the observed difference of $|\bar{y} - \bar{x}| = 0.44 \,\text{cm}$. Therefore, the difference is statistically significant. However, it is very small in magnitude, and therefore we can expect it to be of little practical relevance.

In many cases, the **effect size** can be gauged well by specifying the effect in natural units. In the above example, we chose metric units of length. Another possibility is to specify it in units corresponding to a multiple of the standard deviation.

For two samples $x = (x_1, \ldots, x_{N_x})$ and $y = (y_1, \ldots, y_{N_y})$, **Cohen's $d$** is a measure of the practical relevance of a statistical effect, defined as follows:

$$d(y, x) = \frac{\bar{y} - \bar{x}}{s_{\text{pool}}(x, y)}$$

with the **pooled variance** [8, p. 67]:

$$s_{\text{pool}}^2(x, y) = \frac{N_x s^2(x) + N_y s^2(y)}{N_x + N_y - 2}$$

**Example.** The difference $|\bar{y} - \bar{x}| = 0.44$ cm observed in the example above corresponds to a value of $d(y, x) = 0.05$ for Cohen's $d$. When we compare the average height of respondents in Puerto Rico $\bar{z}$ with those in New York, we get $d(y, z) = 0.50$, corresponding to a difference in metric units of $\bar{y} - \bar{z} = 4.1$ cm.

Rules of thumb for interpreting values of Cohen's $d$ are noted in the following table [9]:

| $|d(y, x)|$ | 0.01 | 0.2 | 0.5 | 0.8 | 1.2 | 2.0 |
|---|---|---|---|---|---|---|
| effect size | negligible | small | medium | large | very large | huge |

**Table 4.3.** *Effect size according to Cohen's d*

## 4.4 Parameter and density estimation

For a sequence of numerical observations $x = (x_1, \ldots, x_N)$, the **empirical cumulative distribution function** is given by the proportion of data points below a certain value:

$$\hat{F} \colon \mathbb{R} \to [0, 1], \; \hat{F}(u) = \frac{1}{N} \cdot |\{m \in \{1, \ldots, N\} | x_m \leq u\}|$$

The concept is closely related to the percentage rank (see Sect. 2.5.2): $\hat{F}(x_n) = \%\text{-rg}(x_n)$ holds for all $n \in \{1, \ldots, N\}$.

The histogram, which is based on a division of the real number line into intervals $]u_k, u_{k+1}] \subset \mathbb{R}$, $k \in \mathbb{Z}$, can be expressed using the empirical cumulative distribution function:

$$\hat{p} \colon \mathbb{R} \to [0, \infty[, \; \hat{p}(u) = \frac{\hat{F}(u_{k+1}) - \hat{F}(u_k)}{u_{k+1} - u_k} \text{ for all } u \in ]u_k, u_{k+1}]$$

The **Glivenko–Cantelli theorem** is another important law of large numbers (see [10, Theorem 8.2.4] for details). For a sequence of independent and identically distributed random variables, the empirical cumulative distribution function converges uniformly against the true distribution function.

Under certain conditions, with the bin width as a function of the sample size, the histogram is a consistent estimator of the underlying probability density

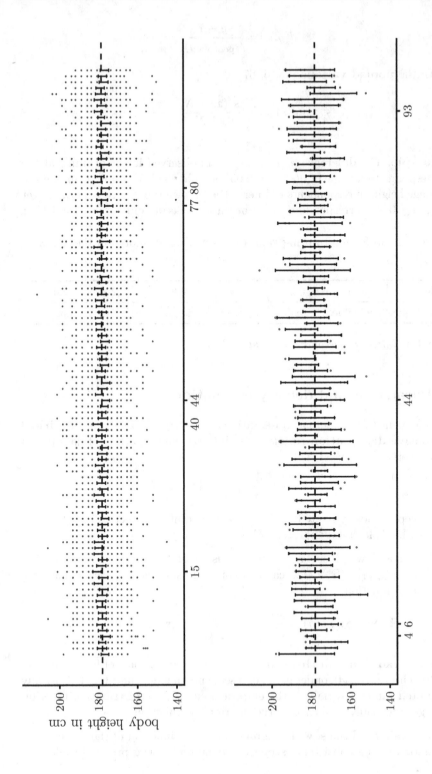

**Fig. 4.4.** *95% confidence intervals for sample size N = 50 (top) and N = 5 (bottom)*

function (see [11, Sect. 6.2]). We can therefore regard the histogram to be an empirical density function. In what follows, we will present other methods of density estimation. When these methods are based on the assumption that the observations are produced by a particular statistical model, the task of density estimation becomes one of determining the optimal parameters from the observed data.

### 4.4.1 Maximum likelihood estimation

Suppose we have good reason to believe that a statistical variable is distributed according to some specific parametric model. We want to determine the parameters of the model in such a way that it matches the observed distribution as closely as possible.

In order to accomplish this, for a sequence of observed values $x = (x_1, \ldots, x_N)$ and a statistical model $p(\,\cdot\,|\theta_1, \ldots, \theta_K)$, we consider the **likelihood function**:

$$L(\theta_1, \ldots, \theta_K | x_1, \ldots, x_N) = p(x_1|\theta_1, \ldots, \theta_K) \cdots p(x_N|\theta_1, \ldots, \theta_K)$$

$$= \prod_{n=1}^{N} p(x_n|\theta_1, \ldots, \theta_K)$$

As usual, we assume that each observation is produced by a random variable. We also assume that the random variables have equal distribution and are mutually independent. Under these assumptions, the likelihood function is the joint probability function of the random variables, evaluated at the point realized by the sample.

A useful criterion for selecting optimal parameters $\hat{\theta}_1, \ldots, \hat{\theta}_K$ is to maximize the likelihood function, which are the parameters that make the observations most likely under the model assumption.

For many common models, the calculation simplifies if we use the **log-likelihood function** instead of the likelihood function, which assumes its maximum at the same point:

$$\ell(\theta_1, \ldots, \theta_K | x_1, \ldots, x_N) = \ln(L(\theta_1, \ldots, \theta_K | x_1, \ldots, x_N))$$

$$= \ln \left( \prod_{n=1}^{N} p(x_n|\theta_1, \ldots, \theta_K) \right)$$

$$= \sum_{n=1}^{N} \ln(p(x_n|\theta_1, \ldots, \theta_K))$$

To simplify the notation, we will frequently omit the dependence on the observations and assume those values as fixed: $\ell(\theta_1, \ldots, \theta_K) = \ell(\theta_1, \ldots, \theta_K | x_1, \ldots, x_N)$.

In summary, the procedure can be described as follows.

Suppose $p(\,\cdot\,|\theta_1,\ldots,\theta_K)$ is a statistical model and $x = (x_1,\ldots,x_N)$ is a sample. The **maximum likelihood estimate** $\hat{\theta}_1,\ldots,\hat{\theta}_K$ of the parameters is given by the maximum point of the log-likelihood function:

$$\ell(\theta_1,\ldots,\theta_K) = \sum_{n=1}^{N} \ln(p(x_n|\theta_1,\ldots,\theta_K))$$

Consequently, the estimated density is the function $\hat{p}(\,\cdot\,) = p(\,\cdot\,|\hat{\theta}_1,\ldots,\hat{\theta}_K)$.

If the likelihood of an observation vanishes, i.e., $p(x_n|\theta_1,\ldots,\theta_K) = 0$ for at least one $n \in \{1,\ldots,N\}$, we may formally set $\ell(\theta_1,\ldots,\theta_K) = -\infty$.

As an example, let us calculate the log-likelihood for the Pareto distribution. We can assume that all observed values are at least as large as the parameter $x_{\min}$ because the likelihood vanishes otherwise:

$$\ell(x_{\min},\alpha) = \sum_{n=1}^{N} \ln\left(\mathrm{Par}(x_n|x_{\min},\alpha)\right) = \sum_{n=1}^{N} \ln\left(\frac{\alpha x_{\min}^{\alpha}}{x_n^{\alpha+1}}\right)$$

$$= N \ln\alpha + N\alpha \ln x_{\min} - (\alpha+1)\sum_{n=1}^{N} \ln x_n$$

The log-likelihood grows with $x_{\min}$. Therefore, we want to choose the parameter that is just large enough to agree with what we learn from the sample:

$$\hat{x}_{\min} = \min_{n\in\{1,\ldots,N\}} \{x_n\}$$

The partial derivative with respect to the second parameter $\alpha$ can be calculated as follows:

$$\frac{\partial\ell}{\partial\alpha}(\hat{x}_{\min},\alpha) = \frac{N}{\alpha} + N\ln\hat{x}_{\min} - \sum_{n=1}^{N} \ln x_n$$

The derivative is monotonically decreasing with $\alpha$. Therefore, the maximum point of the log-likelihood function is given by the zero of the derivative:

$$\hat{\alpha} = \frac{N}{\sum_{n=1}^{N} \ln x_n - N\ln\hat{x}_{\min}} = \left(\left\langle \ln\frac{x}{\hat{x}_{\min}} \right\rangle\right)^{-1}$$

where $\langle\,\cdot\,\rangle$ denotes the arithmetic mean.

**Example.** We believe that all higher incomes (more than 2700 EUR) of German households according to the ALLBUS survey can be modeled by a Pareto distribution. After plugging the surveyed income values into the above formula, we get $x_{\min} = 2750\,\mathrm{EUR}$ and $\alpha = 2.36$ for the optimal parameters.

The following figure shows a histogram of the data compared to the Pareto density.

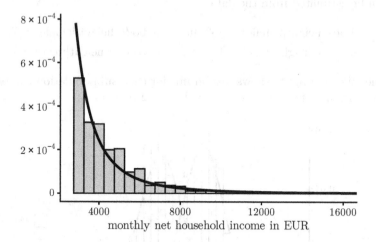

**Fig. 4.5.** *Fit of a Pareto distribution to higher incomes in Germany*

Now, given a sequence of numeric observations $x = (x_1, \ldots, x_N)$, suppose that we want to fit a normal distribution. The log-likelihood as a function of the location parameter $\mu$ and the scale parameter $\sigma$ computes as follows:

$$\ell(\mu, \sigma) = \sum_{n=1}^{N} \ln(\mathcal{N}(x_n | \mu, \sigma^2)) = \sum_{n=1}^{N} \ln\left(\frac{1}{\sigma\sqrt{2\pi}} \cdot e^{-\frac{1}{2} \cdot \frac{(x_n - \mu)^2}{\sigma^2}}\right)$$

$$= -\frac{1}{2\sigma^2} \sum_{n=1}^{N} (x_n - \mu)^2 - N \ln(\sigma) - \frac{N}{2} \ln(2\pi)$$

The gradient of this function:

$$\text{grad}\, \ell(\mu, \sigma) = \begin{pmatrix} \frac{\partial \ell}{\partial \mu}(\mu, \sigma) \\ \frac{\partial \ell}{\partial \sigma}(\mu, \sigma) \end{pmatrix} = \begin{pmatrix} \frac{1}{\sigma^2} \sum_{n=1}^{N} (x_n - \mu) \\ \frac{1}{\sigma^3} \sum_{n=1}^{N} (x_n - \mu)^2 - \frac{N}{\sigma} \end{pmatrix}$$

Any maximum $(\hat{\mu}, \hat{\sigma})$ must be a stationary point, i.e., a point where the gradient vanishes:

$$\hat{\mu} = \frac{1}{N} \sum_{n=1}^{N} x_n = \bar{x}, \quad \hat{\sigma}^2 = \frac{1}{N} \sum_{n=1}^{N} (x_n - \hat{\mu})^2 = s^2(x)$$

By examining the Hessian matrix of second derivatives (see Sect. B.3.3 in the appendix of this book or [12, Sect. 7.6]), we may confirm that these parameters do indeed represent a maximum point. Thus, according to the maximum likelihood paradigm, we simply need to compute the empirical mean and the variance of the data, and then we plug these values into the model as parameters.

**Example.** We believe that the body height of female and male respondents in the CDC survey are normally distributed. The mean and standard deviation can be estimated from the data:

$$\hat{\mu}(\text{body height}|\text{female}) = 163\,\text{cm}, \qquad \hat{\sigma}(\text{body height}|\text{female}) = 7.3\,\text{cm}$$
$$\hat{\mu}(\text{body height}|\text{male}) = 178\,\text{cm}, \qquad \hat{\sigma}(\text{body height}|\text{male}) = 7.8\,\text{cm}$$

The following figure shows histograms for each subpopulation compared to the normal distributions with the above parameters:

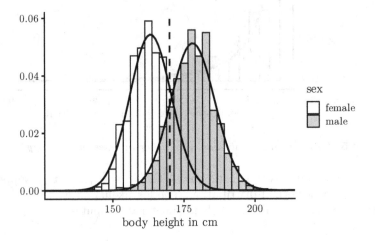

**Fig. 4.6.** *Normal fit to distribution of body height*

The vertical dashed line denotes a body height of 170 cm. This point serves as an optimal division of the subpopulations based on body height. In Sect. 6.3.3, we will explain how to compute this so-called *decision boundary*.

The maximum likelihood method can also be applied to models of discrete/categorical variables, as we will illustrate with the following example. We participate in a lottery in which colored balls are drawn randomly from an opaque urn with replacement. The number of balls is unknown, as is the number of colors. However, we have reason to believe that each color appears with equal probability. If we number the colors from 1 to $K$, this assumption translates to the following uniform distribution:

$$\Pr(\{\text{color no. } k \text{ drawn}\}) = \mathcal{U}(k|K) = \begin{cases} \frac{1}{K} & \text{if } k \in \{1, \dots, K\} \\ 0 & \text{otherwise} \end{cases}$$

Suppose we observe $N$ lottery draws and note down the outcomes $k_1, \dots, k_N \in \{1, 2, \dots\}$. Each previously unobserved color $k_{n+1}$ is denoted by the next consecutive number $\max_n\{k_n\} + 1$. For example:

| $n$ | 1 | 2 | 3 | 4 | 5 | 6 | 7 | 8 $\cdots$ |
|---|---|---|---|---|---|---|---|---|
| color | red | red | yellow | blue | red | yellow | green | green $\cdots$ |
| $k_n$ | 1 | 1 | 2 | 3 | 1 | 2 | 4 | 4 $\cdots$ |

**Table 4.4.** *Outcomes of a lottery*

The log-likelihood function is given as follows:

$$
\ell(K) = \sum_{n=1}^{N} \ln(\mathcal{U}(k_n|K))
$$

$$
= \begin{cases} -\infty & \text{if } k_n > K \text{ for at least one } n \in \{1, \ldots, N\} \\ -N \ln K & \text{otherwise} \end{cases}
$$

The case $\max_n\{k_n\} > K$ corresponds to the situation where more than $K$ different colors would be drawn. Such a model would not be consistent with the observed data, so we can assume $k_n \le K$ right away.

The term $\ell(K) = -N \ln K$ decreases strictly monotonically with $K$. Therefore, we must choose a $K$ value as small as possible without contradicting the condition $k_n \le K$. This leaves only one choice as the maximum likelihood estimate for $K$:

$$
\hat{K} = \max_{n \in \{1, \ldots, N\}} \{k_n\}
$$

At the same time, $\hat{K}$ is also an estimator for the number of different colors present in the urn. If we observe that $\hat{K}$ different colors were drawn, then the optimal assumption according to the maximum likelihood paradigm is that there are a total of $\hat{K}$ different colors in the urn.

However, note that this estimator is biased. Let $\hat{K} = \hat{K}(N)$ be a function in the random variables $X_1, \ldots, X_N$ that produce the lottery's outcomes. We assume that these variables are mutually independent and have a uniform distribution $\mathcal{U}(\cdot|K_0)$, where $K_0$ denotes the true number of different colors in the urn. The probability that the $k$-th color appears in a sample of size $N$ is given by:

$$
1 - \Pr(X_1 \ne k, \ldots, X_N \ne k) = 1 - \prod_{n=1}^{N} \Pr(X_n \ne k)
$$

$$
= 1 - \prod_{n=1}^{N} (1 - \Pr(X_n = k)) = 1 - \left(1 - \frac{1}{K_0}\right)^N
$$

The expected value of the number of colors drawn is therefore:

$$
E[\hat{K}(N)] = \sum_{k=1}^{K_0} \left(1 - \left(1 - \frac{1}{K_0}\right)^N\right) = K_0 - K_0 \cdot \left(1 - \frac{1}{K_0}\right)^N
$$

Our estimator is asymptotically unbiased, $\lim_{N \to \infty} E[\hat{K}(N)] = K_0$. Still, $\hat{K}$ always yields a result that is smaller than the true value $K_0$. However, since the difference depends on the unknown value $K_0$, in practice, we cannot simply add it to the estimate in order to correct it.

### 4.4.1.1 Power transforms

Suppose that we apply a strictly monotonically increasing function $f \colon I \to \mathbb{R}$, $I \subseteq \mathbb{R}$ to a sequence of numeric observations $x_1, \ldots, x_N \in I$, and then we study the transformed data $f(x_1), \ldots, f(x_N)$ instead. For many applications, such a transformation does not change essential characteristics of the data. For example, the rank statistics stay the same. Furthermore, we can invert the transformation to retrieve the original observations so that no information is lost.

However, the distribution of the data will change. We can use this fact to our advantage and transform the data to fit some desired distribution. Two popular choices of transformations for this purpose are the families of **Box–Cox transforms** [13] $g_\lambda(\,\cdot\,)$ and **Yeo–Johnson transforms** [14] $h_\lambda(\,\cdot\,)$ :

$$g_\lambda \colon \left]0, \infty\right[ \to \mathbb{R}, \ g_\lambda(u) = \begin{cases} \lambda^{-1} \cdot (u^\lambda - 1) & \text{if } \lambda \neq 0 \\ \ln(u) & \text{if } \lambda = 0 \end{cases}$$

$$h_\lambda \colon \mathbb{R} \to \mathbb{R}, \ h_\lambda(u) = \begin{cases} \lambda^{-1} \cdot ((1+u)^\lambda - 1) & \text{if } \lambda \neq 0 \text{ and } u \geq 0 \\ \ln(1+u) & \text{if } \lambda = 0 \text{ and } u \geq 0 \\ -(2-\lambda)^{-1} \cdot ((1-u)^{2-\lambda} - 1) & \text{if } \lambda \neq 2 \text{ and } u < 0 \\ -\ln(1-u) & \text{if } \lambda = 2 \text{ and } u < 0 \end{cases}$$

The following figure shows the graph of the Yeo–Johnson transform for different values of the power parameter $\lambda$. Box–Cox transforms show similar characteristics, but their domain is restricted to positive values.

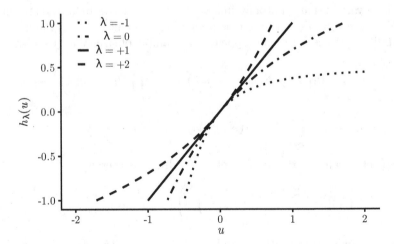

**Fig. 4.7.** *Yeo–Johnson transforms*

Given a family of transformations $f_\lambda(\cdot)$ like the above, and assuming that the transformed data follow the parametric model $p(\cdot | \theta_1, \ldots, \theta_K)$, the log-likelihood in terms of the original data is given by:

$$\ell(\lambda, \theta_1, \ldots, \theta_K) = \sum_{n=1}^{N} \ln(p(f_\lambda(x_n) | \theta_1, \ldots, \theta_K)) + \sum_{n=1}^{N} \ln\left(f_\lambda'(x_n)\right)$$

Maximizing this function with respect to the parameters, including $\lambda$, selects the optimal data transformation.

**Example.** According to the 2018 ALLBUS survey, the monthly net income in German households follows a skewed distribution, see Fig. 2.9. After applying a Box–Cox transform with power parameter $\lambda = 0.277$, the distribution resembles a normal distribution, as seen in the following figure.

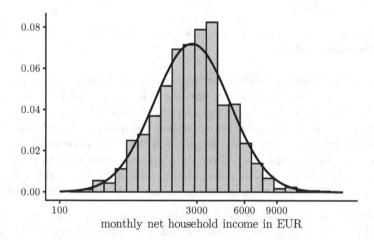

**Fig. 4.8.** *Data resembling a normal distribution after applying a Box–Cox transform*

### 4.4.2 Bayesian parameter estimation

The maximum likelihood method assumes that the distribution of observed values is determined by a statistical model with a specific set of parameters. Conversely, we can reconstruct those model parameters from the observations (by maximizing the likelihood function). Due to limited data, the model parameters may only be estimated with some finite accuracy: given different samples, estimations may vary. Still, the method assumes that there exists an underlying true set of parameters, the values of which may be unknown to the statistician but nevertheless are fixed.

Thus, in maximum likelihood estimation, the model parameters are not random variables. Conceptually, we cannot speak of a probability with which the parameters take on particular values. This is a view common in frequentist statistics.

On the other hand, **Bayesian statistics** present us with a different philosophy: we assume that the models and model parameters themselves follow some probability distribution which gauges our incomplete knowledge about their value. These distributions may in turn depend on so-called *hyperparameters*.

In order to apply this idea in a meaningful way, we modify the likelihood function as follows.

> The **posterior probability distribution** of the parameters of a statistical model $p(\,\cdot\,|\theta_1,\ldots,\theta_K)$ is proportional to the product of the likelihood function and the **prior probability distribution** $p_{\mathrm{prior}}(\,\cdot\,|\alpha_1,\ldots,\alpha_L)$:
>
> $$p_{\mathrm{post}}(\theta_1,\ldots,\theta_K|x_1,\ldots,x_N;\alpha_1,\ldots,\alpha_L)$$
> $$\propto \left(\prod_{n=1}^{N} p(x_n|\theta_1,\ldots,\theta_K)\right)\cdot p_{\mathrm{prior}}(\theta_1,\ldots,\theta_K|\alpha_1,\ldots,\alpha_L)$$
>
> where $x = (x_1,\ldots,x_N)$ is a sequence of observations and $\alpha_1,\ldots,\alpha_L$ are **hyperparameters**.

The prior/posterior probability distributions are often simply called the prior/the posterior. The right-hand side of the above formula must be multiplied by a suitable constant so that the integral (or sum) over the parameters is equal to one, thus satisfying the usual normalization condition for a probability density (or mass) function. The prior distribution does not necessarily have to be normalized. In fact, the integral (or sum) does not even need to exist—as long as the right-hand side can be normalized to yield a proper posterior probability distribution. If the prior cannot be normalized, it is called an **improper prior**.

The final result of the procedure is not a point estimate $\hat{\theta}_1,\ldots,\hat{\theta}_K$ of the optimal model parameters. Rather, the final result is a joint distribution over possible values $\theta_1,\ldots,\theta_K$.

The prior distribution models information that may already be available about the parameters even before the data are collected. For example, a player betting on coin tosses might be convinced that the coin has a high probability of being fair even before the game starts. In mathematical terms, it is assumed a priori that with high probability the parameter of a Bernoulli distribution that produces the sequence of coin tosses will differ little from $p = 50\%$. Accordingly, for a Bayesian analysis of their chances of winning, the player would apply a prior with expectation $\langle p \rangle = 50\%$ and a variance reflecting their initial uncertainty about this information. Maybe the person that tosses the coin is particularly trustworthy and the coin does not look or feel unusual or loaded in any way. In that case, the player is quite convinced that the coin is fair, implying a small variance of potential parameter values. By multiplying the initial prior distribution by the likelihood function that depends on the actual data (i.e., the observed sequence of coin tosses), the player may then update their belief about the value for $p$. That updated belief is modeled by the posterior distribution.

This example illustrates why Bayesian approaches are of particular importance to the field of machine learning. These methods provide a framework that can be interpreted and implemented as a learning process, continuously updating the machine's model of the world by observation and processing of new data.

Analogously to the maximum likelihood procedure, we can determine the mode of the posterior distribution (i.e., its maximum point) in order to obtain a point estimate for the model parameters, the **maximum a posteriori estimate** (MAP).

If there is no a priori information available about the parameters to be estimated, we choose a constant function as the (improper) prior. Before data are collected, all parameter values are equally likely. Such a prior is also called a **noninformative prior**. In that case, the posterior is proportional to the likelihood, and the results of maximum a posteriori and maximum likelihood estimation coincide.

Let us illustrate the Bayesian method of parameter estimation by computing an example (see also [15, Sect. 3.4.1]). Let $x = (x_1, \ldots, x_N)$ be a sequence of numeric observations that we have reason to believe are distributed normally. In order to keep this example simple, we further assume that we already know the value of the scale parameter of that normal distribution: $\sigma = \sigma_0$.

Thus, we only need to estimate the location parameter $\mu$, which determines the likelihood of a single observation $x_n$ as follows:

$$p(x_n|\mu) = \mathcal{N}(x_n|\mu, \sigma_0^2)$$

We assume that prior to collecting the data, the model parameter $\mu$ is distributed according to a normal distribution with mean $\mu_0$ and standard deviation $\Delta_0$:

$$p_{\text{prior}}(\mu|\mu_0, \Delta_0) = \mathcal{N}(\mu|\mu_0, \Delta_0)$$

Therefore, the posterior distribution computes as follows:

$$p_{\text{post}}(\mu|x; \mu_0, \Delta_0) \propto \left( \prod_{n=1}^{N} p(x_n|\mu) \right) \cdot p_{\text{prior}}(\mu|\mu_0, \Delta_0)$$

$$= \left( \prod_{n=1}^{N} \mathcal{N}(x_n|\mu, \sigma_0^2) \right) \cdot \mathcal{N}(\mu|\mu_0, \Delta_0^2)$$

$$\propto e^{-\frac{(\mu-\mu_0)^2}{2(\Delta_0)^2}} \cdot \prod_{n=1}^{N} e^{-\frac{(x_n-\mu)^2}{2(\sigma_0)^2}}$$

We can see from the functional form that this must also be a normal distribution:

$$p_{\text{post}}(\mu|x; \mu_0, \Delta_0) = \frac{1}{\sqrt{2\pi}\Delta_N} \cdot e^{-\frac{(\mu-\hat{\mu}_N)}{2\Delta_N^2}}$$

A somewhat longer calculation that we omit here yields the location and scale parameters of this distribution:

$$\hat{\mu}_N = \frac{N\Delta_0^2 \cdot \bar{x} + \sigma_0^2 \cdot \mu_0}{N\Delta_0^2 + \sigma_0^2}, \quad \Delta_N = \frac{\Delta_0 \sigma_0}{\sqrt{N\Delta_0^2 + \sigma_0^2}}.$$

The maximum a posteriori estimate of $\mu$ is given by $\hat{\mu}_N$. In addition to this point estimate, we can also provide the following interval estimate:

$$[\bar{x}]_{\text{MAP},\gamma} = [\hat{\mu}_N - z(\gamma) \cdot \Delta_N, \hat{\mu}_N + z(\gamma) \cdot \Delta_N]$$

with $0 < \gamma < 1$ and, for example, $z(0.95) = 1.96$. With probability $\gamma$, the parameter is contained in this interval. In Bayesian statistics, such an interval is called a **credibility interval** in order to distinguish the term from the confidence interval of frequentist statistics. The interpretation of the latter is that estimates may vary within the interval with each sample or experiment, but the true parameter has a fixed, deterministic value.

Notwithstanding the various interpretations, for large samples, Bayesian interval estimates are consistent with the frequentist estimation described in Sect. 4.3.1. This follows, firstly, from the fact that the maximum likelihood estimate ($=$ arithmetic mean $\bar{x}$) and the maximum a posteriori estimate $\hat{\mu}_N$ coincide for large samples:

$$\lim_{N \to \infty} \hat{\mu}_N = \bar{x}$$

Secondly, for large samples, credibility and confidence intervals coincide as well:

$$\lim_{N \to \infty} \frac{\sigma_0/\sqrt{N}}{\Delta_N} = 1$$

**Example.** We assume that the height of male survey participants in the CDC study is normally distributed with known standard deviation $\sigma_0 = 7.8\,\text{cm}$, but unknown mean $\mu$.

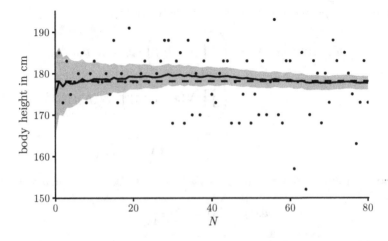

**Fig. 4.9.** *Maximum a posteriori estimate of body height with increasing sample size*

The above figure illustrates how an a priori estimate of $\mu_0 = 175\,\text{cm}$ can be improved by observing new data, with increasing sample size $N$. The grey area indicates the 95% credibility interval based on $\Delta_0 = 5.0\,\text{cm}$. The dashed line indicates the "true" value of 178 cm, determined from the entire sample.

### 4.4.3 Kernel density estimation

Another estimator for the shape of a probability density function is the following.

Let $x_1, \ldots, x_N$ be a sequence of numeric observations. The **kernel density estimate** with **bandwidths** $h_1, \ldots, h_N \in\; ]0, \infty[$ is given as follows, for all $u \in \mathbb{R}$:

$$p_{h_1, \ldots, h_N}(u) = \sum_{n=1}^{N} \frac{1}{Nh_n} K\left(\frac{u - x_n}{h_n}\right)$$

The **kernel** $K\colon \mathbb{R} \to [0, \infty[$ is a nonnegative function that satisfies the following conditions:

1. it is even, i.e., $K(u) = K(-u)$ for all $u \in \mathbb{R}$, and

2. $\int_{-\infty}^{\infty} K(\xi)\, d\xi = 1$.

A popular choice for the kernel is a Gaussian function:

$$K(u) = \frac{1}{\sqrt{2\pi}} e^{-\frac{1}{2}u^2}$$

We can think of the procedure as placing a small Gaussian bell curve at each location of a data point. The total density will then be the sum of all those small heaps of probability mass. In regions with high density, where the data points tend to cluster, this sum will be large:

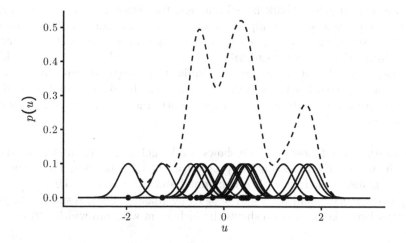

**Fig. 4.10.** *Principle behind kernel density estimation*

By construction, $p_{h_1,\ldots,h_N}(u) \geq 0$ holds for all $u \in \mathbb{R}$, and moreover:

$$
\int_{-\infty}^{\infty} p_{h_1,\ldots,h_N}(\xi)\, d\xi = \frac{1}{N} \sum_{n=1}^{N} \frac{1}{h_n} \int_{-\infty}^{\infty} K\left(\frac{\xi - x_n}{h_n}\right) d\xi
$$

$$
= \frac{1}{N} \sum_{n=1}^{N} \frac{1}{h_n} \int_{-\infty}^{\infty} h_n \cdot K(y_n)\, dy_n
$$

$$
= \frac{1}{N} \sum_{n=1}^{N} \int_{-\infty}^{\infty} K(y_n)\, dy_n = 1
$$

Here, the substitution $y_n = \frac{\xi - x_n}{h_n}$ was made in the integral for each summand. Therefore, $p_{h_1,\ldots,h_N}(\cdot)$ is indeed a probability density.

Since we also assume that the kernel is an even function, $\int_{-\infty}^{\infty} \xi \cdot K(\xi)\, d\xi = 0$ holds and a similar calculation shows:

$$
\int_{-\infty}^{\infty} \xi \cdot p_{h_1,\ldots,h_N}(\xi)\, d\xi = \bar{x}
$$

Thus, the estimated density function reproduces the arithmetic mean: $E[X \sim p_{h_1,\ldots,h_N}(\cdot)] = \bar{x}$. However, the variance obtained from the kernel density estimate always differs from the empirical variance:

$$
\sigma^2[X \sim p_{h_1,\ldots,h_N}(\cdot)] = s^2(x) + \int_{-\infty}^{\infty} \xi^2 \cdot K(\xi)\, d\xi \cdot \frac{1}{N} \sum_{n=1}^{N} h_n^2
$$

The parameters $h_1, \ldots, h_N$ are hyperparameters in the sense that they cannot be readily determined by maximizing the likelihood function. As a result, this approach would lead to vanishing bandwidths.

Given a smooth differentiable kernel function, the kernel density estimation can be considered as a smooth approximation to the histogram, a discontinuous step function. Consequently, the method is sometimes referred to as **kernel smoothing**. Kernel density estimates that have small values for the bandwidth yield large local variations in density, similar to histograms that have a small bin width. On the other hand, higher values for the bandwidth imply a need to average across wider intervals, and consequently the need for a higher amount of smoothing.

**Example.** The following figure shows two kernel density estimates based on the frequency distribution of body weight in the CDC dataset. The estimates use a Gaussian kernel with a constant bandwidth of $h = 5.0\,\mathrm{kg} = h_1 = h_2 = \cdots = h_N$ (solid line) and $h = 1.2\,\mathrm{kg}$ (dashed line), respectively. For comparison, the figure also shows the histogram with bin width $2.27\,\mathrm{kg}$.

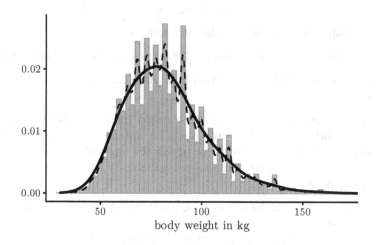

**Fig. 4.11.** *Kernel density estimates of different bandwidth*

## 4.5 Regression analysis

In the previous section, we studied individual statistical/random variables and modeled their distribution through parameterized families of probability mass and density functions. Now, we will use regression analysis to study such variables and their functional dependencies. Consequently, we want to model their joint and/or conditional distribution so that we can understand these relationships.

### 4.5.1 Simple linear regression

Let us consider a sequence of paired observations $(x_1, y_1), \ldots, (x_N, y_N)$. As usual, we think of these as realizations of independent and identically distributed random variables $X_1, X_2, \ldots, X_N$ and $Y_1, Y_2, \ldots, Y_N$, respectively.

An unseen future observation may be thought of as being produced by random variables $X_* = X_{N+1}$ and $Y_* = Y_{N+1}$. Our goal is to make a **prediction** $\hat{y}_* = \hat{f}(x_*)$ for $y_*$ given the information that $X_* = x_*$.

To make this prediction, we must assume that essential characteristics of the distribution of $Y_n$ are determined upon conditioning on the corresponding observation $X_n = x_n$ for any trial, $n \in \{1, \ldots, N, *, \ldots\}$. More specifically, in **simple linear regression**[2], we assume that there exist constants $m, c \in \mathbb{R}$ and normally distributed, mutually independent random variables $\varepsilon_n \sim \mathcal{N}(\cdot \,|\, 0, \sigma^2)$ such that:

---

[2] The term "simple" refers to the fact that only a single independent variable $X$ is used to make predictions. Therefore, we may also speak of univariate linear regression.

$$Y_n = mx_n + c + \varepsilon_n$$

In other words, we assume the following scenario:

- The values $y_*$ of $Y_*$ are normally distributed around some mean values,

- and these mean values lie on a straight line $f(x_*) = mx_* + c$, depending on the observations $x_*$ of $X_*$.

In regression models, the variable $Y_*$ is called the **dependent variable, response variable**, or **target variable**. The variable $X_*$ is the **independent variable, explanatory variable**, or **predictor variable**. In machine learning, we may call $X_*$ a **feature**. The random variable $\varepsilon_*$ is called the **error term**, or **disturbance**.

This model leads to the following conditional probability density to explain the observations:

$$p(y_n|x_n; m, c, \sigma) = \mathcal{N}(y_n|mx_n + c, \sigma^2)$$

for all $n \in \{1, \ldots, N, *, \ldots\}$.

The log-likelihood function derived from this model is the following:

$$\ell(m, c, \sigma) = \sum_{n=1}^{N} \ln(\mathcal{N}(y_n|mx_n + c, \sigma^2))$$

$$= -\frac{1}{2\sigma^2} \sum_{n=1}^{N} (y_n - mx_n - c)^2 - N\ln(\sigma) - \frac{N}{2}\ln(2\pi)$$

A calculation of the maximum point $(\hat{m}, \hat{c}, \hat{\sigma})$ yields the optimal model parameters in a procedure summarized as follows.

> **Simple linear regression.** The slope and intercept of a **regression line**
>
> $$\hat{f}: \mathbb{R} \to \mathbb{R}, \; \hat{f}(x_*) = \hat{m}x_* + \hat{c}$$
>
> fitting a sequence of data points $(x_1, y_1), \ldots, (x_N, y_N) \in \mathbb{R}^2$ are determined as follows:
>
> $$\hat{m} = \frac{\sum_{n=1}^{N}(x_n - \bar{x}) \cdot (y_n - \bar{y})}{\sum_{n=1}^{N}(x_n - \bar{x})^2} = \frac{s(x, y)}{s^2(x)},$$
>
> $$\hat{c} = \frac{1}{N} \sum_{n=1}^{N} (y_n - \hat{m}x_n) = \bar{y} - \hat{m}\bar{x}$$

The **mean squared error** (MSE)

$$\hat{\sigma}^2 = \frac{1}{N} \sum_{n=1}^{N} (y_n - \hat{y}_n)^2$$

with $\hat{y}_n = \hat{f}(x_n)$ and the **root mean squared error** (RMSE) $\hat{\sigma} = \sqrt{\hat{\sigma}^2}$ can be interpreted as measures of accuracy for the predictions from a regression model. In simple linear regression, the (root) mean squared error indicates how widely the data points are dispersed around the regression line. If $\hat{\sigma} = 0$ were to hold, the data points would lie exactly on a straight line, indicating a perfect fit. The very similar quantity, shown by the below formula,

$$N\hat{\sigma}^2 = \sum_{n=1}^{N}(y_n - \hat{y}_n)^2 = \sum_{n=1}^{N}(y_n - \hat{m}x_n - \hat{c})^2$$

is called the **residual sum of squares** (RSS). If we omit the estimation of $\sigma$, then maximizing the likelihood is equivalent to minimizing the residual sum of squares. Here, we show it written as a function of the model parameters:

$$R(m, c) = \sum_{n=1}^{N}(y_n - mx_n - c)^2$$

This view on linear regression is called the **method of least squares**. In fact, many statistical methods and machine learning techniques rely on solving some optimization problem, minimizing or maximizing some **objective function**.

Another measure for goodness of fit is the **coefficient of determination**, which equals one if the regression line fits the data perfectly:

$$r^2 = 1 - \frac{\sum_{n=1}^{N}(y_n - \hat{y}_n)^2}{\sum_{n=1}^{N}(y_n - \bar{y})^2} = 1 - \frac{\hat{\sigma}^2}{s^2(y)}$$

The coefficient of determination is equal to zero if the regression is not a better fit than the horizontal line passing through the mean $\bar{y}$. Even though the measure is written as "R squared," in settings more general than simple linear regression, it may be negative.

**Example.** The Berkeley Earth Surface Temperature project provides measurement data to quantify global warming [16]. The figure below shows the time series of average global air temperatures since 1850 along with a line fitted to the data points through linear regression.

The slope of the regression line is given by $\hat{m} = 0.006\,{}^{K}/{a}$, which corresponds to a warming of 0.6 Kelvin per century. In general, the difference between the actual observation and the prediction is called the **residual**. Geometrically, the residual is the distance of the data point from the regression line along the $y$-axis. In this figure, the residual for each observation is indicated by the size of the data point, which visually highlights any outliers. The root mean squared error is given by $\hat{\sigma} = 0.17\,\mathrm{K}$ and the coefficient of determination is given by $r^2 = 0.77$.

The analysis also shows that measurement data from the 20th century lie below the regression line, while most data points from before 1900 or after

2000 lie above the line. This association indicates that the true functional relationship is a convex function, i.e., the increase in temperature is not uniform but *accelerated*. To account for this, we will later fit a second-order polynomial instead of a regression line (see Fig. 6.3).

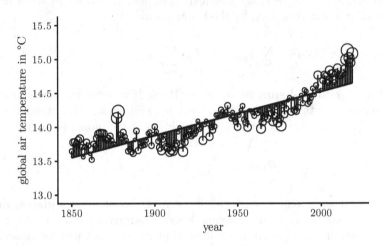

**Fig. 4.12.** *Time series of average global air temperature with regression line*

Alternatively, the equation for the regression line can be written as follows:

$$\hat{f} \colon \mathbb{R} \to \mathbb{R},\ \hat{f}(x_*) = \hat{\beta}(x_* - \bar{x}) + \hat{\alpha}$$

with $\hat{\beta} = \hat{m}$ and $\hat{\alpha} = \bar{y}$. In particular, we recognize that the regression line always passes through the centroid $(\bar{x}, \bar{y})$ of the data.

Statistical estimation theory allows us to compute a $\gamma$ confidence interval instead of just a point estimate $\hat{f}(x_*)$. We omit the derivation. Given some predictor value $x_* \in \mathbb{R}$, for sufficiently large samples, we get the following:

$$[\hat{f}(x_*)]_\gamma = \left[ \hat{f}(x_*) \pm z(\gamma) \cdot \frac{\hat{\sigma}}{\sqrt{N}} \cdot \sqrt{1 + \frac{(x_* - \bar{x})^2}{s^2(x)}} \right]$$

with, e.g., $z(0.95) = 1.96$. With a varying $x_*$, the above formula defines a **confidence band** around the regression line. The confidence band reflects our uncertainty about the position and orientation of the regression line that is a result of having limited data. The width of the confidence band increases with the distance from the centroid $(\bar{x}, \bar{y})$.

The confidence band is not to be confused with a **prediction band**. A prediction band denotes a region where a certain proportion $0 < \delta \leq 1$ of the data points can be expected. For sufficiently large samples, a $\delta$-prediction band can be given as follows:

$$\left[ \hat{f}(x_*) \pm z(\delta) \cdot \hat{\sigma} \cdot \sqrt{1 + \frac{1}{N} \cdot \left( 1 + \frac{(x_* - \bar{x})^2}{s^2(x)} \right)} \right]$$

For $x$-values that are not too far away from the mean $\bar{x}$, the prediction band is approximately $[\hat{f}(x_*) \pm z(\delta) \cdot \hat{\sigma}]$.

The following figure once more shows measurement data on global air temperatures with a regression line fitted to the data points. The 95% confidence band is shown as a narrow gray ribbon around the regression line, and the dashed curves indicate the borders of the 95% prediction band.

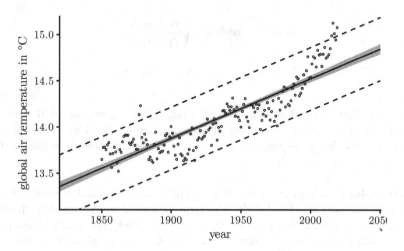

**Fig. 4.13.** *Confidence and prediction bands for linear regression*

Finally, armed with what we have just learned, we want to take another look at the Bravais–Pearson correlation coefficient. The slope of the regression line is given by

$$\hat{m}(y, x) = \frac{s(x, y)}{s^2(x)}$$

where $x = (x_1, \ldots, x_N)$ is the sequence of independent values and $y = (y_1, \ldots, y_N)$ is the sequence of dependent values. If we swap the roles of dependent and independent values, we get:

$$\hat{m}(x, y) = \frac{s(y, x)}{s^2(y)} = \frac{s(x, y)}{s^2(y)}$$

First, we note that $\mathrm{sgn}(\hat{m}(y, x)) = \mathrm{sgn}(\hat{m}(x, y))$ holds: the slopes of both regression lines have the same sign. Consequently, we can derive the following formula for the correlation coefficient:

$$r(x, y) = \frac{s(x, y)}{s(x) \cdot s(y)} = \mathrm{sgn}(\hat{m}(y, x)) \cdot \sqrt{\hat{m}(y, x) \cdot \hat{m}(x, y)}$$

In magnitude, the Pearson coefficient is the geometric mean of the slopes of the regression lines through the data points. This further emphasizes the interpretation of this measure as one of linear correlation.

### 4.5.2 Theil–Sen regression

The following procedure presents an alternative to simple linear regression.

Let $(x_1, y_1), \ldots, (x_N, y_N) \in \mathbb{R}^2$ be a sequence of data points/paired numeric obervations. The **Theil–Sen estimator** determines the slope and intercept of a regression line:

$$\hat{m}_{\mathrm{TS}} = \underset{\substack{k,l \in \{1,\ldots,N\} \\ x_k \neq x_l}}{\mathrm{median}} \left( \frac{y_l - y_k}{x_l - x_k} \right),$$

$$\hat{c}_{\mathrm{TS}} = \underset{k \in \{1,\ldots,N\}}{\mathrm{median}} \left( y_k - \hat{m}_{\mathrm{TS}} \cdot x_k \right)$$

The Theil–Sen method is more robust than linear regression, as it uses the median so that outliers have less of an influence on the line of best fit. In contrast, with linear regression, even a few outliers in the data can lead to completely different results.

**Example.** In addition to monthly net income, the ALLBUS survey also records respondents' total daily television viewing time. We want to find out if there is a relationship between those two variables. The figure below plots television viewing time against income for the cohort of male respondents who live alone and work full time.

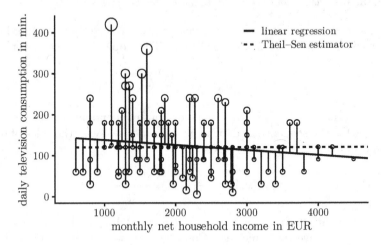

**Fig. 4.14.** *Theil–Sen regression vs. simple linear regression*

The regression line, determined by linear regression, has a slightly negative slope. This negative slope might suggest that, in the selected cohort, the people with higher income watch less television than the people not of higher income. More precisely, $\hat{m} = -0.012 \frac{\text{min.}}{\text{EUR}}$, i.e., a reduction in television viewing time of 12 minutes per 1000 EUR of additional income.

However, $\hat{\sigma} = 68\,\text{min}$, which makes the predictive power of this statement questionable. Furthermore, the regression line obtained through the more robust Theil–Sen estimation has a vanishing slope: $\hat{m}_{\text{TS}} = 0$. As a result, this analysis reaches a different conclusion—on average, each person watches 120 minutes of television daily, regardless of income.

### 4.5.3 Simple logistic regression

In linear regression, the target variable $Y_*$ is a continuous random variable, potentially assuming arbitrary values. In logistic regression, the target variable is Bernoulli distributed: only the values $Y_* = 0$ and $Y_* = 1$ are possible.

The logistic regression model makes the following assumption, where the $x_n$ represent realizations of the explanatory variable:

$$Y_n = \begin{cases} 1 & \text{if } mx_n + c + \varepsilon_n > 0 \\ 0 & \text{otherwise} \end{cases}$$

Furthermore, it is assumed that the error terms $\varepsilon_n$ follow a **logistic distribution**: $\varepsilon_n \sim \text{Logist}(\,\cdot\,|0, 1)$. The logistic distribution is defined as follows:

$$\text{Logist}(u|\mu, s) = \frac{1}{4s}\text{sech}^2\left(\frac{u - \mu}{2s}\right)$$

with the *hyperbolic secant*

$$\text{sech}(u) = \frac{2}{e^u + e^{-u}}.$$

We can think of the approach as follows. To explain the observations produced by the discrete random variables $Y_n$, we introduce the auxiliary continuous random variable $Y_n^* = mx_n + c + \varepsilon_n$, which causes observation $Y_n = 1$ whenever the condition $Y_n^* > 0$ is met. In this context, $Y_n^*$ is called a **latent variable**; its realizations do not correspond to any actual observation.

This approach yields the following probabilities for the two possible values of the response variable $Y_n$ conditioned on the observed values for the explanatory variable $X_n$:

$$\Pr(Y_n = 1|X_n = x_n) = \Pr(Y_n^* > 0|X_n = x_n)$$
$$= \int_0^\infty \text{Logist}(\xi|mx_n + c, 1)\,\mathrm{d}\xi$$
$$= \frac{1}{1 + e^{-mx_n - c}}$$

or

$$\Pr(Y_n = 0 | X_n = x_n) = 1 - \Pr(Y_n = 1 | X_n = x_n)$$
$$= 1 - \frac{1}{1 + e^{-mx_n - c}}$$
$$= \frac{1}{1 + e^{mx_n + c}}$$

In summary, we can describe the situation by the following statistical model:

$$p(y_n | x_n; m, c) = \frac{1}{1 + \exp\left((-1)^{y_n} \cdot (mx_n + c)\right)}$$

for all $n \in \{1, \ldots, N, \ldots\}$.

The following formula shows the corresponding log-likelihood function for simple logistic regression:

$$\ell(m, c) = -\sum_{n=1}^{N} \ln\left(1 + \exp\left((-1)^{y_n} \cdot (mx_n + c)\right)\right)$$

The maximum point $(\hat{m}, \hat{c})$ of this function cannot be calculated in closed form. Instead, numerical methods need to be used.

The estimated parameters determine the **decision boundary** $\hat{x}$, where we have $\hat{m}\hat{x} + \hat{c} = 0$. For observed values $x_*$ with $\hat{m}x_* + \hat{c} > 0$, according to the logistic model, it is more likely to find $Y_* = 1$: $\Pr(Y_* = 1 | X_* = x_*) > \Pr(Y_* = 0 | X_* = x_*)$. On the other side of the decision boundary, we expect that the occurrence of $Y_* = 0$ is more likely.

For simple (i.e., univariate) logistic regression, the decision boundary consists of only a single point, since we consider only one independent variable. In practice, univariate logistic regression is of little importance. However, building on these ideas, we introduce multivariate logistic regression in Sect. 6.3.1.

# References

[1]   CDC Population Health Surveillance Branch. *Behavioral Risk Factor Surveillance System (BRFSS) Survey Data 2018*. Accessed Feb. 1, 2020. URL: https://www.cdc.gov/brfss/.

[2]   UN Department of Economic and Social Affairs. *World Population Prospects 2019*. Accessed July 10, 2020. URL: https://population.un.org/wpp/.

[3]   Rick Durrett. *Probability: Theory and Examples*. 5th ed. Cambridge University Press, May 2019.

[4]   GESIS – Leibniz-Institut für Sozialwissenschaften. *Allgemeine Bevölkerungsumfrage der Sozialwissenschaften ALLBUS 2018*. 2019. DOI: 10.4232/1.13250.

[5]   Andrew M. Penner et al. "Within-job gender pay inequality in 15 countries". In: *Nature Human Behaviour* (Nov. 2022). DOI: 10.1038/s41562-022-01470-z.

[6]   Eugene Lukacs. "A Characterization of the Normal Distribution". In: *Ann. Math. Statist.* 13.1 (Mar. 1942), pp. 91–93. DOI: 10.1214/aoms/1177731647.

[7]   Radha G. Laha. "On an extension of Geary's theorem". In: *Biometrika* 40.1-2 (1953), pp. 228–229. DOI: 10.1093/biomet/40.1-2.228.

[8]   Jacob Cohen. *Statistical power analysis for the behavioral sciences*. 2nd ed. New Jersey, USA: Lawrence Earlbaum Associates, 1988.

[9]   Shlomo S. Sawilowsky. "New Effect Size Rules of Thumb". In: *Journal of Modern Applied Statistical Methods* 8.2 (Nov. 2009), pp. 597–599. DOI: 10.22237/jmasm/1257035100.

[10]   Krishna B. Athreya and Soumendra N. Lahiri. *Measure Theory and Probability Theory*. Springer New York, 2006. DOI: 10.1007/978-0-387-35434-7.

[11]   Larry Wasserman. *All of Nonparametric Statistics*. Springer New York, 2006. DOI: 10.1007/0-387-30623-4.

[12]   Jan R. Magnus. *Matrix Differential Calculus with Applications in Statistics and Econometrics*. 3rd ed. Wiley, Feb. 2019. DOI: 10.1002/9781119541219.

[13]   George E. P. Box and David R. Cox. "An Analysis of Transformations". In: *Journal of the Royal Statistical Society: Series B (Methodological)* 26.2 (July 1964), pp. 211–243. DOI: 10.1111/j.2517-6161.1964.tb00553.x.

[14]   In-Kwon Yeo and Richard A. Johnson. "A new family of power transformations to improve normality or symmetry". In: *Biometrika* 87.4 (Dec. 2000), pp. 954–959. DOI: 10.1093/biomet/87.4.954.

[15]   Richard O. Duda, Peter E. Hart, and David G. Stork. *Pattern Classification*. 2nd ed. Wiley, 2000.

[16]   Berkeley Earth. *Time Series Data – Monthly Global Average Temperature (Annual Summary)*. Accessed Feb. 1, 2020. URL: http://berkeleyearth.org/data/.

# 5

---

# Multivariate statistics

When we learned about association measures and regression methods in the previous chapters, we gained a first glimpse into multivariate statistics. Multivariate methods allow for the joint study of statistical variables and their relationships between each other with the goal of capturing a complete picture of the data. In machine learning, this approach is essential, as predictions are based on a large number of features and their statistical characteristics.

## 5.1 Data matrices

Let us consider a sample of size $N$ where each observation collects the values for a total of $D$ statistical variables/features. To simplify the discussion, we will assume that the data are quantitative/numeric, unless stated otherwise. Thus, each of the $N$ observations is a $D$-dimensional data point/feature vector $x_n \in \mathbb{R}^D$, $n \in \{1, \ldots, N\}$. In order to work with tools from linear algebra, such as matrix multiplication, we need to agree on conventions for the format of the vectors and matrices involved. One option is to write the feature vectors $x_1, \ldots, x_N$ as the columns of a matrix, like so: $x = (x_1, \ldots, x_N)$. In general, this matrix has the format $D \times N$, consisting of $D$ rows and $N$ columns. For example, four observations of both body height (in meters) and body weight (in kilograms) can be summarized as a matrix as follows:

$$x = \begin{pmatrix} 1.63 & 1.66 & 1.47 & 1.79 \\ 59 & 91 & 64 & 86 \end{pmatrix}$$

This convention is especially helpful when we want to use tools of multivariate calculus. Otherwise, we would have to break with common conventions, so that, e.g., the gradient becomes a row vector. In statistics, however, it is also common to write the feature vectors as rows in a way that each column contains a sequence of observed values, resulting in a matrix with $N$ rows and $D$ columns:

© Springer-Verlag GmbH Germany, part of Springer Nature 2023
M. Plaue, *Data Science*, https://doi.org/10.1007/978-3-662-67882-4_5

$$x^T = \begin{pmatrix} 1.63 & 59 \\ 1.66 & 91 \\ 1.47 & 64 \\ 1.79 & 86 \end{pmatrix}$$

We simply call a matrix constructed in this way a **data matrix**. It has the same format as the data tables that we already encountered in previous chapters. We give the data matrix its own notation: $X = x^T$. Despite this convention, we always write data points/feature vectors as column vectors. Similarly, when we refer to a *random vector*, i.e., a tuple of random variables, we always mean column vectors. We may also write column vectors as transposed rows, e.g., $u = (u_1, u_2, u_3)^T$.

Sequences of observed values, on the other hand, are usually to be understood as row vectors. If a different convention is used, we may indicate it in bold print for the sake of clear notation: $\boldsymbol{y} = y^T = (y_1, \ldots, y_N)^T$.

We will also combine, for example, model parameters to vectors and matrices. We then explicitly point out the format if necessary.

In general, a data matrix with $N$ rows/observations and $D$ columns/features has the following form:

$$X = \begin{pmatrix} x_{11} & x_{12} & \cdots & x_{1D} \\ x_{21} & x_{22} & \cdots & x_{2D} \\ \vdots & \vdots & & \vdots \\ x_{N1} & x_{N2} & \cdots & x_{ND} \end{pmatrix} = (x_{nd})_{\substack{n \in \{1,\ldots,N\} \\ d \in \{1,\ldots,D\}}}$$

The **dimensionality** of the data is given by $D$. The rows $X_{1\bullet}, \ldots, X_{N\bullet}$ of a data matrix $X$ are the transposed feature vectors:

$$X_{1\bullet} = x_1^T, \ X_{2\bullet} = x_2^T, \ \ldots, \ X_{N\bullet} = x_N^T$$

We can denote the columns—i.e., the sequences of observed univariate values for each feature—by $X_{\bullet 1}, \ldots, X_{\bullet D}$. If we subtract the respective arithmetic mean, we get the **mean-centered data matrix**:

$$X_{\bullet d} \mapsto X_{\bullet d} - \overline{X_{\bullet d}}$$

for all $d \in \{1, \ldots, D\}$; the minus sign here means that the mean is to be subtracted from each observation in the sequence.

If, in addition to mean-centering, we divide by the respective standard deviation, we get the **standardized data matrix** that consists of the so-called **standard scores**, or **z-scores**:

$$X_{\bullet d} \mapsto Z_{\bullet d} = \frac{X_{\bullet d} - \overline{X_{\bullet d}}}{s\left(X_{\bullet d}\right)}$$

If the variable has a vanishing variance, i.e., $s\left(\boldsymbol{X}_{\bullet d}\right) = 0$ holds, we may assume $\boldsymbol{Z}_{\bullet d} = 0$. By construction, arithmetic means of variables/features in mean-centered or standardized data are always zero. Moreover, the sample variance of a $z$-score is always equal to one (unless it vanishes).

Standardization is particularly advisable when the variables have different units (e.g., meters vs. kilograms) or operate on a different scale (e.g., body height vs. daily walking distance). For example, the above data matrix of height and weight (using unbiased sample variance) becomes:

$$\boldsymbol{Z} = \begin{pmatrix} -0.05 & -1.00 \\ 0.17 & 1.00 \\ -1.27 & -0.69 \\ 1.16 & 0.69 \end{pmatrix}$$

Another popular form of normalization is **min–max scaling**:

$$\boldsymbol{X}_{\bullet d} \mapsto \frac{\boldsymbol{X}_{\bullet d} - \boldsymbol{X}_{\min,d}}{\boldsymbol{X}_{\max,d} - \boldsymbol{X}_{\min,d}}$$

where

$$\boldsymbol{X}_{\min,d} = \min_{n \in \{1,\ldots,N\}} \{x_{nd}\} \text{ and } \boldsymbol{X}_{\max,d} = \max_{n \in \{1,\ldots,N\}} \{x_{nd}\}.$$

Min–max scaled variables take values between zero and one.

For some applications, it is convenient to add a zeroth column to the data matrix, the entries of which are all equal to one:

$$\boldsymbol{X}_{\bullet 0} = \left.\begin{pmatrix} 1 \\ \vdots \\ 1 \end{pmatrix}\right\} N \text{ times}$$

This extended data matrix is especially useful in the discussion of (multivariate) regression, where we will call it the **regressor matrix**. Other common names are the **design matrix** or the **model matrix**.

## 5.2 Distance and similarity measures

In order to compare two observations $u, v \in \mathbb{R}$ of a univariate, numeric statistical variable, we can compute their distance via the absolute value of their difference: $\delta(u, v) = |u - v|$. For ordinal variables, we may use the difference in rank. For a categorical variable with domain $\{m_1, \ldots, m_K\}$, the discrete metric is the standard choice to compare values:

$$\delta(m_k, m_l) = \begin{cases} 0 & \text{if } k = l \\ 1 & \text{if } k \neq l \end{cases}$$

For multivariate variables, when we want to numerically compare two observations, there are a number of useful measures that can be interpreted as a distance or similarity.

### 5.2.1 Distance and similarity measures for numeric variables

First, we define various measures for the length of a single feature vector. In mathematical parlance, such a measure is generally called a **norm**.

For a numeric vector $u = (u_1, \ldots, u_D)^T \in \mathbb{R}^D$ and fixed $p \in \mathbb{N}$, $p \geq 1$, we define its **Minkowski norm of order $p$** (or **$p$-norm** for short):

$$\|u\|_p = \left( \sum_{d=1}^{D} |u_d|^p \right)^{\frac{1}{p}}$$

The **maximum norm** is the following:

$$\|u\|_\infty = \max_{d \in \{1, \ldots, D\}} \{|u_d|\}$$

The maximum norm can be interpreted as the limit of the $p$-norm for large values of $p$, since for all $u \in \mathbb{R}^D$ we have the following:

$$\lim_{p \to \infty} \|u\|_p = \|u\|_\infty$$

The 2-norm is the ordinary Euclidean norm, which we fall back to if no particular $p$ is specified: $\|u\| = \|u\|_2$. The Euclidean norm is in turn related to the **standard Euclidean inner product**

$$\langle u, v \rangle = u \bullet v = \sum_{d=1}^{D} u_d \cdot v_d$$

as follows:

$$\|u\| = \sqrt{\langle u, u \rangle}$$

For two column vectors $u, v \in \mathbb{R}^D$, the inner product can be written as a matrix multiplication as follows:

$$\langle u, v \rangle = u^T \cdot v$$

The distance between two data points can be computed via the length of the vector that connects them:

For two numeric data points $u, v \in \mathbb{R}^D$, we define the **Euclidean distance**

$$\delta_2(u, v) = \|u - v\|_2,$$

the **Manhattan distance**

$$\delta_1(u, v) = \|u - v\|_1,$$

and the **Chebyshev distance**, or **maximum distance**,

$$\delta_\infty(u, v) = \|u - v\|_\infty.$$

The above list represents the **Minkowski distances** most commonly used in data analysis. These distance measures can be defined in that way for any Minkowski norm. Unless specified otherwise, we will use the Euclidean distance.

Let's calculate an example:

$$u = \begin{pmatrix} -0.5 \\ 0.5 \end{pmatrix}, v = \begin{pmatrix} 1.0 \\ 1.0 \end{pmatrix}$$

The following values result from the various distance measures:

$$\delta_1(u, v) = |-0.5 - 1.0| + |0.5 - 1.0| = 2.0$$
$$\delta_2(u, v) = \sqrt{(-0.5 - 1.0)^2 + (0.5 - 1.0)^2} \approx 1.6$$
$$\delta_\infty(u, v) = \max\{|-0.5 - 1.0|, |0.5 - 1.0|\} = 1.5$$

**Example.** We consider three plants of the genus *Iris*, which were selected from a larger dataset of 150 specimens [1]:

| $n$ | type | sepal length | sepal width | petal length | petal width |
|-----|------|--------------|-------------|--------------|-------------|
| 1 | *Iris setosa* | 5.1 cm | 3.5 cm | 1.4 cm | 0.2 cm |
| 2 | *Iris setosa* | 4.9 cm | 3.0 cm | 1.4 cm | 0.2 cm |
| 101 | *Iris virginica* | 6.3 cm | 3.3 cm | 6.0 cm | 2.5 cm |

**Table 5.1.** *Measurements of sepal and petal of iris flowers*

We can combine the lengths of sepal and petal to data points $x_1, x_2, x_{101} \in \mathbb{R}^4$. The Chebyshev distance between specimens of the species *Iris setosa* is $\delta_\infty(x_1, x_2) = 0.5$ cm. The distance of these specimens to the *Iris virginica* specimen is much larger due to the significantly longer petal: $\delta_\infty(x_1, x_{101}) = \delta_\infty(x_2, x_{101}) = 4.6$ cm.

If the data come with units of measurement, those units must be identical in order to compare them in a meaningful way. For example, in the calculation above, we chose centimeters to measure and compare lengths. In many cases, a prior standardization of the data (i.e., computing the $z$-scores) is recommended so that the variables are unitless and have an equal variance.

The concept dual to distance is *similarity*: when two information objects or observations are associated with feature vectors that have a small distance, we can think of them as being similar.

For two numeric feature vectors with non-negative components $u, v \in [0, \infty[^D$, we define their **cosine similarity**:

$$\sigma_{\cos}(u, v) = \frac{\langle u, v \rangle}{\|u\| \cdot \|v\|} = \cos \sphericalangle(u, v)$$

unless $u = 0$ or $v = 0$, in which case one may either leave the measure undefined or set $\sigma_{\cos}(u, v) = 0$.

The **Tanimoto similarity** is given as follows:

$$\sigma_{\text{Tanim}}(u, v) = \frac{\langle u, v \rangle}{\|u\|^2 + \|v\|^2 - \langle u, v \rangle}$$

unless $u = v = 0$, in which case the Tanimoto similarity is undefined.

Conventions that fill the gaps in the above similarity measures' domains are not set in stone. For example, some may want to argue for $\sigma_{\text{Tanim}}(0, 0) = 0$.

If the compared vectors are orthogonal, both the cosine and the Tanimoto similarity vanish. If they are colinear, the cosine similarity takes its maximum possible value of one: It holds $\sigma_{\cos}(u, v) = 1$ precisely when $v = \lambda u$ for some $\lambda > 0$. On the other hand, the maximum Tanimoto similarity $\sigma_{\text{Tanim}}(u, v) = 1$ is given if and only if $u = v$.

> **Example.** We can also compare the dimensions of the sepal and petal of irises from the above example using similarity measures. The Tanimoto similarity of the two *Iris setosa* specimens in the above example is given by $\sigma_{\text{Tanim}}(x_1, x_2) \approx 0.99$. The similarity with the specimen of the species *Iris virginica* is much lower and given by $\sigma_{\text{Tanim}}(x_1, x_{101}) \approx 0.65$ and $\sigma_{\text{Tanim}}(x_2, x_{101}) \approx 0.64$, respectively.

### 5.2.2 Distance and similarity measures for categorical variables

Suppose we are given two ordered lists of equal length that contain symbols or other objects where we can discern (in-)equality. An obvious idea to determine a measure of their distance/similarity is to compare each position in the list and count instances of different/identical symbols or objects. For example, the two strings ABCDE and ABXDY would have a distance of two, since two characters do not match (or a similarity of three, since three characters do match). See also the edit distances that we introduced in Sect. 1.4.5.1 for the purpose of data deduplication.

This idea gives rise to the following formal definition.

> For two data tuples $u, v$ of length $D$ that contain values for the same categorical variables, we define their **Hamming distance**:
>
> $$\delta_{\text{Hamm}}(u, v) = |\{d \in \{1, \ldots, D\} | u_d \neq v_d\}|$$
>
> The **normalized Hamming distance** is given by: $1/D \cdot \delta_{\text{Hamm}}(u, v)$

The Hamming distance simply counts the number of features that have a different value. An everyday example would be plant identification. Two plants with identical characteristics (e.g. growth habit, leaf shape, leaf arrangement, etc.) have a Hamming distance of zero, and thus they can be presumed to belong to the same species.

For binary variables, the Hamming distance is the number of distinct bits. Let $x_1 = (1, 1, 0, 1)^T$ and $x_2 = (0, 1, 0, 1)^T$ be two binary feature vectors. For easier comparison, we can also write them on top of each other as a data matrix:

$$X = \begin{pmatrix} 1 & 1 & 0 & 1 \\ 0 & 1 & 0 & 1 \end{pmatrix}.$$

Only the first entries are different, so $\delta_{\text{Hamm}}(x_1, x_2) = 1$ holds. The normalized Hamming distance is given by 0.25 since there are four entries in total.

The following similarity measures are specifically designed to compare lists of binary variable values.

For two data tuples $u, v \in \{0, 1\}^D$ of binary values, we define their **Jaccard similarity** as follows:

$$\sigma_{\text{Jacc}}(u, v) = \frac{|\{d | u_d = 1 \text{ and } v_d = 1\}}{|\{d | u_d = 1 \text{ or } v_d = 1\}|}$$

If $u = v = 0$ holds, this measure is not defined.

The **Szymkiewicz–Simpson similarity**, or **overlap coefficient**, is defined as follows:

$$\sigma_{\text{overlap}}(u, v) = \frac{|\{d | u_d = 1 \text{ and } v_d = 1\}|}{\min\{|\{d | u_d = 1\}|, |\{d | v_d = 1\}|\}}$$

If $u = 0$ or $v = 0$ hold, this measure is not defined.

In Sect. 2.5.3, we introduced the Jaccard index as a measure of association between two binary variables as a method to compare sequences of observations. Here, we apply the same formula to measure the similarity between two observations of multiple categorical variables. Instead of the binary vectors $u$ and $v$, we can consider the sets $U$ and $V$ for the variables which have been assigned a value of one. The Jaccard similarity and overlap coefficient prove to be useful measures for the size of those sets' intersection:

$$\sigma_{\text{Jacc}}(U, V) = \frac{|U \cap V|}{|U \cup V|} \quad \text{and} \quad \sigma_{\text{overlap}}(U, V) = \frac{|U \cap V|}{\min\{|U|, |V|\}}$$

Both measures take values between zero and one, where $\sigma_{\text{Jacc}}(U, V) = \sigma_{\text{overlap}}(U, V) = 0$ corresponds to a vanishing overlap: $U \cap V = \emptyset$. The maximum values are characterized as follows:

$$\sigma_{\text{Jacc}}(U, V) = 1 \Leftrightarrow U = V,$$
$$\sigma_{\text{overlap}}(U, V) = 1 \Leftrightarrow U \subseteq V \text{ or } U \supseteq V$$

An alternative formula for the Jaccard similarity that is often convenient is the following:

$$\sigma_{\text{Jacc}}(U, V) = \frac{|U \cap V|}{|U| + |V| - |U \cap V|}$$

Let us compute an example and consider the two binary vectors $u = (1, 1, 0, 1)^T$ and $v = (0, 1, 0, 1)^T$. If the features associated with each entry of $u$ and $v$ are named by letters from $a$ to $d$, these binary vectors correspond to the sets $U = \{a, b, d\}$ and $V = \{b, d\}$, thus $U \cap V = \{b, d\}$. The similarity measures are calculated as follows:

$$\sigma_{\text{Jacc}}(u, v) = \sigma_{\text{Jacc}}(U, V) = \frac{2}{3 + 2 - 2} \approx 0.67,$$

$$\sigma_{\text{overlap}}(u, v) = \sigma_{\text{overlap}}(U, V) = \frac{2}{\min\{3, 2\}} = 1.00$$

### 5.2.3 Distance and similarity matrices

In the last sections, we learned about examples of distance and similarity measures. We now approach a general definition.

Let $\Omega$ be any set, e.g., some feature space like $\Omega \subseteq \mathbb{R}^D$ or $\Omega = \{0, 1\}^D$. Let $\delta \colon \Omega \times \Omega \to [0, \infty[$ be a map for which we consider the following conditions, for all $u, v, w \in \Omega$:

$$
\begin{array}{rl}
(1) & \delta(u, u) = 0 \\
(2) & \delta(u, v) = 0 \Rightarrow u = v \\
\textbf{symmetry:} & \delta(u, v) = \delta(v, u) \\
\textbf{triangle inequality:} & \delta(u, w) \leq \delta(u, v) + \delta(v, w)
\end{array}
$$

If at least condition (1) is satisfied, we call $\delta(\,\cdot\,, \cdot\,)$ a **premetric**[1]. If *all* of the conditions above are satisfied, the map is called a **metric**.

All distance measures induced by the Minkowski norms $\| \cdot \|_p$ are metrics, as is the Hamming distance.

We can construct the **Jaccard distance** from the Jaccard similarity as follows:

$$\delta_{\text{Jacc}}(u, v) := 1 - \sigma_{\text{Jacc}}(u, v)$$

for all $u, v \in \{0, 1\}^D$. One can prove that this distance measure is a metric.

The following is an example of a symmetric premetric that is not a metric (neither the triangle inequality nor condition (2) are satisfied):

---

[1] The definition given is not a standard definition. Some authors may mean something different when they talk about a "premetric."

$$\delta_{\text{overlap}}(u, v) := 1 - \sigma_{\text{overlap}}(u, v)$$

for all $u, v \in \{0, 1\}^D$.

Many distance measures commonly used in data science are at least symmetric premetrics, and similarity measures can often be constructed from those. This observation gives rise to the following conditions, all or some of which can be reasonably imposed on similarity measures:

$$
\begin{aligned}
&(1) && \sigma(u, u) = \sigma(v, v) \\
&(2) && \sigma(u, v) < \sigma(v, v) \Leftarrow u \neq v \\
&\textbf{symmetry:} && \sigma(u, v) = \sigma(v, u)
\end{aligned}
$$

for all $u, v \in \Omega$. In particular, if $\delta(\cdot, \cdot)$ is some metric, then all of the above properties are satisfied by the following similarity measure:

$$\sigma(u, v) := a^2 \cdot e^{-\frac{1}{2}\left(\frac{\delta(u,v)}{h}\right)^q}$$

with real numbers $a, h, q > 0$; $q = 1$ or $q = 2$ are common choices.

In some cases, the following multiplicative triangle inequality is also a desirable property:

$$\sigma(u, w) \geq \sigma(u, v) \cdot \sigma(v, w)$$

Let $x_1, \ldots, x_N$ be a sequence of observations that take values in some domain $\Omega$ with distance measure $\delta(\cdot, \cdot)$. We can define the **distance matrix** as follows:

$$
\Delta(x) = \begin{pmatrix}
\delta(x_1, x_1) & \delta(x_1, x_2) & \cdots & \delta(x_1, x_N) \\
\delta(x_2, x_1) & \delta(x_2, x_2) & \cdots & \delta(x_2, x_N) \\
\vdots & \vdots & & \vdots \\
\delta(x_N, x_1) & \delta(x_N, x_2) & \cdots & \delta(x_N, x_N)
\end{pmatrix}
$$

For a similarity measure $\sigma(\cdot, \cdot)$, the **similarity matrix** $\Sigma(x)$ is defined analogously.

The distance/similarity matrix consists of all pairwise distances/similarities between observations. Given $N$ observations, it is a square matrix of format $N \times N$. If the distance measure is a symmetric premetric, the distance matrix is symmetric and has vanishing diagonal entries.

Let us calculate a real-life example and consider the pairwise geographic distances (measured in kilometers) of the German cities Berlin, Hamburg, Munich, Cologne, and Frankfurt. These can be summarized in a distance matrix as follows:

$$
\Delta((\text{Berlin}, \text{Hamburg}, \text{Munich}, \text{Cologne}, \text{Frankfurt})) = \begin{pmatrix}
0 & 282 & 508 & 534 & 459 \\
282 & 0 & 615 & 377 & 396 \\
508 & 615 & 0 & 470 & 312 \\
534 & 377 & 470 & 0 & 163 \\
459 & 396 & 312 & 163 & 0
\end{pmatrix}
$$

As another example, let us consider the standardized data matrix of height and weight from Sect. 5.1, which is composed of four data points:

$$z_1 = \begin{pmatrix} -0.05 \\ -1.00 \end{pmatrix}, \; z_2 = \begin{pmatrix} 0.17 \\ 1.00 \end{pmatrix}, \; z_3 = \begin{pmatrix} -1.27 \\ -0.69 \end{pmatrix}, \; z_4 = \begin{pmatrix} 1.16 \\ 0.69 \end{pmatrix}$$

The Euclidean distance matrix is given as follows:

$$\Delta(z) = \begin{pmatrix} 0 & 2.01 & 1.26 & 2.08 \\ 2.01 & 0 & 2.22 & 1.04 \\ 1.26 & 2.22 & 0 & 2.79 \\ 2.08 & 1.04 & 2.79 & 0 \end{pmatrix}$$

Finally, we can use similarity measures to characterize, for example, cooking recipes by ingredients. Each ingredient defines a binary variable: for each dish, the ingredient is either needed or not needed. Examples:

$U_1$ = spaghetti aglio e olio = {chili, garlic, olive oil, spaghetti},

$U_2$ = curry = {broccoli, chili, coconut milk, curry powder, garlic, tofu},

$U_3$ = spaghetti al pomodoro = {basil, olive oil, onions, spaghetti, tomatoes},

$U_4$ = ratatouille = {aubergines, basil, bell pepper, garlic, olive oil,

onions, rosemary, tomatoes, zucchini}

The pairwise computation of the Jaccard similarity and overlap coefficient for these sets of ingredients leads to the following similarity matrices:

$$\Sigma_{\text{Jacc}}(U) = \begin{pmatrix} 1 & 0.25 & 0.29 & 0.18 \\ 0.25 & 1 & 0 & 0.07 \\ 0.29 & 0 & 1 & 0.40 \\ 0.18 & 0.07 & 0.40 & 1 \end{pmatrix},$$

$$\Sigma_{\text{overlap}}(U) = \begin{pmatrix} 1 & 0.5 & 0.5 & 0.5 \\ 0.5 & 1 & 0 & 0.17 \\ 0.5 & 0 & 1 & 0.8 \\ 0.5 & 0.17 & 0.8 & 1 \end{pmatrix}$$

## 5.3 Multivariate measures of central tendency and variation

Suppose that we are given a sequence of univariate observations $x_1, \ldots, x_N \in \mathbb{R}$. We want to find the location $\hat{u} \in \mathbb{R}$ on the number line that minimizes the sum of squared distances to those observations. That is, $\hat{u}$ is the minimum point of the following function:

$$\ell_2 : \mathbb{R} \to [0, \infty[, \; \ell_2(u) = \sum_{n=1}^{N} (u - x_n)^2$$

The derivative of this function is given as follows:

$$\frac{d}{du}\ell_2(u) = \sum_{n=1}^{N} 2(u - x_n) = 2N \cdot u - 2 \sum_{n=1}^{N} x_n$$

The second derivative is positive, so the uniquely determined minimum point is at the zero of the derivative: $\hat{u} = \frac{1}{N} \sum_{n=1}^{N} x_n$. We recognize this number to be the arithmetic mean of the observations. Similarly, it can be shown that the sample median is a minimum point of the sum of the distances without squaring them, i.e., a minimum point of the following function:

$$\ell_1 \colon \mathbb{R} \to [0, \infty[, \ \ell_1(u) = \sum_{n=1}^{N} |u - x_n|$$

The mean and the median can be generalized to multivariate measures of central tendency by applying the above idea to more general distance measures, such as the Euclidean distance instead of the distance along the number line.

We also want to describe how multidimensional data points disperse, in which case it is important to note that this dispersion may vary when computed along different directions in feature space.

### 5.3.1 Centroid and geometric median, medoid

The centroid is the multivariate analogue of the arithmetic mean.

Given a sequence of data points/feature vectors $x_1, \ldots, x_N$ with $x_n \in \mathbb{R}^D$ for all $n \in \{1, \ldots, N\}$, the **sample mean vector**, or **centroid**, is defined as follows:

$$\bar{x}_{\text{centroid}} = \frac{1}{N} \sum_{n=1}^{N} x_n$$

It is not difficult to show that the centroid is characterized by minimizing the sum of squared Euclidean distances to the data points; i.e., it is the uniquely determined minimum point of the following function:

$$\ell_2 \colon \mathbb{R}^D \to [0, \infty[, \ \ell_2(u) = \sum_{n=1}^{N} \|u - x_n\|^2$$

Furthermore, the centroid can be characterized as the column-wise arithmetic mean of the data matrix:

$$\bar{x}_{\text{centroid}} = \begin{pmatrix} \overline{X_{\bullet 1}} \\ \overline{X_{\bullet 2}} \\ \vdots \\ \overline{X_{\bullet D}} \end{pmatrix}$$

In other words, the centroid is the vector composed of the arithmetic means of the individual univariate variables. In particular, we see that mean-centering is equivalent to the operation of subtracting the centroid from each data point. Therefore, the centroid of a set of mean-centered data points is always the coordinate origin.

We may use the alternative notations $\mu(x)$, $\bar{x}$ for the centroid, which coincide with the arithmetic mean.

Given a sequence of data points $x_1, \ldots, x_N$ with $x_n \in \mathbb{R}^D$ for all $n \in \{1, \ldots, N\}$, a **geometric median** is any point $\bar{x}_{\text{median}} \in \mathbb{R}^D$ that is a minimum point of the following function:

$$\ell_1 : \mathbb{R}^D \to [0, \infty[, \ \ell_1(u) = \sum_{n=1}^{N} \|u - x_n\|$$

Thus, a geometric median minimizes the sum of the distances to the data points in the sample. If the data points $x_1, \ldots, x_N$ do not happen to all lie on a single line, their geometric median is uniquely determined.

In contrast to the centroid, there is no closed formula for the calculation of the geometric median. In particular, the geometric median is *not* given by the *sample median vector*, i.e., the vector comprised of the univariate medians for each variable. The geometric median can be calculated numerically using **Weiszfeld's algorithm** [2, 3]. This algorithm applies the following iteration:

$$y_{k+1} = \left( \sum_{n=1}^{N} \frac{1}{\|x_n - y_k\|} \right)^{-1} \cdot \sum_{n=1}^{N} \frac{x_n}{\|x_n - y_k\|}$$

with a suitable starting value $y_0$, for example $y_0 = \bar{x}_{\text{centroid}}$. If the geometric median is unique, and if one of the intermediate results $y_k$ does not happen to lie exactly on one of the data points $x_n$, then the sequence $y_0, y_1, \ldots$ converges to $\bar{x}_{\text{median}}$.

> **Example.** Fig. 5.2 shows the scatter plot of the width and length of the petal leaf of iris flower plants, along with the geometric median and the centroid of the data points.

The geometric median and the centroid minimize the sum of the Euclidean distances and squared Euclidean distances to the data points $x_1, \ldots, x_N$, respectively. We can generalize this idea to metrics $\delta(\cdot, \cdot)$ defined on some space $\Omega$: the **generalized geometric median** and the **Fréchet mean** are the minimum points of the function

$$\ell_\alpha : \Omega \to [0, \infty[, \ \ell(u) = \sum_{n=1}^{N} (\delta(u, x_n))^\alpha$$

with $\alpha = 1$ or $\alpha = 2$, respectively [4, 5].

Computing these minima is in general a hard task. The following measure of central tendency is more relevant in practice, as it is easier to calculate.

Let $x_1, \ldots, x_N$ be data points that can be compared by some symmetric premetric $\delta(\,\cdot\,,\,\cdot\,)$. A **medoid** is a data point $\bar{x}_{\text{medoid}} \in \{x_1, \ldots, x_N\}$ that minimizes the sum of the distances to the other data points. Thus, it is a minimum point of the following function:

$$f\colon \{x_1, \ldots, x_N\} \to [0, \infty[\,, \ f(u) = \sum_{n=1}^{N} \delta(u, x_n)$$

Unlike the geometric median, the medoid is selected from the set of observed data points.

Given the distance matrix $\Delta$, a medoid $x_i$ corresponds to an index $i \in \{1, \ldots, N\}$ with minimum row or column sum:

$$\Delta_{\bullet i} = \sum_{n=1}^{N} \Delta_{ni} = \sum_{n=1}^{N} \Delta_{in} = \Delta_{i\bullet}$$

### 5.3.2 Sample covariance and correlation matrix

We can calculate pairwise measures of association between a collection of statistical variables and write them into a matrix.

Let $x_1, \ldots, x_N \in \mathbb{R}^D$ be data points and $\boldsymbol{X}$ be the corresponding data matrix. The **sample covariance matrix** is given by the pairwise sample covariances of its columns:

$$S(x) = \begin{pmatrix} s(\boldsymbol{X}_{\bullet 1}, \boldsymbol{X}_{\bullet 1}) & s(\boldsymbol{X}_{\bullet 1}, \boldsymbol{X}_{\bullet 2}) & \cdots & s(\boldsymbol{X}_{\bullet 1}, \boldsymbol{X}_{\bullet D}) \\ s(\boldsymbol{X}_{\bullet 2}, \boldsymbol{X}_{\bullet 1}) & s(\boldsymbol{X}_{\bullet 2}, \boldsymbol{X}_{\bullet 2}) & \cdots & s(\boldsymbol{X}_{\bullet 2}, \boldsymbol{X}_{\bullet D}) \\ \vdots & \vdots & & \vdots \\ s(\boldsymbol{X}_{\bullet D}, \boldsymbol{X}_{\bullet 1}) & s(\boldsymbol{X}_{\bullet D}, \boldsymbol{X}_{\bullet 2}) & \cdots & s(\boldsymbol{X}_{\bullet D}, \boldsymbol{X}_{\bullet D}) \end{pmatrix}$$

Similarly, the **sample correlation matrix** $R(x)$ is the matrix of Bravais–Pearson correlation coefficients.

The covariance matrix and the correlation matrix are both symmetric matrices. If the data matrix $\boldsymbol{X}$ is mean-centered, the covariance matrix can be expressed succinctly as a matrix product:

$$S(x) = \frac{1}{N} \boldsymbol{X}^T \cdot \boldsymbol{X},$$

or $S_{\text{cor}}(x) = \frac{1}{N-1} \boldsymbol{X}^T \cdot \boldsymbol{X}$ if a Bessel-corrected estimation is to be used. If the data matrix has been standardized, then the covariance matrix and correlation matrix are identical.

**Example.** We refer once more to the *Iris* flower dataset and analyze the length and width of the petal leaves. First, we standardize the data—the resulting $z$-scores have unit variance. The covariance/correlation matrix of those data computes to the following:

$$R(x) = \begin{pmatrix} 1 & 0.96 \\ 0.96 & 1 \end{pmatrix}$$

Length and width are thus strongly correlated; the bottom chart in Fig. 5.2 shows the scatter plot of $z$-scores. The figure also shows the **covariance error ellipse**. This ellipse is centered around the centroid of the data points and its semi-axes point in the directions of the smallest and largest variances, respectively. The lengths of the semi-axes scale with the respective standard deviation. The directions of the semi-axes are given by the eigenvectors of the covariance matrix: further details can be found in Sect. 7.2.1 on principal component analysis.

## 5.4 Random vectors and matrices

Now that we have generalized the basic notions from descriptive statistics to the multivariate case, let us do the same for probability theory and inferential statistics.

Just as we can think of univariate observations as realizations of a random variable, data points/feature vectors can be thought of as realizations of a **random vector**. Most conveniently, we may view a random vector $X$ as a (column) vector, the entries of which are random variables, that is:

$$X = \begin{pmatrix} X_1 \\ \vdots \\ X_D \end{pmatrix}$$

where $X_1, \ldots, X_D$ are random variables.

Similarly, **random matrices** are matrices, the entries of which are given by random variables.

### 5.4.1 Expectation vector and covariance matrix

We can generalize the notion of expected values for random vectors by computing the expectation of each of the components and then writing those as a vector again.

In a similar fashion, we can compute the covariances of the random vector's components and then summarize them as a matrix.

Let $X = (X_1, \ldots, X_D)^T$ be a random vector. The **expectation vector** or **mean vector** of $X$ is given as follows:

$$E[X] = \begin{pmatrix} E[X_1] \\ \vdots \\ E[X_D] \end{pmatrix}$$

The **covariance matrix** is the symmetric matrix of pairwise covariances:

$$\Sigma[X] = \begin{pmatrix} \sigma[X_1, X_1] & \sigma[X_1, X_2] & \cdots & \sigma[X_1, X_D] \\ \sigma[X_2, X_1] & \sigma[X_2, X_2] & \cdots & \sigma[X_2, X_D] \\ \vdots & & & \vdots \\ \sigma[X_D, X_1] & \sigma[X_D, X_2] & \cdots & \sigma[X_D, X_D] \end{pmatrix}$$

We may omit the word "vector" and simply talk about the expectation, mean, or expected value of a random vector. The expected value of a random matrix is defined analogously.

The linearity of the univariate expected value carries over:

$$E[A \cdot X + B \cdot Y + C] = A \cdot E[X] + B \cdot E[Y] + C$$

where $X, Y$ are arbitrary random vectors with $D$ entries, while $A, B$ are matrices of format $K \times D$, and $C$ is a column vector of length $K$.

We recall that the following relationship exists between the covariance and variance: $\sigma[X_i, X_i] = \sigma^2[X_i]$. Therefore, the diagonal of the covariance matrix contains the variances of the components of the random vector.

The covariance matrix can also be written more succinctly as the expected value of a random matrix. Remember that we write $X$ as a column vector:

$$\Sigma[X] = E[(X - E[X]) \cdot (X - E[X])^T]$$

A square matrix $A$ of format $D \times D$ is *positive semidefinite* if it does not reverse the direction of vectors. Accordingly, for all vectors $v \in \mathbb{R}^D$ with $v \neq 0$ and scalars $\lambda \in \mathbb{R}$, the following holds:

$$A \cdot v = \lambda v \Rightarrow \lambda \geq 0$$

In other words, the matrix has no negative eigenvalues (see also Sect. B.2.4 in the appendix). Another equivalent characterization is the following [6, Sect. 1.6]:

$$\langle v, A \cdot v \rangle \geq 0$$

for all $v \in \mathbb{R}^D$.

The covariance matrix is always positive semidefinite. This is not difficult to show, setting $Y = X - E[X]$:

$$\langle v, \Sigma[X] \cdot v \rangle = v^T \cdot \Sigma[X] \cdot v = v^T \cdot E[Y \cdot Y^T] \cdot v$$
$$= E[v^T \cdot Y \cdot Y^T \cdot v] = E[(v^T \cdot Y)^2] \geq 0$$

where we used the linearity and monotonicity of the expected value.

### 5.4.2 Multivariate normal distributions

The **probability density function** of a random vector $X = (X_1, \ldots, X_D)^T$, the components of which are continuous random variables, is the joint probability density function of these components:

$$p_X \colon \mathbb{R}^D \to [0, \infty[ \, , \, p_X(u) = p_{X_1, \ldots, X_D}(u_1, \ldots, u_D)$$

One of the most important families of multivariate density functions is given by the generalization of the univariate normal distribution.

---

The **$D$-dimensional normal distribution** is the following (multivariate) probability density function:

$$\mathcal{N}(u|\mu, \Sigma) = \mathcal{N}(u_1, \ldots, u_D|\mu, \Sigma)$$
$$= \frac{1}{\sqrt{(2\pi)^D \cdot \det(\Sigma)}} \cdot e^{-\frac{1}{2}(u-\mu)^T \cdot \Sigma^{-1} \cdot (u-\mu)}$$

where $\mu \in \mathbb{R}^D$, and $\Sigma$ is a square matrix of format $D \times D$ that is symmetric and positive definite.

---

The expectation and covariance matrix of a normally distributed random vector $X \sim \mathcal{N}(\cdot \,|\mu, \Sigma)$ are given by $E[X] = \mu$ and $\Sigma[X] = \Sigma$, respectively.

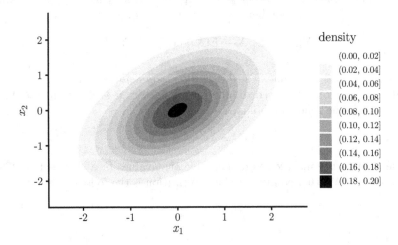

**Fig. 5.1.** *Probability density function of a bivariate normal distribution*

The figure above shows the density of a bivariate normal distribution with a mean vector and covariance matrix given by:

$$\mu = \begin{pmatrix} 0 \\ 0 \end{pmatrix}, \quad \Sigma = \begin{pmatrix} 1 & \frac{1}{2} \\ \frac{1}{2} & 1 \end{pmatrix}.$$

The squared Euclidean norm of a $D$-dimensional normally distributed random vector $X$ centered at the origin, the covariance matrix of which is given by the identity matrix $\Sigma = \mathrm{diag}(1, \ldots, 1)$, follows a chi-squared distribution with $D$ degrees of freedom (see Sect. 3.5.1):

$$\|X\|^2 \sim \chi_D^2$$

The following theorems can be shown by way of cumbersome calculations, see for example [7, Chap. VIII, Sect. 9]. In what follows, let $X = (X_1, \ldots, X_D)^T$ be a normally distributed random vector: $X \sim \mathcal{N}(\cdot \,|\, \mu, \Sigma)$.

**Linear transformations of a normally distributed random vector.**
Let $A$ be a matrix of format $K \times D$, $K \le D$, of rank $K$. Then, the random vector $Y := A \cdot X$ is also normally distributed, with mean $A \cdot \mu$ and covariance $A \cdot \Sigma \cdot A^T$:

$$Y \sim \mathcal{N}(\cdot \,|\, A \cdot \mu, A \cdot \Sigma \cdot A^T)$$

As a special case of the above theorem, we can derive the formula for a linear combination $Y = \sum_{n=1}^N a_n X_n$ of independent univariate normally distributed random variables $X_1, \ldots, X_N$ with expected values $\mu_1, \ldots, \mu_N$ and variances $\sigma_1^2, \ldots, \sigma_N^2$. Plugging in the values

$$\Sigma = \mathrm{diag}(\sigma_1^2, \ldots, \sigma_N^2) \text{ and } A = (a_1, \ldots, a_N),$$

we then get this formula:

$$Y \sim \mathcal{N}(\cdot \,|\, \mu_Y, \sigma_Y^2) \text{ with } \mu_Y = \sum_{n=1}^N a_n \mu_n \text{ and } \sigma_Y^2 = \sum_{n=1}^N (a_n \sigma_n)^2$$

We now want to provide formulas for computing the marginal densities. For this purpose, we imagine $X$ divided into two random vectors of lengths $K$ and $D - K$, respectively, where $1 \le K < D$:

$$X^{(0)} = \begin{pmatrix} X_1 \\ \vdots \\ X_K \end{pmatrix} \text{ and } X^{(1)} = \begin{pmatrix} X_{K+1} \\ \vdots \\ X_D \end{pmatrix}$$

We divide the mean vector the same way:

$$\mu^{(0)} = \begin{pmatrix} \mu_1 \\ \vdots \\ \mu_K \end{pmatrix} \text{ and } \mu^{(1)} = \begin{pmatrix} \mu_{K+1} \\ \vdots \\ \mu_D \end{pmatrix}$$

Finally, we write the covariance matrix into the following block form:

$$\Sigma = \begin{pmatrix} \Sigma^{(00)} & \Sigma^{(01)} \\ \Sigma^{(10)} & \Sigma^{(11)} \end{pmatrix}$$

Deleting the last $D - K$ columns and rows from $\Sigma$ yields the $K \times K$-matrix $\Sigma^{(00)}$, and deleting the first $K$ columns and rows yields $\Sigma^{(11)}$, etc.

> **Marginal distributions of a normally distributed random vector.** The random vector $X^{(0)}$ is also normally distributed:
>
> $$X^{(0)} \sim \mathcal{N}(\cdot \,|\mu^{(0)}, \Sigma^{(00)})$$

For reasons of symmetry, this theorem can be applied to $X^{(1)}$ or to any other selection of components. In particular, each individual component $X_d$, $d \in \{1, \ldots, D\}$ of a normally distributed random vector follows a one-dimensional normal distribution with mean $\mu_d$ and variance $\Sigma_{dd}$.

> **Conditional distributions of components of a normally distributed random vector.** Let $x^{(0)} \in \mathbb{R}^K$. The conditional distribution of $X^{(1)}$ under the condition $X^{(0)} = x^{(0)}$ is a normal distribution:
>
> $$p_{X^{(1)}|X^{(0)}}\left(\cdot \,\middle|\, x^{(0)}\right) = \mathcal{N}\left(\cdot \,\middle|\, \mu^{(1|0)}, \Sigma^{(1|0)}\right)$$
>
> with
>
> $$\mu^{(1|0)} = \mu^{(1)} + \Sigma^{(10)} \cdot \left(\Sigma^{(00)}\right)^{-1} \cdot \left(x^{(0)} - \mu^{(0)}\right),$$
>
> $$\Sigma^{(1|0)} = \Sigma^{(11)} - \Sigma^{(10)} \cdot \left(\Sigma^{(00)}\right)^{-1} \cdot \Sigma^{(01)}$$

As an example, let us consider a normally distributed random vector $X = (X_1, X_2)^T$ with the following mean vector and covariance matrix:

$$\mu = \begin{pmatrix} 0 \\ 0 \end{pmatrix}, \quad \Sigma = \begin{pmatrix} 1 & \frac{1}{2} \\ \frac{1}{2} & 1 \end{pmatrix}.$$

The marginal densities are standard normal distributions: $X_1, X_2 \sim \mathcal{N}(\cdot \,|0, 1)$.

The distribution of $X_2$ under the condition $X_1 = -1$ is a normal distribution with mean $\mu_{2|1} = -\frac{1}{2}$ and variance $\sigma_{2|1}^2 = \frac{3}{4}$.

We can interpret this result as follows. The random variables $X_1$ and $X_2$ are positively correlated: $\sigma[X_1, X_2] = \Sigma_{12} = \frac{1}{2}$. Upon learning that $X_1 = -1$, we also know more about $X_2$. Since there is a positive correlation, we no longer expect the value to be close to the prior mean $\mu_2 = 0$, as we will find that the value is more likely to be found near $\mu_{2|1} = -\frac{1}{2}$. After including this new information of what we know about the value of $X_2$, we are less uncertain of it: $\sigma_{2|1} \approx 0.87 < 1 = \sigma_2$.

### 5.4.3 Multinomial distributions

The multivariate **probability mass function** of a random vector $X = (X_1, \ldots, X_D)^T$, the components of which are all discrete random variables, is given by the joint probability mass function of these components:

$$p_X \colon \operatorname{supp}(X_1) \times \cdots \times \operatorname{supp}(X_D) \to [0, \infty[, \, p_X(u) = p_{X_1, \ldots, X_D}(u_1, \ldots, u_D)$$

The following distribution can be considered a multivariate generalization of the binomial distribution (Sect. 4.1.1).

> The **multinomial distribution** is given by the following (multivariate) probability mass function:
>
> $$\mathcal{M}(\cdot \,|p, N) \colon \mathbb{N} \to [0, 1],$$
> $$\mathcal{M}(k|p, N) = \mathcal{M}(k_1, \ldots, k_D|p_1, \ldots, p_D, N)$$
> $$= \begin{cases} \frac{N!}{\prod_{d=1}^{D} k_d!} \cdot \prod_{d=1}^{D} p_d^{k_d} & \text{if } \sum_{d=1}^{D} k_d = N \\ 0 & \text{otherwise} \end{cases}$$
>
> where $p_1, \ldots, p_D \in [0, 1]$ with $\sum_{d=1}^{D} p_d = 1$, and $N \in \mathbb{N}$.

The expectation vector and the covariance matrix of a multinomially distributed random vector $X \sim \mathcal{M}(\cdot \,|p, N)$ are given as follows:

$$E[X] = N \cdot p, \, \Sigma[X] = N \cdot (\Sigma_p - p \cdot p^T)$$

where $p = (p_1, \ldots, p_D)^T$, and $\Sigma_p = \operatorname{diag}(p_1, \ldots, p_D)$ is the matrix with diagonal entries $p_1, \ldots, p_D$ and otherwise vanishing entries.

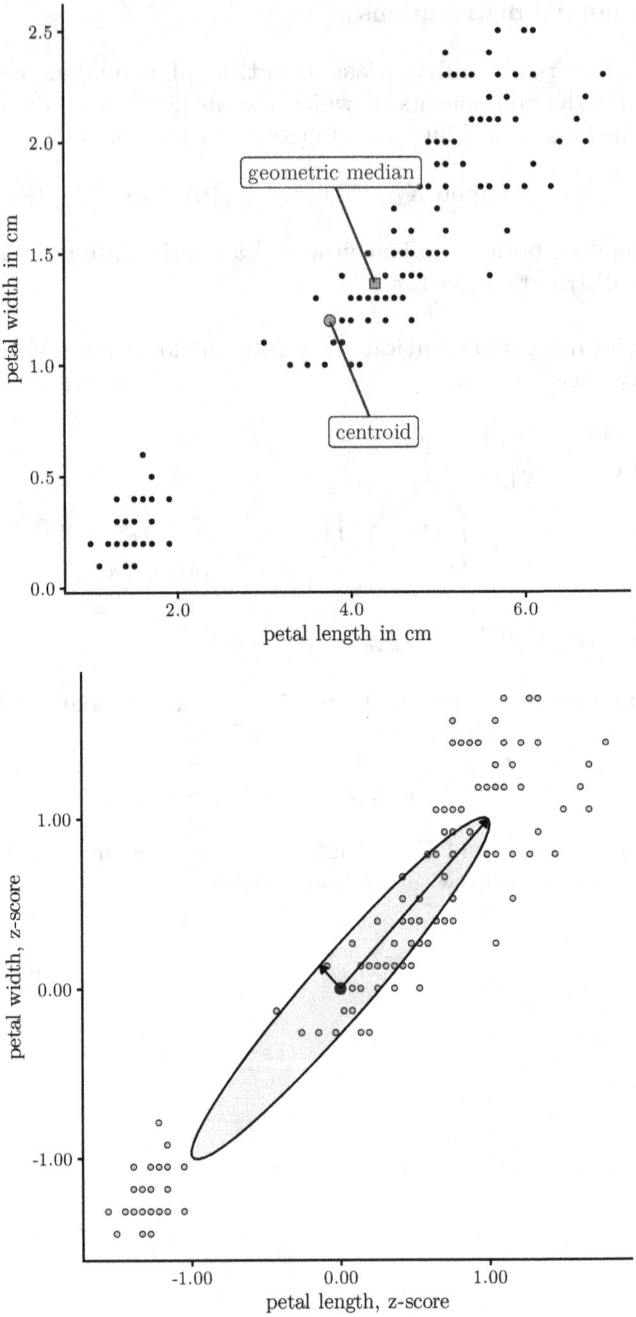

**Fig. 5.2.** *Centroid and geometric median (top); covariance error ellipse (bottom)*

# References

[1]  Ronald Aylmer Fisher. "The use of multiple measurements in taxonomic problems". In: *Annals of Eugenics* 7.2 (Sept. 1936), pp. 179–188. DOI: 10.1111/j.1469-1809.1936.tb02137.x.

[2]  Endre Weiszfeld. "Sur le point pour lequel la somme des distances de $n$ points donnés est minimum". In: *Tohoku Mathematical Journal* 43 (1937), pp. 355–386.

[3]  Amir Beck and Shoham Sabach. "Weiszfeld's Method: Old and New Results". In: *Journal of Optimization Theory and Applications* 164.1 (May 2014), pp. 1–40. DOI: 10.1007/s10957-014-0586-7.

[4]  Maurice René Fréchet. "Les éléments aléatoires de nature quelconque dans un espace distancié". In: *Annales de l'Institut Henri Poincaré* 10.4 (1948), pp. 215–310.

[5]  Miroslav Bacák. "Computing Medians and Means in Hadamard Spaces". In: *SIAM J. Optim.* 24 (2014), pp. 1542–1566. arXiv:1210.2145.

[6]  Jan R. Magnus. *Matrix Differential Calculus with Applications in Statistics and Econometrics*. 3rd ed. Wiley, Feb. 2019. DOI: 10.1002/9781119541219.

[7]  Richard von Mises. *Mathematical Theory of Probability and Statistics*. Ed. by Hilda Geiringer. 1st ed. Academic Press, 1964.

# Part III

## Machine learning

# 6

# Supervised machine learning

According to the International Organization for Standardization (ISO), an **algorithm** [1] is a "finite ordered set of well-defined rules for the solution of a problem."

For example, an algorithm can sort a list of arbitrary strings—such as a list of words or names—in alphabetical order. If such an algorithm was presented with the tuple (David, Robert, Anna, Carl), the correct output would be the tuple (Anna, Carl, David, Robert). A simple algorithm for sorting lists is **bubble sort**. Bubble sort steps through the input list and puts adjacent entries into the correct order. This process is repeated until no more swaps are necessary and the list has become fully sorted:

$N :=$ length of the input tuple $x$
swapped := true
**while** *swapped = true* **do**
    swapped := false
    **for** $i = 1$ *to* $N - 1$ **do**
        **if** $x_i > x_{i+1}$ **then**
            swap $x_i$ with $x_{i+1}$
            swapped := true
        **end**
    **end**
**end**
Output: $x$

Here, $x_i > x_{i+1}$ stands for "$x_i$ comes after $x_{i+1}$ in alphabetical order."

Algorithms are powerful tools because they can produce correct results for any given input. The bubble sort algorithm defines instructions according to a set of rules explicitly specified by the programmer. Any modern computer can execute these instructions in a very short amount of time.

However, a purely rule-based approach is not always practical or even feasible. For example, it would be very difficult to define explicit rules that were to instruct an algorithm to perform the following image classification task: "arrange photos into categories depending on whether the photo shows a landscape, a portrait, or something other than a landscape or a portrait." **Machine learning** provides a set of tools that are able to solve such problems.

In this chapter, we will deal with **supervised machine learning**. Supervised methods are based on the statistical evaluation of a sample where each observation comes with an already known assignment of a **label** that the algorithm is ultimately supposed to predict for yet unseen data. That sample is called the **training dataset**. Keeping with the image classification example, a training dataset would consist of a (large) number of photographs, each of which has been (manually) annotated with one of the labels: *landscape, portrait*, etc. Ideally, the learning algorithm is then able to recognize patterns that characterize and distinguish between landscape and portrait photographs. More concretely, these patterns are statistical variations of **features**. For digital photographs, the raw features are given by the color values of each pixel. From these statistical patterns, rules are generated that are able to categorize new, yet to be seen photos that were not contained in the training dataset. These rules are not explicitly specified by the programmer but are "learned" by the machine on the basis of the training dataset.

Some typical characteristics of machine learning tasks and problems are the following:

- How to extract useful features is not always obvious; typically, some form of **feature engineering** is required. For example, unstructured text data cannot be used in their raw form but must be represented numerically in order to make them accessible to computational and statistical analysis.

- Datasets are often **high-dimensional**, i.e., there are a great number of features that need to be processed. For example, digital photographs with a typical resolution of $1280 \times 720$ pixels and three color channels are represented by $D = 1280 \cdot 720 \cdot 3 = 2.764.800$ color values per image. Therefore, it may be useful or even necessary to apply **feature selection** and other procedures to reduce the dimensionality of the data.

- It is not uncommon for datasets used for machine learning tasks to also consist of a large number of records. ImageNet, for example, is a publicly available dataset that consists of more than 14 million digital images assigned to a total of 1000 different categories [2]. JFT-300M is an example of a dataset used for machine learning that consists of 300 million images [3]. Datasets in **natural language processing** may consist of millions of sentences or paragraphs as well. **Machine learning engineering**—applying machine learning algorithms to massive datasets at scale—is an important topic that we will, however, not address in this book (see, for example, [4] instead).

- Machine learning is primarily used for the **prediction** of features associated with instances that have yet to be seen by the algorithm, instances that were *not* included in the training dataset. The classification algorithm should be able to recognize a large variety of landscapes, portraits, etc., given any new input image. Consequently, the purpose of such an algorithm is not necessarily or primarily to produce a statistical model that describes the training dataset well: it should adapt and *generalize* well to new situations.

- For some tasks that humans master quite effortlessly, computers have only been able to perform through machine learning. These tasks include, for example: image recognition—especially face recognition—speech recognition, natural language translation [5], or playing the Japanese board game Go above amateur level [6]. Consequently, machine learning provides an important set of tools in the field of **artificial intelligence** (AI).

## 6.1 Elements of supervised learning

An important task of intelligent computer systems is to predict the value of a statistical variable. If this prediction is made from patterns and correlations previously observed in a so-called training dataset, then it is referred to as supervised learning.

**Classification algorithms** are used to predict the values of a categorical variable. If the target variable is a numeric variable, we speak of a **regression algorithm**[1]. For example, predicting stock prices based on historical trends would fall into the realm of regression. The automatic assignment of digital photographs to categories (e.g., landscapes, portraits, etc.) belongs to the field of classification. Another example of a classification algorithm is an automated email filter that assigns each email to one of the categories *spam* or *not spam/ham*.

Formally, the situation can be described as follows. We are given a feature space $\mathcal{X}$ that represents the range of values for the explanatory variables that the prediction is based on. For example, $\mathcal{X} = \mathbb{R}^D$, so $D$ numeric features would be used for prediction. These could be, for example, the color values of a digital photograph.

The target variable, the values of which we want to predict, has some range $\mathcal{Y}$. For regression algorithms, $\mathcal{Y} = \mathbb{R}$. For classification algorithms, $\mathcal{Y}$ consists of a finite number of categories, classes, or **labels**: for example, $\mathcal{Y} = \{\text{landscape}, \text{portrait}, \text{other}\}$. We can always assume $\mathcal{Y} \subset \mathbb{N}$ by appropriate numbering of the categories.

Binary classification, $\mathcal{Y} = \{0, 1\}$, is especially useful for illustration and visualization. Moreover, binary classification can be used to solve **multi-label classification** tasks, i.e., when an instance can be assigned multiple of $K$ nonexclusive labels $l_1, \ldots, l_K$. A multi-label classification task can be transformed

---

[1] In statistical learning theory, *both* classification and regression algorithms are based on the idea of regression analysis. Consequently, the terminology may appear somewhat inconsistent.

into a series of binary classification tasks, each aimed at making a prediction of the form "does belong to class $l_k$" versus "does not belong to class $l_k$."

In Sections 6.2 and 6.3, we present a number of regression and classification procedures. The result of such a procedure is a map $\hat{f}\colon \mathcal{X} \to \mathcal{Y}$ learned from the training data that is called a **decision rule**, or **hypothesis**. Such a decision rule assigns, for example, a photo to an image category based on the color values of the pixels.

In classification, decision rules are also called classification rules, or **classifiers**. Decision rules in regression come in the form of **regression functions**. Any classifier or regression function is selected from some **hypothesis space** $\mathcal{F} \subset \{f\colon \mathcal{X} \to \mathcal{Y}\}$. In many cases, the hypothesis space is a parameterized family of functions:

$$\mathcal{F} = \{f(\,\cdot\,; \theta)\colon \mathcal{X} \to \mathcal{Y} | \theta \in \mathcal{P}\}$$

where $\mathcal{P}$ is the space of **model parameters** to be determined during training, typically $\mathcal{P} \subseteq \mathbb{R}^K$. The optimal hypothesis[2] is selected by estimating the optimal model parameters $\hat{\theta} \in \mathcal{P}$ from the training dataset.

Often, we consider not just a single hypothesis space but a family of such spaces where a member is identified by a set of **hyperparameters**.

Formally, we can summarize this concept as follows:

Let $\mathcal{X}$ be the feature space and $\mathcal{Y}$ the range of possible predictions. Furthermore, let $\mathcal{S}_N = (\mathcal{X} \times \mathcal{Y})^N$ denote all possible training datasets of size $N$.

A family of **methods of supervised learning** parameterized by hyperparameters $\alpha \in \mathcal{H}$ consists, firstly, of a family of hypothesis spaces:

$$\mathcal{F}_\alpha = \{f_\alpha(\,\cdot\,; \theta)\colon \mathcal{X} \to \mathcal{Y} | \theta \in \mathcal{P}_\alpha\}_{\alpha \in \mathcal{H}}$$

Secondly, it is defined by a procedure that selects a set of model parameters given a training dataset (of arbitrary size $N$):

$$\hat{\theta}_\alpha\colon \mathcal{S}_N \to \mathcal{P}_\alpha$$

A supervised algorithm processes a training dataset $(x, y) \cong ((x_1, y_1), \ldots, (x_N, y_N))$ with the goal of learning a decision rule:

$$\hat{f}(\,\cdot\,) = f_\alpha(\,\cdot\,; \hat{\theta}_\alpha(x, y))$$

Given yet unseen data $x_*$, this decision rule can be used to make a prediction $\hat{y}_* = \hat{f}(x_*)$.

---

[2] We skip a few mathematical details. For example, decision rules $f\colon \mathcal{X} \to \mathcal{Y}$ should at least be measurable maps so that, for example, expected values like $E[f(X)]$ are well-defined, given a random variable $X\colon \Omega \to \mathcal{X}$.

The hyperparameters $\alpha$ are determined from a priori assumptions, or alternatively they may be calculated through a validation procedure—a concept we discuss in more detail in Sect. 6.1.3.

A binary classification rule $f\colon \mathcal{X} \to \{0,1\}$ is equivalent to dividing the feature space into two disjoint regions:

$$\mathcal{X} = f^{-1}(\{0\}) \cup f^{-1}(\{1\}) \text{ with } f^{-1}(\{0\}) \cap f^{-1}(\{1\}) = \emptyset$$

These two regions are separated by a **decision boundary**. In some sense, along this boundary, the classifier is actually "undecided." The **linear classifiers** are an important family of decision rules with domain $\mathcal{X} = \mathbb{R}^D$. Linear classifiers are characterized by their decision boundary being an (affine) hyperplane:

$$\mathcal{F} = \left\{ f(u; w) = \begin{cases} 1 \text{ if } \sum_{d=1}^{D} w_d u_d < w_0 \\ 0 \text{ otherwise} \end{cases} \middle| w = (w_0, \ldots, w_D) \in \mathbb{R}^{D+1} \right\}$$

The logistic regression described in Sect. 6.3.1 is an example of a procedure that learns a linear classifier.

Let us illustrate how a procedure that we already know fits into the above characterization. Simple linear regression (Sect. 4.5.1) can be interpreted as a regression algorithm with $\mathcal{X} = \mathbb{R}$ and $\mathcal{Y} = \mathbb{R}$. The hypothesis space is given by the space of regression lines:

$$\mathcal{F}_1 = \{f\colon \mathbb{R} \to \mathbb{R}, \ f(u; m, c) = mu + c \mid m, c \in \mathbb{R}\}$$

The space of model parameters consists of all possible values for the slope and the intercept of the regression line: $\mathcal{P} = \{(m, c) \mid m, c \in \mathbb{R}\} = \mathbb{R}^2$. There are no hyperparameters.

The training consists of computing the minimum point of the residual sum of squares. Given a training dataset $(x, y) \cong ((x_1, y_1), \ldots, (x_N, y_N))$, we recall that this minimum point is given as follows:

$$\hat{\theta}(x, y) = \left( \frac{s(x, y)}{s^2(x)}, \bar{y} - \frac{s(x, y)}{s^2(x)} \cdot \bar{x} \right) = (\hat{m}, \hat{c})$$

The decision rule $\hat{f}\colon \mathbb{R} \to \mathbb{R}$ thus learned is the function $\hat{f}(x_*) = \hat{m}x_* + \hat{c}$, which makes a prediction for any $x_* \in \mathbb{R}$.

In Sect. 6.2.1, we will see how we can fit polynomials of higher degree to a sample of data points, and contrary to the previous methods shown it will not just be a straight line. For each degree of a polynomial $K$, which we interpret as a hyperparameter, we can associate the following hypothesis space:

$$\mathcal{F}_K = \left\{ f_K(u; w) = \sum_{k=0}^{K} w_k u^k \middle| w = (w_0, \ldots, w_K) \in \mathbb{R}^{K+1} \right\}_{K \in \mathbb{N}}$$

### 6.1.1 Loss functions and empirical risk minimization

We can numerically evaluate the goodness of a prediction by introducing a **loss function**:

$$\lambda \colon \mathcal{Y} \times \mathcal{Y} \to [0, \infty[, \ (y, \hat{y}) \mapsto \lambda(y, \hat{y})$$

The loss function measures the cost or damage caused by predicting $\hat{y}$ when the true value of the target variable is in fact $y$.

A popular loss function for regression algorithms is the **quadratic loss**, or $L_2$ **loss**:

$$\lambda_2(y, \hat{y}) = (y - \hat{y})^2$$

for all $y, \hat{y} \in \mathbb{R}$. Another possible choice is the $L_1$ **loss**: $\lambda_1(y, \hat{y}) = |y - \hat{y}|$. More generally, we may consider the family of $L_p$ **loss functions** $\lambda_p(y, \hat{y}) = |y - \hat{y}|^p$ with $p \geq 1$.

For classification tasks, the **zero–one loss function** is a particularly simple choice. When using a zero–one loss, if the predicted category is correct, the loss is given by $\lambda = 0$. If the prediction is incorrect, the loss is given by $\lambda = 1$. In general, the loss function for a classification task can be described by a **cost matrix**, which tabulates the loss for each possible combination of true and predicted values.

The cost matrix for identifying spam emails might look like this ($a, b \in \mathbb{R}$ with $a, b > 0$):

| prediction true class | no spam | spam |
|---|---|---|
| **no spam** | 0 | $a$ |
| **spam** | $b$ | 0 |

**Table 6.1.** *Cost matrix of a spam filter*

If the message is classified correctly, the loss is always zero. If the message is incorrectly classified as spam, the cost incurred by the misclassification is given by $\lambda(\text{not spam}, \text{spam}) = a$. If the spam filter lets an unwanted message through, the cost incurred is $\lambda(\text{spam}, \text{no spam}) = b$.

Thus, if the user of the email software were to consider it a greater harm for a relevant message to accidentally end up in the spam folder than for an unwanted message to remain in the inbox, then $a > b$ should be chosen when designing the algorithm.

The **empirical risk** or **training error** of a decision rule $f \colon \mathcal{X} \to \mathcal{Y}$ with respect to a training dataset $((x_1, y_1), \ldots, (x_N, y_N)) \in (\mathcal{X} \times \mathcal{Y})^N$ and a loss function $\lambda \colon \mathcal{Y} \times \mathcal{Y} \to [0, \infty[$ is given by the average loss over the training examples:

$$\hat{R}[f] = \frac{1}{N} \sum_{n=1}^{N} \lambda(y_n, f(x_n))$$

If the decision rule is selected from a hypothesis space $\mathcal{F}_\alpha = \{f_\alpha(\,\cdot\,;\theta)\}$, the training error can be interpreted as a function $R_\alpha(\cdot)$ in the model parameters:

$$R_\alpha(\theta) = \hat{R}[f_\alpha(\,\cdot\,;\theta)] = \frac{1}{N} \sum_{n=1}^{N} \lambda(y_n, f_\alpha(x_n;\theta))$$

The **empirical risk minimization** (ERM) paradigm is a simple and popular approach to supervised machine leaning:

- Choose a hypothesis space and hyperparameters, i.e., a set of classifiers/regression functions to select from,

- choose a loss function that evaluates goodness of fit,

- given the training dataset, minimize the training error to determine the optimal classifier/regression function.

For example, simple linear regression corresponds to minimizing the following empirical risk with respect to the quadratic loss function, with $m, c \in \mathbb{R}$:

$$R(m, c) = \frac{1}{N} \sum_{n=1}^{N} (y_n - m x_n - c)^2$$

If we ignore the constant factor $1/N$, then this formula is just the residual sum of squares.

However, minimizing empirical risk alone is not sufficient to ensure the **generalizability** of the learned decision rule: the quality of predictions for *new* data may be severely limited even if the classifier or regression function is very well adapted to the training data. This fact can be illustrated by the following extreme example. Given a training dataset $((x_1, y_1), \ldots, (x_N, y_N)) \in (\mathbb{R} \times \mathbb{R})^N$, we define the following decision rule:

$$\hat{f}(x_*) = \begin{cases} y_i & \text{if } x_* = x_i \text{ for some } i \in \{1, \ldots, N\} \\ 10^{42} & \text{otherwise} \end{cases}$$

Here, we assume that the $x_i$ are pairwise distinct, or they at least uniquely determine the corresponding $y_i$. In other words, the procedure looks up the input in the training dataset and then assigns the corresponding response. If it can't find the input, the prediction is just the arbitrary value $\hat{y} = 10^{42}$. For any reasonable loss function, e.g., the quadratic loss function, the empirical risk vanishes for this procedure: the decision rule is always an optimal fit to the training data—all pairs of values are reproduced perfectly. Nevertheless, the algorithm is no better than random guessing, since for *almost all* input values the prediction can produce an arbitrarily high loss.

The example shows that minimizing the empirical risk/training error $\hat{R}[f]$ does not imply minimization of the true **expected risk** $R[f]$, also called the **test error** or **generalization error**. At first glance, this may seem like a contradiction, since for some *fixed* decision rule $f \colon \mathcal{X} \to \mathcal{Y}$ and all $\varepsilon > 0$ the arithmetic mean is a consistent estimator after all:

$$\lim_{N \to \infty} \Pr\left(\left|\hat{R}_N[f] - R[f]\right| < \varepsilon\right) = 1$$

with

$$\hat{R}_N[f] = \frac{1}{N} \sum_{n=1}^{N} \lambda(Y_n, f(X_n)), \; R[f] = E[\lambda(Y_*, f(X_*))]$$

where $X_1, \ldots, X_N, X_*$ and $Y_1, \ldots, Y_N, Y_*$ are independent and identically distributed random variables with range $\mathcal{X}$ and $\mathcal{Y}$, respectively. However, when applying a machine learning procedure, the decision rule is not fixed but varies over the hypothesis space $\mathcal{F}$, depending on the training data. In order to guarantee that the expected risk is always minimized, we would have to make sure that *uniform* convergence holds, for all $\varepsilon > 0$:

$$\lim_{N \to \infty} \Pr\left(\sup_{f \in \mathcal{F}} \left|\hat{R}_N[f] - R[f]\right| < \varepsilon\right) = 1$$

In that case, the empirical risk cannot arbitrarily deviate from the expected risk. An essential task of the theory of statistical learning is to investigate in which cases and to what extent such requirements on the generalizability of the procedures can be guaranteed.

### 6.1.2 Overfitting and underfitting

The figure below shows polynomials of varying degree, fitted to the time series of measurements of the global average temperature of the Earth [7]. The higher the degree of the polynomial, the better it fits the sample/training data. In other words, the empirical risk—the average squared deviation between the regression curve and the data points—is minimized.

This observation points to a more general phenomenon: the more parameters a model has (in this case, the coefficients of the polynomial), the more "wiggle room" there is to fit it to the training data.

At first glance, it seems to make sense to use models that are as complex as possible, because then the empirical risk is minimized. Within the interval where there is sufficient training data, the 20th degree polynomial is the best fit, accounting also for small variations in temperature. However, we also clearly see that this model is poorly suited for predictions. For instance, for the early and late times that are beyond what was included in the training data, the regression curve drops off very quickly. The decision rule predicts a "sudden ice age" for Earth's past and future.

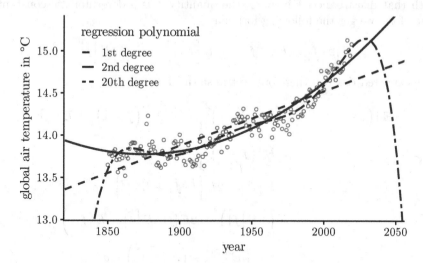

**Fig. 6.1.** *Average global air temperature and regression polynomials of varying degree*

In order to investigate in more detail the impact of model complexity on the generalization error, we consider regression algorithms that operate under the following assumptions:

- The target variable is distributed according to $Y_n = f(x_n) + \varepsilon_n$, where the independent and identically distributed error terms $\varepsilon_n$ have a vanishing expected value and finite variance $\sigma^2 > 0$.

- The training dataset $((x_1, y_1), \ldots, (x_N, y_N))$ and a point in the test dataset $(x_*, y_*)$ are realizations of independent and identically distributed random variables $(X_1, \ldots, X_N, X_*)$ and $(Y_1, \ldots, Y_N, Y_*)$, respectively.

- The loss function is the quadratic loss function $\lambda(y, \hat{y}) = (y - \hat{y})^2$ for all $y, \hat{y} \in \mathbb{R}$.

Let $x_* \in \mathcal{X}$ be a value of the predictor variable observed in the test dataset. Then, the test error (subject to the condition $X_* = x_*$) is given as follows:

$$R\left[\hat{f}_\alpha \Big| X_* = x_*\right] = E\left[\left(f(x_*) + \varepsilon_* - \hat{f}_\alpha(x_*)\right)^2\right]$$

with the prediction $\hat{f}_\alpha(x_*) = f_\alpha(x_*; \hat{\theta}_\alpha)$. In doing so, we interpret the prediction as a statistical estimator, i.e., a function in the random variables that produce the distribution of the training data:

$$\hat{\theta}_\alpha = \hat{\theta}_\alpha((X_1, Y_1), \ldots, (X_N, Y_N))$$

For clarity, we introduce short notation: $\hat{f} = \hat{f}_\alpha(x_*)$, $f = f(x_*)$ and $\varepsilon = \varepsilon_*$. The disturbance $\varepsilon$ with $E[\varepsilon] = 0$ and $E[\varepsilon^2] = \sigma^2$ generates variations in the test dataset that cannot be explained by any observation in the training dataset. Since the estimator $\hat{f}$ was learned using the training dataset, it is uncorrelated with that disturbance. Given $x_*$, the quantity $f$ is a deterministic constant. Therefore, we get the following formula:

$$E\left[\varepsilon \cdot \hat{f}\right] = E\left[\varepsilon \cdot f\right] = 0, \; E\left[f \cdot \hat{f}\right] = E\left[f\right] \cdot E\left[\hat{f}\right]$$

The expected risk can therefore be transformed as follows:

$$E\left[\left(f - \hat{f} + \varepsilon\right)^2\right] = E\left[\left(f - \hat{f}\right)^2\right] + E\left[2\varepsilon \cdot \left(f - \hat{f}\right)\right] + E\left[\varepsilon^2\right]$$

$$= E\left[\left(f - \hat{f}\right)^2\right] + \sigma^2$$

$$= E\left[f^2\right] - 2E\left[f \cdot \hat{f}\right] + E\left[\hat{f}^2\right] + \sigma^2$$

$$= \left(\left(E\left[\hat{f}\right]\right)^2 - 2E\left[f\right] \cdot E\left[\hat{f}\right] + E\left[f^2\right]\right)$$

$$+ \left(E\left[\hat{f}^2\right] - \left(E\left[\hat{f}\right]\right)^2\right) + \sigma^2$$

$$= \left(E\left[\hat{f} - f\right]\right)^2 + E\left[\left(\hat{f} - E[\hat{f}]\right)^2\right] + \sigma^2$$

**Bias–variance decomposition.** The expected risk of a regression algorithm with quadratic loss can be decomposed as follows:

$$R\left[\hat{f}_\alpha \middle| X_* = x_*\right]$$

$$= \left(E\left[\hat{f}_\alpha(x_*) - f(x_*)\right]\right)^2 + E\left[\left(\hat{f}_\alpha(x_*) - E\left[\hat{f}_\alpha(x_*)\right]\right)^2\right] + \sigma^2$$

$$= \left(\text{bias}\left[\hat{f}_\alpha \middle| X_* = x_*\right]\right)^2 + \text{variance}\left[\hat{f}_\alpha \middle| X_* = x_*\right] + \sigma^2$$

The generalization error is comprised of the following components:

- Due to the random fluctuations of the target quantity $Y_*$ around the true value $f(x_*)$, a prediction with vanishing error based on observation of data is not possible: at least the **irreducible error** $\sigma^2$ is to be expected.

- The **bias** $E[\hat{f}_\alpha(x_*) - f(x_*)]$ is the expected deviation of the estimate from the true value.

- The **variance** $E[(\hat{f}_\alpha(x_*) - E[\hat{f}_\alpha(x_*)])^2]$ indicates the variation that occurs when the procedure is repeatedly applied to different training datasets, even if those are drawn from the same distribution.

**Model selection**—i.e., the optimal choice of hyperparameters $\alpha$—implies a suitable **bias–variance tradeoff** such that the model minimizes the expected risk. A complex model that can be fitted to training data, and therefore has low empirical risk and bias, often comes at the price of a high variance. For instance, such a model may be too sensitive to random fluctuations in the training data. In this case, we say that the model is **overfitting** (the data). On the other hand, models that are too simple and exhibit a high bias are said to be **underfitting**.

The following decomposition of expected risk is independent of a specific loss function and applies to classification methods as well.

> **Approximation and estimation error.** The expected risk of a supervised machine learning procedure can be decomposed as follows:
>
> $$R[\hat{f}_\alpha] = \left( \inf_{f \in \mathcal{F}_\alpha} \{R[f]\} - R[f_{\text{Bayes}}] \right) + \left( R[\hat{f}_\alpha] - \inf_{f \in \mathcal{F}_\alpha} \{R[f]\} \right) + R[f_{\text{Bayes}}]$$
>
> $$= \text{approximation error} + \text{estimation error} + \text{Bayes error}$$
>
> Here, $R[f_{\text{Bayes}}]$ is the minimum expected risk under all decision rules that are allowed in principle without restriction to the hypothesis space.

In more detail, the terms have the following meaning:

- The **Bayes error** is the irreducible error that corresponds to the minimum expected risk, produced by an optimal decision rule $f_{\text{Bayes}}$.

- The **approximation error** $\inf\{R[f]\} - R[f_{\text{Bayes}}]$ is analogous to the bias and indicates the excess risk generated by selecting the optimal decision rule from a limited hypothesis space.

- The **estimation error** $R[\hat{f}_\alpha] - \inf\{R[f]\}$ results from estimating the model parameters from a limited training dataset: Even if the algorithm minimizes empirical risk, $\hat{f}_\alpha$ does not necessarily minimize expected risk. It is analogous to the variance.

A classification model with a large estimation error is overfitting, and a model with high approximation error is underfitting the data.

### 6.1.2.1 Regularization

One approach to avoid overfitting is the method of **regularization**, where a regularizer term $\mathcal{R}[f]$ is added to the empirical risk to be minimized:

$$\hat{R}[f] = \frac{1}{N} \sum_{n=1}^{N} \lambda(y_n, f(x_n)) + \mathcal{R}[f]$$

The regularizer does not depend on the data, yet it measures the complexity of the classifier/regression function $f$. The idea is to minimize not only the training error but also to put a penalty on decision rules that represent more complex models, such as a "wigglier" polynomial. The hope is to reduce the estimation error/variance and, consequently, the generalization error. Written as a function of model parameters and hyperparameters, the regularized objective function becomes:

$$R_{\alpha,\beta}(\theta) = \frac{1}{N} \sum_{n=1}^{N} \lambda(y_n, f_\alpha(x_n; \theta)) + \mathcal{R}_\beta(\theta)$$

where we introduced regularization parameters $\beta$ that gauge the strength of the regularization. Let us consider the case where we wish to learn a function of the form $f(u; b, w) = \varphi(w^T \cdot u + b)$, $u \in \mathbb{R}^D$, where the model parameters are given by $b \in \mathbb{R}$ and a row vector of coefficients $w = (w_1, \ldots, w_D)$. The function $\varphi \colon \mathbb{R} \to \mathbb{R}$ can be assumed to be monotone. Linear regression (Sect. 6.2.1) and logistic regression (Sect. 6.3.1) are both examples of models that can be written in such a way. Thus, we have the following model:

$$R_\beta(b, w) = \frac{1}{N} \sum_{n=1}^{N} \lambda(y_n, \varphi(w^T \cdot u + b)) + \mathcal{R}_\beta(w)$$

There is no particular motivation to put a penalty on the size of $b$, so we removed it from the regularizer's argument. In fact, we argue that $f(u) = \text{const.}$ is the simplest hypothesis that we may assume a priori, and we want to put a penalty on the size of the coefficients $w_1, \ldots, w_D$. This rationale motivates the use of one of the following regularizers:

$$\mathcal{R}_\beta^{(1)}(w) = \beta \cdot \|w\|_1, \ \mathcal{R}_\beta^{(2)}(w) = \beta \cdot \|w\|^2, \ \mathcal{R}_{\beta_1,\beta_2}^{(1,2)}(w) = \beta_1 \cdot \|w\|_1 + \beta_2 \cdot \|w\|^2$$

where $\| \cdot \|_1$ is the Minkowski 1-norm and $\| \cdot \|$ the usual Euclidean 2-norm. These are called **$L_1$ regularization**, **$L_2$ regularization**, and **elastic net regularization**, respectively.

In a more general sense, regularization in machine learning refers to any technique that aims at reducing the test error but not the training error in order to avoid overfitting. Another example for a regularization technique in that sense is *dropout* in neural networks, presented later in Sect. 6.4.2.1.

### 6.1.3 Training, model validation, and testing

A model that minimizes empirical risk does not necessarily perform well on unseen data, and the true expected risk cannot be estimated from a single training dataset. One solution to this problem is to compare the model's predictions with the true labels of a second, separate dataset, which is called a *validation dataset*. The idea is to test the generalizability of the model by applying it to data that has not been used to train it. In this way, overfitting in particular can be detected and avoided.

> **Training dataset.** Sample of annotated/labeled training examples, based on which a machine learning procedure determines the parameters of a model (by minimizing the training error).

**Validation dataset.** Sample of annotated examples used for **hyperparameter tuning** and model selection. The generalization error is estimated from this dataset to determine the optimal hyperparameters.

**Test dataset.** Annotated examples also used to estimate the generalization error, to evaluate the performance of the final classifier or regression model.

A typical workflow for training, validating, and testing a supervised learning algorithm is the following:

1. The original raw annotated data is partitioned—usually by random selection—to produce a training dataset, a validation dataset, and a test dataset. A typical partition could be 70% of training data in addition to 15% each of validation and test data.

2. The model parameters are determined from the training dataset for a selection of hyperparameters. For example, the selection of hyperparameters can be done manually (**grid search**) or (partially) at random (**random search**).

3. For each selection of hyperparameters, the average loss is determined from the validation dataset, serving as an estimator of the expected risk/test error. Other performance measures can be used for the evaluation as well, some of which we explain in the following sections. For example, the **efficiency**, in terms of data processing speed or required computational resources, may also play an important role in selecting a model in practice.

4. The model/hyperparameters that are considered optimal according to the validation step are finally evaluated using the test dataset.

There is some terminological confusion, as some practitioners may not make a strict distinction between validation and test dataset. In any event, it is *crucial* that the training dataset does not overlap with the validation/test dataset!

The simple validation technique described above is called **holdout validation**. Another technique is **$K$-fold cross-validation**: the initial dataset (after withholding a test dataset) is divided into a total of $K$ non-overlapping validation datasets of equal size. Each of these validation datasets is assigned a training dataset, which is simply the remaining observations that are not included in the respective validation dataset. Training and validation is performed using this sequence of datasets a total of $K$ times, and the model performance scores are averaged over those trials. This procedure is aimed at improving the chances of selecting optimal hyperparameters, compared to using only a single validation dataset.

### 6.1.3.1 Performance measures for regression

In order to evaluate the predictive power of the result of a simple linear regression (Sect. 4.5.1), we computed the mean squared error (MSE), root mean squared error (RMSE), and coefficient of determination ($r^2$) from the whole dataset at our disposal, with confidence that our model was a good description of the relation between predictor and target variable.

In multivariate statistics and machine learning, such confidence can rarely be gained from first principles. In fact, these measures gauge what we now know as the *training error*. However, we can use those measures to gauge the size of the generalization error by applying them to a withheld test dataset.

Suppose we are given a regression function $\hat{f}\colon \mathcal{X} \to \mathbb{R}$ that a regression algorithm learned from some training dataset. We would like to validate or test the model based on test examples $(x_1, y_1), \ldots, (x_M, y_M)$. Writing $\hat{y}_* = \hat{f}(x_*)$ for the model's prediction, we may compute the following measures for goodness of fit, including the **mean absolute error (MAE)**:

$$\text{MSE} = \frac{1}{M} \sum_{m=1}^{M} (y_m - \hat{y}_m)^2, \ \text{RMSE} = \left( \frac{1}{M} \sum_{m=1}^{M} (y_m - \hat{y}_m)^2 \right)^{\frac{1}{2}},$$

$$\text{MAE} = \frac{1}{M} \sum_{m=1}^{M} |y_m - \hat{y}_m|$$

We recognize that the mean squared error is the average quadratic loss, and the mean absolute error is the average $L_1$ loss. In order to optimize performance, those measures need to be minimized. On the other hand, the coefficient of determination needs to be maximized:

$$r^2 = 1 - \frac{\sum_{m=1}^{M} (y_m - \hat{y}_m)^2}{\sum_{m=1}^{M} (y_m - \bar{y})^2}$$

where $\bar{y}$ is the arithmetic mean of the target variable's values in the test dataset.

A useful baseline to compare a given regression function with is the decision rule that assigns the arithmetic mean to every test example: $\hat{y}_* = \bar{y}$. Assuming that $\bar{y}$ is accurately estimated, the mean squared error of this baseline predictor is equal to the variance of the test dataset, the root mean squared error is equal to the standard deviation, and the mean absolute error is equal to the mean absolute deviation from the mean. A regression function with $r^2 < 0$ performs worse than this baseline, while $r^2 = 1$ indicates a perfect prediction for every test example.

### 6.1.3.2 Performance measures for binary classification

Suppose that we are given a binary decision rule $\hat{f}\colon \mathcal{X} \to \{0, 1\}$ and a single test example $(x_*, y_*)$. Writing $\hat{y}_* = \hat{f}(x_*)$ for the prediction, we note that any one of the following cases may occur:

|  | $\hat{y}_* = 0$ | $\hat{y}_* = 1$ |
|---|---|---|
| $y_* = 0$ | true negative | false positive |
| $y_* = 1$ | false negative | true positive |

**Table 6.2.** *True/false positive/negative results*

In the literature, the convention $y \in \{-1, +1\}$ for binary classification problems is also quite common. That convention is more consistent with the manner preferred when speaking about negative and positive results.

Given $M$ test examples and noting down the absolute frequencies of correct and incorrect classification, we obtain a contingency table which is called the **confusion matrix**:

|  | $\hat{y} = 0$ | $\hat{y} = 1$ | $\sum$ |
|---|---|---|---|
| $y = 0$ | TN | FP | N |
| $y = 1$ | FN | TP | P |
| $\sum$ | $\hat{N}$ | $\hat{P}$ | $M$ |

In general, the rates of misclassification FP and FN should be kept as low as possible. Given a cost matrix of the form

| $\lambda(\cdot, \cdot)$ | $\hat{y} = 0$ | $\hat{y} = 1$ |
|---|---|---|
| $y = 0$ | 0 | $a$ |
| $y = 1$ | $b$ | 0 |

the empirical test error can be computed as follows: $\hat{R}[f] = a \cdot \frac{FP}{M} + b \cdot \frac{FN}{M}$.

Instead of the test error, the following measures are also common to assess the learned classification rule. Unlike the test error, these metrics should be *maximized*.

For a binary classifier, the following performance measures can be computed from the entries of the confusion matrix:

$$\textbf{accuracy} = \frac{TN + TP}{M},$$

$$\textbf{specificity} = \frac{TN}{N},$$

$$\textbf{recall, sensitivity} = \frac{TP}{P},$$

$$\textbf{precision} = \frac{TP}{\hat{P}}$$

Precision refers to the fraction of correct positive labels out of all examples that the classifier labeled as positive. Recall refers to the fraction of examples that the classifier labeled as positive among all examples that are truly positive.

When viewed alone, precision and recall are not sufficient for assessing a classifier's goodness of fit. For example, a classifier that would assign a positive class label to every observation would yield a recall of 100%, but overall it would perform poorly. For many algorithms, we can increase precision at the cost of lowering recall, or vice versa. However, to help balance this trade-off, we can use the following combination of precision and recall.

Given a fixed weighting parameter $\beta > 0$, the $F_\beta$-score is defined as follows:

$$F_\beta = \frac{(1 + \beta^2) \cdot \text{TP}}{(1 + \beta^2) \cdot \text{TP} + \beta^2 \cdot \text{FN} + \text{FP}}$$

$$= (1 + \beta^2) \cdot \frac{\text{precision} \cdot \text{recall}}{\beta^2 \cdot \text{precision} + \text{recall}}$$

The $F_\beta$-score for $\beta = 1$ is just the harmonic mean of precision and recall. If the value for $\beta$ is small, more importance is attributed to precision. A large value for $\beta$ corresponds to a higher weighting of recall.

Accuracy refers to the fraction of correct labels among all test examples. Using a zero–one loss function, accuracy equals one minus the empirical test error. If the data are highly imbalanced, the accuracy may be a misleading measure for goodness of fit. For example, if 95% of the test examples have a true positive class assignment, a decision rule that would assign a positive class label to every observation would achieve an accuracy of 95%.

The following measures are considered to be suitable for imbalanced data as well as when positive and negative class labels are to be treated on an equal footing.

The **balanced accuracy** is given by the arithmetic mean of specificity and sensitivity:

$$\text{balanced accuracy} = \frac{1}{2} \cdot \left( \frac{\text{TN}}{\text{N}} + \frac{\text{TP}}{\text{P}} \right)$$

**Matthews' correlation coefficient** (MCC) is defined as follows [8]:

$$\text{MCC} = \frac{\text{TN} \cdot \text{TP} - \text{FN} \cdot \text{FP}}{\sqrt{\hat{\text{P}} \cdot \text{P} \cdot \hat{\text{N}} \cdot \text{N}}}$$

**Youden's index** is a simple rescaling of the balanced accuracy [9]:

$$\text{Youden's index} = 2 \times \text{balanced accuracy} - 1 = \frac{\text{TN}}{\text{N}} + \frac{\text{TP}}{\text{P}} - 1$$

$$= \frac{\text{TN} \cdot \text{TP} - \text{FN} \cdot \text{FP}}{\text{P} \cdot \text{N}}$$

If Matthews' correlation coefficient or Youden's index are equal to one, the classifier only makes correct predictions. A classifier with MCC or Youden's index equal to minus one produces a false prediction every time.

As an example, imagine that we want to automatically detect and filter spam email. First, we train a classifier, and then we run it on a test dataset of size $M = 1620$, yielding the following confusion matrix:

| prediction<br>true class | not spam | spam | $\sum$ |
|---|---|---|---|
| **not spam** | 400 | 20 | 420 |
| **spam** | 200 | 1000 | 1200 |
| $\sum$ | 600 | 1020 | 1620 |

**Table 6.3.** *Confusion matrix for spam detection*

The accuracy of this classifier is given by $\frac{400+1000}{1620} \approx 86\%$, the balanced accuracy is $\frac{1}{2} \cdot \left( \frac{400}{420} + \frac{1000}{1200} \right) \approx 89\%$. The Matthews correlation coefficient computes to MCC $= 0.71$.

Moreover, we get the following results:

$$\text{precision for detecting spam} = \frac{1000}{1020} \approx 98\%$$

$$\text{recall for detecting spam} = \frac{1000}{1200} \approx 83\%$$

In reference to the *error rates*: 2% of the emails moved to the spam folder should not have been; while 17% of unwanted emails slipped through the spam filter.

Instead of detecting spam emails, we could have just as well declared the identification of relevant, non-spam messages as the positive class assignment. In that case, we would get the following results:

$$\text{precision for identifying relevant email} = \frac{400}{600} \approx 67\%$$

$$\text{recall for identifying relevant email} = \frac{400}{420} \approx 95\%$$

In terms of the error rates, 33% of emails in the inbox are in fact spam, and 5% of actually relevant emails have been moved to the spam folder.

In terms of $F_\beta$-scores, the classifier for identifying relevant emails performs as follows:

$$F_{0.5} \approx 70\%, \ F_1 \approx 78\%, \ F_2 \approx 88\%$$

Precision and recall are also used for the evaluation of **information retrieval** systems like search engines. A search engine takes a user query as input and then outputs those data records from a dataset that are a relevant match to the query. For this use case, the performance measures are to be interpreted as follows:

$$\text{precision} = \frac{\text{nb. of relevant results}}{\text{total nb. of results}},$$

$$\text{recall} = \frac{\text{nb. of relevant results}}{\text{total nb. of relevant data records}}$$

Precision indicates the quality of results presented to the user: how many records in the result set are a relevant match to the query? Recall, on the other hand, can be interpreted as a measure for result coverage: how many out of all relevant records did the system match and retrieve successfully?

### 6.1.4 Numerical optimization

Given a training dataset and fixed hyperparameters, the empirical risk is a real-valued function in the model parameters $\theta_1, \ldots, \theta_K$. Training via empirical risk minimization means determining the minimum point of that function. As a rule, this minimum point cannot be given by a closed formula—it must be calculated by means of numerical analysis.

In fact, many statistical learning procedures, including unsupervised learning, lead to the numerical computation of extrema of some **objective function** $R \colon \mathbb{R}^K \to \mathbb{R}$. The prototypical algorithm for solving this problem is called **gradient descent**, also called **steepest descent**.

Let us sketch the rationale behind gradient descent. The derivative of $R$ in the direction $h \in \mathbb{R}^K$, $\|h\| = 1$ can be expressed via the gradient of $R$ as follows:

$$D_h R(\theta) = \lim_{\alpha \to 0} \frac{R(\theta + \alpha h) - R(\theta)}{\alpha} = \langle h, \operatorname{grad} R(\theta) \rangle$$

If $\theta$ does not happen to be a stationary point (i.e., a point with vanishing gradient), and if $h$ points in the opposite direction of the gradient, this directional derivative is negative:

$$D_h R(\theta) = -\|\operatorname{grad} R(\theta)\|^{-1} \cdot \langle \operatorname{grad} R(\theta), \operatorname{grad} R(\theta) \rangle < 0$$

if $h = -\|\operatorname{grad} R(\theta)\|^{-1} \cdot \operatorname{grad} R(\theta)$ and $\operatorname{grad} R(\theta) \neq 0$. If we choose some $\alpha > 0$ that is not too large, we can approximate the directional derivative by the difference quotient:

$$D_h R(\theta) \approx \frac{R(\theta + \alpha h) - R(\theta)}{\alpha}$$

Sufficiently far away from a stationary point, this approximation will also have a negative value. Thus, given all those conditions, we have $R(\theta) > R(\theta + \alpha h)$. We can take advantage of this fact and define an iterative method

$$\theta^{(j+1)} = \theta^{(j)} + \alpha h^{(j)}$$

that produces descending function values:

$$R\left(\theta^{(0)}\right) > R\left(\theta^{(1)}\right) > \dots$$

Once the norm of the gradient $\|\operatorname{grad} R(\theta^{(j)})\|$ reaches some small enough threshold, the argument $\theta^{(j)}$ is near a local minimum and the iteration can be terminated.

Let $R\colon \mathbb{R}^K \to \mathbb{R}$, $\theta \mapsto R(\theta)$ be a (continuously differentiable) function. **Simple gradient descent** computes a local minimum point $\hat{\theta}$ of $R(\cdot)$ by running the following iteration:

> initialize $\theta \in \mathbb{R}^K$,
> converged := false, $j := 0$
> **while** *converged = false and $j < j_{\max}$* **do**
> > update $j \leftarrow j + 1$
> > update $\theta \leftarrow \theta + \alpha \cdot h$, where $h := -\|\operatorname{grad} R(\theta)\|^{-1} \cdot \operatorname{grad} R(\theta)$
> > **if** $\|\operatorname{grad} R(\theta)\| \leq \tau$ **then**
> > > converged := true
> >
> > **end**
> >
> **end**
> output: converged, $\hat{\theta} = \theta$

Here, $\alpha > 0$ and $\tau > 0$ represent step size and the tolerance parameter, respectively, which are chosen to have small values. The number $j_{\max}$ specifies the maximum number of iteration steps to be performed.

In order to find a *maximum* point of $R$ instead of a minimum, we can simply apply the algorithm to $-R$. In the context of machine learning, the step size $\alpha$ is also called the **learning rate**.

A common variant of the above algorithm is given by using the non-normalized gradient: "$h = -\operatorname{grad} R(\theta)$." If we insist on the convention $\|h\| = 1$, such a procedure can also be interpreted as a continuously adjusted step size:

$$\alpha^{(j)} = \alpha^{(0)} \cdot \|\operatorname{grad} R(\theta^{(j)})\|$$

Such a control of the step size/learning rate is quite reasonable: that way, we avoid overshooting the target. A more sophisticated method for adjusting the step size is the **Barzilai–Borwein method** [10], which we will just state here without further comment:

$$\alpha^{(j)} = \alpha^{(j-1)} \cdot \frac{\|g^{(j)}\|}{\|g^{(j-1)}\|} \cdot \frac{\langle g^{(j-1)}, g^{(j)} - g^{(j-1)} \rangle}{\|g^{(j)} - g^{(j-1)}\|^2}$$

with $g^{(j)} := \operatorname{grad} R\left(\theta^{(j)}\right)$.

Fig. 6.2 (top) illustrates a gradient descent for an objective function with two arguments, $R(\theta) = R(\theta_1, \theta_2)$. The non-normalized gradient was used, and the arrows indicate the difference of successive pairs of values:

$$\alpha^{(j)} h^{(j)} = -\alpha^{(0)} \cdot \operatorname{grad} R\left(\theta^{(j)}\right) = \theta^{(j+1)} - \theta^{(j)}$$

These vectors are perpendicular to the contour lines of the objective function and point in the direction of its minimum point.

In statistical learning, objective functions are often of the following form:

$$R(\theta) = \frac{1}{N} \sum_{n=1}^{N} R_n(\theta)$$

For example, the empirical risk is of this form, in which case the loss is summed over the observations in the training dataset. In order to minimize that risk, we need to compute the direction of steepest descent many times:

$$h \propto - \operatorname{grad} R(\theta) = -\frac{1}{N} \sum_{n=1}^{N} \operatorname{grad} R_n(\theta)$$

Often, there are also a very large number of observations/summands, since training datasets can be quite large. Thus, ordinary gradient descent can become expensive and possibly hog computational resources that are not available. Instead, we can use **stochastic gradient descent**: with each iteration step, only a *single* summand with index $n_1$ is selected, and only for this training example we determine the gradient:

$$h_{\text{stoch}} \propto - \operatorname{grad} R_{n_1}(\theta)$$

The index $n_1 \in \{1, \dots, N\}$ is usually selected at random. Another variant of gradient descent is **mini-batch gradient descent**, where we randomly select up to $1 < M \ll N$ training examples, the mini-batch:

$$h_{\text{batch}} \propto -\frac{1}{M} \sum_{k=1}^{M} \operatorname{grad} R_{n_k}(\theta)$$

The random selection is performed without replacement: after a total of $\lceil N/M \rceil$ iteration steps, which make up a so-called **epoch**, all training data have been processed. For the subsequent epoch, all training examples are made available again and processed in mini-batches.

In this context, ordinary gradient descent is also called **full-batch gradient descent**, or just **batch gradient descent**.

Fig. 6.2 (bottom) shows a stochastic gradient descent. Notice that the direction of descent is no longer the steepest and may not be orthogonal to the contour lines of the function to be minimized. Nevertheless, the general direction of descent—averaged over many iteration steps—remains the correct one with high probability.

Another numerical optimization algorithm is the Broyden–Fletcher–Goldfarb–Shanno method (BFGS method for short) [11, 12, 13, 14]. A number of examples in this book have been calculated through this method, so we briefly present it here.

The method's rationale is similar to that of gradient descent except that it considers the second-order Taylor polynomial instead of just a linear approximation:

$$R(\theta + \alpha h) \approx R(\theta) + \alpha \cdot \langle h, \mathrm{grad}\, R(\theta) \rangle + \frac{\alpha^2}{2} \langle h, \mathrm{Hess}\, R(\theta) \cdot h \rangle$$

The search direction $h$ should, once again, be chosen in such a way that the function values become smaller with each iteration step: i.e., the algorithm moves in the direction of a minimum. Assuming that the Hessian matrix is positive definite, the minimum of the Taylor polynomial lies in the following direction:

$$h \propto - (\mathrm{Hess}\, R(\theta))^{-1} \cdot \mathrm{grad}\, R(\theta)$$

The step size $\alpha$ is adaptively controlled by approximately minimizing the one-dimensional function $g(\alpha) = R(\theta + \alpha h)$ by means of a so-called **backtracking line search**:

> set parameters $c_1 \in\, ]0, 1[$, $c_2 \in\, ]0, 1[$,
> intialize $\alpha > 0$,
> compute $m := \langle \mathrm{grad}\, R(\theta), h \rangle$
> **while** $R(\theta + \alpha h) \geq R(\theta) + c_1 m \alpha$ **do**
> | update $\alpha \leftarrow c_2 \alpha$
> **end**
> output: $\alpha$

The termination condition is also called **Armijo rule** [15]; usually a very small value is chosen for $c_1$.

Finally, an essential component of the BFGS algorithm is that the Hessian matrix is not computed exactly, rather an approximation $H$ is used. This approximation arises from the following **quasi-Newton condition** in the $j$-th iteration step [16, Sect. 3.2]. For the sake of clarity, the iteration index is notated as a subscript:

$$H_{j+1} \cdot h_j = \frac{\mathrm{grad}\, R(\theta_j + \alpha_j h_j) - \mathrm{grad}\, R(\theta_j)}{\alpha_j} =: \frac{v_j}{\alpha_j}$$

If $H_j$ is any symmetric and positive definite matrix, then $H_{j+1}$ by the following definition is also symmetric and positive definite, and it also satisfies the above condition:

$$H_{j+1} = H_j + (\alpha_j \cdot \langle v_j, h_j \rangle)^{-1} \cdot v_j \cdot v_j^T - (\langle h_j, H_j \cdot h_j \rangle)^{-1} \cdot (H_j \cdot h_j) \cdot (H_j \cdot h_j)^T$$

Finally, the final formula for the inverse of $H$ used in the following pseudocode follows from the so-called Sherman–Morrison formula [16, Exercise 3.13].

Let $R\colon \mathbb{R}^K \to \mathbb{R}$, $\theta \mapsto R(\theta)$ be a (twice continuously differentiable) function. The **BFGS method** computes a local minimum point $\hat{\theta}$ of $R(\,\cdot\,)$ by running the following iteration where $\tau > 0$ and $j_{\max}$ are a tolerance parameter and the maximum number of iteration steps, respectively. All vectors are to be interpreted as column vectors.

initialize: converged := false,
$j := 0$, $\theta = (\theta_1, \ldots, \theta_K)^T \in \mathbb{R}^K$,
$H^{-1} :=$ identity matrix of format $K \times K$

**while** *converged = false and $j < j_{\max}$* **do**
  update $j \leftarrow j + 1$
  *# Determine step size and search direction:*
  determine (approximately) a minimum $\alpha \in \mathbb{R}$ of the function
  $g(\alpha) := R(\theta + \alpha \cdot h)$ where $h := -H^{-1} \operatorname{grad} R(\theta)$
  *# Update approximation for function value and Hessian matrix:*

  $\theta \leftarrow \theta + \alpha \cdot h$,

  $$H^{-1} \leftarrow H^{-1} + \left(1 + \alpha \frac{v^T H^{-1} v}{h^T v}\right) \cdot \frac{h h^T}{h^T v} - \frac{h v^T H^{-1} + H^{-1} v h^T}{h^T v}$$

  where $v = \operatorname{grad} R(\theta + \alpha \cdot h) - \operatorname{grad} R(\theta)$
  **if** $\| \operatorname{grad} R(\theta) \| \leq \tau$ **then**
    | converged := true
  **end**
**end**
output: converged, $\hat{\theta} = \theta$

## 6.2 Regression algorithms

In the following sections, we present regression algorithms that process a training dataset $(x, y) \cong ((x_1, y_1), \ldots, (x_N, y_N))$ where $x_1, \ldots, x_N \in \mathbb{R}^D$ are feature vectors, and $y_1, \ldots, y_N \in \mathbb{R}$ are realizations of the target variable. The goal is to learn a regression function:

$$\hat{f}\colon \mathbb{R}^D \to \mathbb{R}$$

This regression function can then be used to make a prediction $\hat{y}_* = \hat{f}(x_*)$ given some new, yet unseen feature vector $x_* \in \mathbb{R}^D$.

### 6.2.1 Linear regression

In Sect. 4.5.1, we learned about simple linear regression, which examines a linear relationship between a *univariate* predictor variable and some response variable.

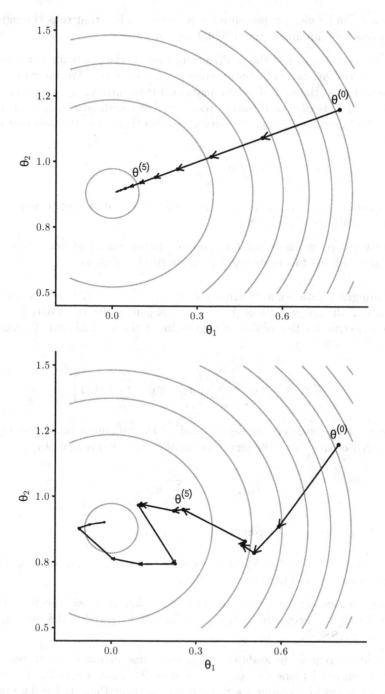

**Fig. 6.2.** *Ordinary, full-batch gradient descent (top) and stochastic gradient descent (bottom)*

The result can be visually presented as a regression line that runs through the data points and minimizes the residual sum of squares.

We want to generalize to the multivariate case. If the predictor is a random vector, its realizations are feature vectors $x_1, \ldots, x_N \in \mathbb{R}^D$. We can write these feature vectors as the rows of a data matrix with the entries $x_{nd}$, $n \in \{1, \ldots, N\}$, $d \in \{1, \ldots, D\}$. Hence, each feature vector $x_n$ has coordinates $x_{n1}, \ldots, x_{nD}$. In multivariate linear regression, the target variable then takes the following form:

$$Y_n = w_0 + \sum_{d=1}^{D} w_d \cdot x_{nd} + \varepsilon_n$$

with constants $w_0, w_1, \ldots, w_D$ and normally distributed, independent error terms $\varepsilon_n \sim \mathcal{N}(\cdot \,|\, 0, \sigma^2)$, $\sigma > 0$.

The least squares method—that is, minimizing the empirical risk with respect to the squared loss function—can be summarized as follows.

We are given data points/feature vectors $x_1, \ldots, x_N \in \mathbb{R}^D$ and the observations of the target variable $y_1, \ldots, y_N \in \mathbb{R}$ paired with those data points. The objective function of **linear regression** is the residual sum of squares $R: \mathbb{R}^{D+1} \to [0, \infty[$:

$$R(w_0, \ldots, w_D) = \sum_{n=1}^{N} \left( y_n - w_0 - \sum_{d=1}^{D} w_d x_{nd} \right)^2$$

The model parameters are determined by the minimum point $\hat{w} = (\hat{w}_0, \ldots, \hat{w}_D)$ of $R(\cdot)$, and the learned regression function is given by:

$$\hat{f}(x_*) = \hat{w}_0 + \sum_{d=1}^{D} \hat{w}_d x_{*d}$$

for all $x_* \in \mathbb{R}^D$ with coordinates $x_{*1}, \ldots, x_{*D}$.

It is possible to give a closed-form expression for the minimum point, which we present in more detail in the next section.

$L_2$ regularization can be applied to linear regression, in which case the method is called **ridge regression**. When including an $L_1$ penalty, the technique is known as **LASSO**[3].

Simple linear regression leads to a regression line fitting the data points in the 2-dimensional plane $(x_n, y_n) \in \mathbb{R} \times \mathbb{R} = \mathbb{R}^2$. More generally, the linear regression algorithm computes a regression *hyperplane* that fits the data points $(x_n, y_n) \in \mathbb{R}^D \times \mathbb{R} \cong \mathbb{R}^{D+1}$ in $(D+1)$-dimensional space.

---

[3] LASSO stands for Least Absolute Shrinkage and Selection Operator.

An important application of the multivariate model is, of course, to include more than just a single feature to make a prediction. Another important application of multivariate linear regression, however, is the modeling of nonlinear relationships. For example, we can include quadratic terms of a single predictor variable as follows:

$$R(w_0, w_1, w_2) = \sum_{n=1}^{N} \left( y_n - w_0 - w_1 \cdot x_n - w_2 \cdot x_n^2 \right)^2$$

For this model, the learned parameters are the coefficients of a quadratic polynomial that we want to fit through the data points. Despite the regression function being nonlinear, we still speak of linear regression: we have merely added a new feature.

**Example.** In Sect. 4.5.1, we determined a regression line to model the time series of global air temperatures on Earth [7] (see Fig. 4.12). If we add a quadratic term to the model, then we get a regression polynomial instead:

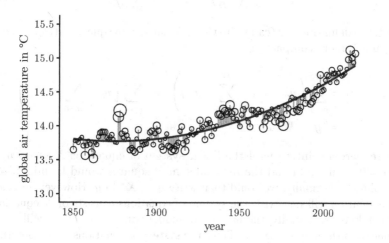

**Fig. 6.3.** *Time series of average global air temperature with quadratic regression polynomial*

The regression curve is given by the following equation:

$$\hat{f}(x_*) = 13.8\,°C + 5.39 \cdot 10^{-5}\,\frac{K}{a^2} \cdot (x_* - 1875\,a)^2$$

Thus, according to this model, the average global air temperature on Earth shows an *accelerated* increase since 1875. The root mean squared error is given by $\hat{\sigma} = 0.13\,K$ and the coefficient of determination is given by $r^2 = 0.88$.

## 6.2.1.1 Moore–Penrose inverse

In order to apply linear regression to our data, we need to compute the minimum point of the residual sum of squares. In this section, we show that this minimum can be given in a closed form via the so-called Moore–Penrose inverse of a matrix. First, we recall the definition of the regressor matrix, where the feature vectors are augmented by a one in the zeroth entry:

$$
\boldsymbol{X} = \begin{pmatrix} 1 & x_{11} & x_{12} & \cdots & x_{1D} \\ 1 & x_{21} & x_{22} & \cdots & x_{2D} \\ \vdots & & \vdots & & \vdots \\ 1 & x_{N1} & x_{N2} & \cdots & x_{ND} \end{pmatrix} = (x_{nd})_{\substack{n \in \{1,\dots,N\} \\ d \in \{0,1,\dots,D\}}}
$$

Furthermore, we combine the model parameters and the observations of the target variable into column vectors of length $D+1$ and $N$, respectively:

$$
w = \begin{pmatrix} w_0 \\ w_1 \\ \vdots \\ w_D \end{pmatrix} \text{ and } \boldsymbol{y} = \begin{pmatrix} y_1 \\ y_2 \\ \vdots \\ y_N \end{pmatrix}
$$

With these definitions, we can write the residual sum of squares $R(w_0, \dots, w_D) = R(w)$ much more compactly:

$$
R(w) = \sum_{n=1}^{N} \left( y_n - w_0 - \sum_{d=1}^{D} w_d x_{nd} \right)^2 = \sum_{n=1}^{N} \left( y_n - \sum_{d=0}^{D} x_{nd} w_d \right)^2
$$
$$
= \| \boldsymbol{y} - \boldsymbol{X} \cdot w \|^2
$$

If there were a solution $\hat{w}$ of the linear system of equations $\boldsymbol{X} \cdot w = \boldsymbol{y}$, then $R(\hat{w}) = 0$ would hold and the residual sum of squares would be minimized by that solution. Formally, we could then write $\hat{w} = \boldsymbol{X}^{-1} \cdot \boldsymbol{y}$. However, in practice, such a solution will not exist: the system of equations consists of $N$ equations in $D+1$ unknowns. Usually, the number of model parameters $D+1$ will be much smaller than the sample size $N$. Thus, the system of equations is overdetermined. In the context of simple linear regression, this circumstance arises from the fact that there are usually many more than just two data points that determine the slope and intercept of the regression line.

Nevertheless, there always exists a minimum point of the residual sum of squares. This minimum point can be written as $\hat{w} = \boldsymbol{X}^\dagger \cdot \boldsymbol{y}$, where $\boldsymbol{X}^\dagger$ is the **generalized inverse, pseudoinverse**, or **Moore–Penrose inverse** of the regressor matrix $\boldsymbol{X}$.

To determine $\boldsymbol{X}^\dagger$, we first calculate the gradient of the objective function $R(\cdot)$:

$$
\operatorname{grad} R(w) = (DR(w))^T = \frac{\mathrm{d}}{\mathrm{d}w} \| \boldsymbol{y} - \boldsymbol{X} \cdot w \|^2
$$
$$
= \left( 2(\boldsymbol{y} - \boldsymbol{X} \cdot w)^T \cdot (-\boldsymbol{X}) \right)^T = \left( -2\boldsymbol{y}^T \boldsymbol{X} + 2w^T \boldsymbol{X}^T \boldsymbol{X} \right)^T
$$
$$
= -2\boldsymbol{X}^T \cdot \boldsymbol{y} + 2\boldsymbol{X}^T \cdot \boldsymbol{X} \cdot w
$$

Here, we used the chain rule (see [17, Sect. 5.12] or Section B.3.3) as well as these basic differentiation rules:

$$\frac{\mathrm{d}}{\mathrm{d}z}(\|z\|^2) = 2z^T, \quad \frac{\mathrm{d}}{\mathrm{d}z}(\boldsymbol{X} \cdot z) = \boldsymbol{X}$$

Moreover, we recall the matrix calculation rules $(A \cdot B)^T = B^T \cdot A^T$ and $(A^T)^T = A$.

Therefore, the gradient vanishes for $\hat{w} \in \mathbb{R}^{D+1}$ with

$$\boldsymbol{X}^T \cdot \boldsymbol{X} \cdot \hat{w} = \boldsymbol{X}^T \cdot \boldsymbol{y}.$$

Provided that $\boldsymbol{X}^T \cdot \boldsymbol{X}$ is invertible, this equation can be solved directly for $\hat{w}$:

$$\hat{w} = (\boldsymbol{X}^T \cdot \boldsymbol{X})^{-1} \cdot \boldsymbol{X}^T \cdot \boldsymbol{y}$$

This stationary point is the uniquely determined minimum point of $R(\cdot)$, since that function is nonnegative and thus bounded from below. In this case, $\boldsymbol{X}^\dagger = (\boldsymbol{X}^T \cdot \boldsymbol{X})^{-1} \cdot \boldsymbol{X}^T$ is the Moore–Penrose inverse of $\boldsymbol{X}$.

If $\boldsymbol{X}^T \cdot \boldsymbol{X}$ *is not* invertible, we can proceed as follows. First, we notice that the matrix $\Sigma := \boldsymbol{X}^T \cdot \boldsymbol{X}$ has only nonnegative eigenvalues. Moreover, $\Sigma$ is a symmetric matrix, which implies (see Sect. B.2.4 in the appendix) that there exists an orthogonal matrix $V$ and a diagonal matrix $\Lambda$ such that:

$$\Sigma = V \cdot \Lambda \cdot V^T$$

The columns of $V$ are the normalized and pairwise orthogonal eigenvectors $v_0, \ldots, v_D \in \mathbb{R}^{D+1}$ of $\Sigma$. The diagonal entries of $\Lambda$ are the corresponding eigenvalues $\lambda_0, \ldots, \lambda_D$. Alternatively, we can write the diagonal decomposition as follows:

$$\Sigma = \sum_{d=0}^{D} \lambda_d v_d \cdot v_d^T$$

We can see that this representation is correct from the fact that $\Sigma$ acts on the eigenvector basis as desired:

$$\Sigma \cdot v_i = \sum_{d=0}^{D} \lambda_d v_d \cdot v_d^T \cdot v_i = \sum_{d=0}^{D} \lambda_d \langle v_d, v_i \rangle v_d = \lambda_i v_i$$

for all $i \in \{0, \ldots, D\}$.

We then determine the Moore–Penrose inverse via the following formula:

$$\boldsymbol{X}^\dagger = \sum_{d=0}^{D} \lambda_d^\dagger v_d \cdot v_d^T \cdot \boldsymbol{X}^T \text{ with } \lambda_d^\dagger = \begin{cases} \frac{1}{\lambda_d} & \text{if } \lambda_d > 0 \\ 0 & \text{if } \lambda_d = 0 \end{cases}$$

This definition solves our problem, because, as we will show below, it satisfies the above condition of a vanishing gradient: $\boldsymbol{X}^T \boldsymbol{X} \hat{w} = \Sigma \boldsymbol{X}^\dagger \boldsymbol{y} = \boldsymbol{X}^T \boldsymbol{y}$.

First, we note that for any eigenvector $v$ with eigenvalue zero, the vector $v^T X^T y$ must vanish:

$$\Sigma v = 0 \Rightarrow X^T X v = 0 \Rightarrow v^T X^T X v = 0$$
$$\Rightarrow \langle X v, X v \rangle = 0 \Rightarrow X v = 0 \Rightarrow v^T X^T y = 0$$

Furthermore, $\sum_{d=0}^{D} v_d v_d^T v_i = v_i$ holds for any eigenvector $v_i$ and—via linearity—for any vector. Thus, $\sum_{d=0}^{D} v_d \cdot v_d^T$ is the identity matrix.

Finally, we note the following formula:

$$\Sigma X^\dagger y = \Sigma \cdot \left( \sum_{d=0}^{D} \lambda_d^\dagger v_d v_d^T X^T \right) \cdot y = \sum_{d=0}^{D} \lambda_d^\dagger \Sigma v_d v_d^T X^T y$$

$$= \sum_{d=0}^{D} \lambda_d^\dagger \lambda_d v_d v_d^T X^T y = \sum_{d=0}^{D} v_d v_d^T X^T y = X^T y$$

Let us summarize.

---

The residual sum of squares

$$R(w) = \| y^T - X \cdot w \|^2$$

with regressor matrix $X$ of format $N \times (D+1)$ and observations of the target variable $y = (y_1, \ldots, y_N)$ has the minimum point

$$\hat{w} = X^\dagger \cdot y^T$$

where $X^\dagger$ is the **Moore–Penrose inverse** of $X$ described above.

---

### 6.2.2 Gaussian process regression

To derive another regression procedure, let us once again consider a training dataset $(x, y) \cong ((x_1, y_1), \ldots, (x_N, y_N))$ with feature vectors $x_1, \ldots, x_N \in \mathbb{R}^D$ and observed target values $y_1, \ldots, y_N \in \mathbb{R}$. We assume that $\bar{y} = 0$ holds: this circumstance can always be achieved by mean-centering, i.e., subtracting the arithmetic mean from the target variable. Instead of a linear relationship, we write down a more general model assumption:

$$Y_n = f(x_n) + \varepsilon_n$$

with normally distributed and independent error terms $\varepsilon_n \sim \mathcal{N}(\cdot \,|\, 0, \delta_n^2)$ for all $n \in \{1, \ldots, N\}$. We want to allow for the variance $\delta_n > 0$ of the error terms to vary in general and allow for **heteroskedasticity**, i.e., we do not necessarily assume $\delta_1 = \delta_2 = \cdots = \delta_N$.

Next, let us consider the pairs of observed predictor values and the corresponding (unknown, not directly observable) value of the regression function:

$$((x_1, f_1), \ldots, (x_N, f_N)) = ((x_1, f(x_1)), \ldots, (x_N, f(x_N)))$$

Gaussian process regression is a Bayesian method: It is based on the assumption that our ignorance of the values $f_1, \ldots, f_N$ can be modeled by a multivariate normal distribution, with location parameter $m = m(x_1, \ldots, x_N) \in \mathbb{R}^N$ and $N \times N$-covariance matrix $\tilde{\Sigma} = \tilde{\Sigma}(x_1, \ldots, x_N)$:

$$p(f_1, \ldots, f_N | x_1, \ldots, x_N) = p(f|x) = \mathcal{N}(f | m(x), \tilde{\Sigma}(x))$$

We make this assumption for any finite selection of pairs of predictor/predicted values, not just those in the training dataset. A family $x_* \mapsto f_*$ of random variables distributed as such is called a **Gaussian process**. Since $\bar{y} = 0$ holds, we make the additional assumption $m(x) = 0$. The covariance matrix is assumed to be of the following form:

$$\tilde{\Sigma}(x) = \begin{pmatrix} \sigma(x_1, x_1) & \cdots & \sigma(x_1, x_N) \\ \vdots & & \vdots \\ \sigma(x_N, x_1) & \cdots & \sigma(x_N, x_N) \end{pmatrix}$$

where $\sigma \colon \mathbb{R}^D \times \mathbb{R}^D \to \mathbb{R}$ is some suitable similarity measure.

The basic rationale is the following: If two feature vectors $x_m$ and $x_n$ are close to each other, the corresponding values of the target variable should be similar as well. On the other hand, if the feature vectors are separated by a large distance, we may assume that only few common influences determine the corresponding values of the target variable—over large distances in feature space, the target variable becomes decorrelated. Like all model assumptions, this one should also be scrutinized in terms of the use case and the data generation process. For example, a naive application of the method to a time series would not take into account periodic influences that could cause correlations over long time spans.

We further assume that the entries of the covariance matrix are of a Gaussian form:

$$\sigma(u, v) = a^2 \exp\left( -\frac{1}{2} u^T \cdot \Sigma_h^{-1} \cdot v \right)$$

for all column vectors $u, v \in \mathbb{R}^D$. The numbers $h_1, \ldots, h_D, a > 0$ are additional parameters, and $\Sigma_h^{-1}$ is the diagonal matrix with diagonal entries $h_1^{-2}, \ldots, h_D^{-2}$ (all other entries being zero).

In order to obtain the distribution of the actual target observations $y_1, \ldots, y_N$, we still need to take into account the error terms:

$$p(y_1, \ldots, y_N | x_1, \ldots, x_N) = p(y|x) = \mathcal{N}(y | 0, K(x))$$

where $K(x) = \tilde{\Sigma}(x) + \Sigma_\delta$. Here, $\Sigma_\delta = \mathrm{diag}(\delta_1^2, \ldots, \delta_N^2)$ is the diagonal matrix with diagonal entries $\delta_1^2, \ldots, \delta_N^2$.

If we were to explicitly write out all of the parameters, we would have $\sigma(\cdot, \cdot) = \sigma_{h,a}(\cdot, \cdot)$, $\tilde{\Sigma}(x) = \tilde{\Sigma}_{h,a}(x)$ and $K(x) = K_\delta(x; h, a)$. For clarity, we will notationally suppress those parameters.

Given a new, yet unseen data point $x_*$, we want to make the prediction $f_* = f(x_*)$. For this purpose, we determine the posterior distribution that describes our knowledge of this value given the data. The joint density function of the observed feature values and the prediction is also a normal distribution:

$$p(y_1, \ldots, y_N, f_* | x_1, \ldots, x_N, x_*) = p(y, f_* | x, x_*) = \mathcal{N}((y, f_*) | 0, K_*)$$

The covariance matrix $K_*$ is given by the following block matrix of the format $(N+1) \times (N+1)$:

$$K_* = \begin{pmatrix} K(x) & K(x_*, x) \\ K(x_*, x)^T & a^2 \end{pmatrix}$$

with the column vector $K(x_*, x) := (\sigma(x_*, x_1), \ldots, \sigma(x_*, x_N))^T$. We can now use the formula for the conditional probability density of components of a normally distributed random vector (see Sect. 5.4.2) to derive the posterior distribution that we are after:

$$p(f_* | x, x_*, y) = \mathcal{N}(f_* | K(x_*, x)^T K(x)^{-1} y, a^2 - K(x_*, x)^T K(x)^{-1} K(x_*, x))$$

with the column vector of target values $y = (y_1, \ldots, y_N)^T$.

Overall, the model has the following parameters: the variances of the error terms $\delta_1^2, \ldots, \delta_N^2$, the correlation ranges $h_1, \ldots, h_D$, and the scale parameter $a$. The correlation ranges and the scale parameter should be treated as model parameters and be estimated from the data. If all error terms were also estimated from the data, overfitting can easily occur. Alternatively, there are the following possibilities:

- The variance of the disturbance is assumed to be constant. This assumption of **homoskedasticity** drastically reduces the number of model parameters: $\delta_1 = \cdots = \delta_N$.

- The standard deviations $\delta_1, \ldots, \delta_N$ are treated as hyperparameters and are inferred from other sources. For example, they can represent the uncertainty of a physical measurement, as determined by an experimental uncertainty assessment.

In any case, a maximum-a-posteriori estimator and a credibility interval can be obtained from the posterior distribution. We summarize our reasoning, treating $\delta_1, \ldots, \delta_N$ as hyperparameters.

The log-likelihood function of **Gaussian process regression** given feature vectors $x_1, \ldots, x_N \in \mathbb{R}^D$ and mean-centered target values $y = (y_1, \ldots, y_N)^T$ is the following:

$$\ell_\delta(h, a) = \ln(p(y | x)) = \ln\left(\mathcal{N}(y | 0, K_\delta(x; h, a))\right)$$

$$= -\frac{1}{2} y^T (K_\delta(x; h, a))^{-1} y - \frac{1}{2} \ln(\det(K_\delta(x; h, a))) - \frac{N}{2} \ln(2\pi)$$

where $K_\delta(x; h, a) = \tilde{\Sigma}_{h,a}(x) + \Sigma_\delta$ is given in the above derivation.

A maximum point $(\hat{h}_1, \ldots, \hat{h}_D, \hat{a})$ of the log-likelihood function leads to the following regression function:

$$\hat{f} \colon \mathbb{R}^D \to \mathbb{R}, \ \hat{f}(x_*) = \hat{K}(x_*, x)^T \hat{K}(x)^{-1} \boldsymbol{y}$$

with

$$\hat{K}(x) = K_\delta(x; \hat{h}, \hat{a}), \ \hat{K}(x_*, x) = (\sigma_{\hat{h}, \hat{a}}(x_*, x_1), \ldots, \sigma_{\hat{h}, \hat{a}}(x_*, x_N))^T.$$

A $\gamma$-credibility band for the estimate is given by

$$\left[ \hat{f}(x_*) \pm z(\gamma) \cdot \sqrt{a^2 - \hat{K}(x_*, x)^T \hat{K}(x)^{-1} \hat{K}(x_*, x)} \right]$$

with, for example, $z(0.95) = 1.96$.

For a univariate predictor $(D = 1)$, the regression curve consists of the pairs of values $(x_*, f_*)$. In the multivariate case $(D > 1)$, these describe a hypersurface in $(D + 1)$-dimensional space.

**Example.** We apply Gaussian process regression to the time series of measurements of the average global air temperature on Earth. The gray band around the regression curve marks the 95%-credibility interval:

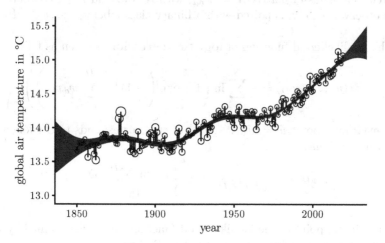

**Fig. 6.4.** *Time series of average global air temperature with Gaussian process regression curve*

Berkley Earth provides uncertainties for the measurements (between 0.03 K and 0.21 K, depending on the year), and those have been incorporated into the model directly as the standard deviations of the disturbance $\delta_1, \ldots, \delta_N$.

With the help of the BFGS procedure, a maximum point of the likelihood function, and thus the remaining parameters, can be determined:

$$\bar{y} = 14.10\,^{\circ}\mathrm{C},\ h = 30\,\mathrm{y},\ a = 0.52\,\mathrm{K}$$

## 6.3 Classification algorithms

In the following sections, we present classification algorithms. These algorithms process a training dataset $(x, y) \cong ((x_1, y_1), \ldots, (x_N, y_N))$ where $x_1, \ldots, x_N \in \mathbb{R}^D$ are feature vectors and $y_1, \ldots, y_N \in \{0, 1, \ldots, K - 1\}$ are the labels. Binary classification corresponds to the case $K = 2$. The goal is to learn a classification rule:

$$\hat{f} \colon \mathbb{R}^D \to \{0, 1, \ldots, K - 1\}.$$

This classification rule can then be used to make a prediction $\hat{y}_* = \hat{f}(x_*)$ given some new, yet unseen feature vector $x_* \in \mathbb{R}^D$.

### 6.3.1 Logistic regression

The simple logistic regression model can be generalized to the multivariate case in a way that is similar to simple linear regression. The training dataset can be written as a regressor matrix $X = (x_{nd})$ with $N$ rows and $D+1$ columns, where each observation/row is paired with a binary class label $y_1, \ldots, y_N \in \{0, 1\}$.

The log-likelihood function of **logistic regression** is given as follows:

$$\ell(w_0, \ldots, w_D) = -\sum_{n=1}^{N} \ln\left(1 + \exp\left((-1)^{y_n} \cdot \sum_{d=0}^{D} w_d x_{nd}\right)\right)$$

A maximum point $\hat{w} = (\hat{w}_0, \ldots, \hat{w}_D)$ of $\ell(\,\cdot\,)$ corresponds to the following learned classifier:

$$\hat{f} \colon \mathbb{R}^{D+1} \to \{0, 1\},\ \hat{f}(x_*) = \begin{cases} 1 & \text{if } \sum_{d=0}^{D} \hat{w}_d x_{*d} > 0 \\ 0 & \text{otherwise} \end{cases}$$

The maximum point of the log-likelihood function can be determined by means of numerical optimization, such as the BFGS algorithm.

To get a first impression of the algorithm's behavior, we apply it to a synthetic training dataset of low dimensionality. The figure below shows the scatter plot of two classes of data points generated using bivariate Gaussian distributions ($N = 1000$). In general, logistic regression learns a linear classifier: the decision boundary is a $(D - 1)$-dimensional hyperplane with normal vector $w^{\perp} = (\hat{w}_1, \ldots, \hat{w}_D)^T$ and the distance $\hat{w}_0 / \|w^{\perp}\|$ from the origin. In this example, the decision boundary is one-dimensional and indicated by the solid line dividing the two-dimensional feature space into regions that correspond to the predictions $\hat{y} = 0$ (light region) and $\hat{y} = 1$ (dark region).

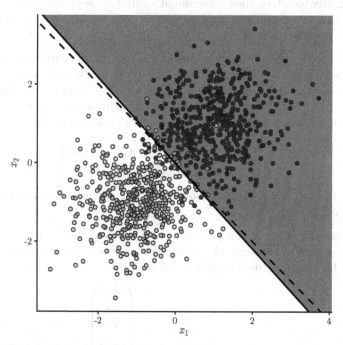

**Fig. 6.5.** *Decision boundary of logistic regression*

Let us use this example to illustrate once more the concepts of empirical and expected risk. The empirical risk—even though it has been minimized by the algorithm—does not vanish, as some of the training data would not be classified correctly by the learned decision rule: they are on the wrong side of the decision boundary.

The dashed line in the above figure indicates the optimal decision boundary corresponding to the best possible classifier, also called the **Bayes classifier**. In practice, this boundary is unknown: here we can only determine it because the dataset is synthetic. For the Bayes classifier, the expected risk is equal to the irreducible Bayes error. Note that even for the optimal classifier, the empirical risk does not vanish, either: some training data is misclassified.

The optimal classifier is linear (again: in practice, we would not know this). Thus, it is contained in the hypothesis space of the applied method. Therefore, the approximation error vanishes: For larger and larger training datasets, it becomes more and more likely that the learned classifier comes very close to the optimal one; the dashed and solid lines coincide. Thus, the deviation between the learned and optimal decision boundaries shown in the figure is due to the estimation error alone.

Fig. 6.6 shows a dataset where logistic regression would suffer from (severe) underfitting if it were to be applied naively: the spiral-shaped regions corresponding to negative/positive class assignment cannot be separated by a decision boundary that is a single straight line. In this case, we say that the classes are not **linearly separable.**

One way to separate those regions using a linear classifier after all is to add higher powers of the predictor variables. This procedure is very similar to using linear regression to fit a polynomial instead of a regression line. For example, we can add terms up to third powers. Instead of a straight line $w_0 + w_1 x_1 + w_2 x_2 = 0$, the decision boundary is now an algebraic curve of degree three:

$$w_0 + w_1 x_1 + w_2 x_2 + w_{11} x_1^2 + w_{12} x_1 x_2 + w_{22} x_2^2$$
$$+ w_{112} x_1^2 x_2 + w_{122} x_1 x_2^2 + w_{111} x_1^3 + w_{222} x_2^3 = 0$$

The result can be seen in Fig. 6.6 above. Note that this model is able to capture the shape of the data.

Another way of looking at the procedure is as follows. The data points are not linearly separated in the original two-dimensional feature space. Thus, we map them to a higher-dimensional space as follows:

$$\Phi \colon \mathbb{R}^2 \to \mathbb{R}^9, \ \begin{pmatrix} x_1 \\ x_2 \end{pmatrix} \mapsto \begin{pmatrix} x_1 \\ x_2 \\ x_1^2 \\ x_1 x_2 \\ x_2^2 \\ x_1^2 x_2 \\ x_1 x_2^2 \\ x_1^3 \\ x_2^3 \end{pmatrix}$$

In this higher-dimensional feature space, the classes are now linearly separated, and we can apply logistic regression successfully. This idea is also at the core of the so-called **kernel trick**, which we describe in the next section.

### 6.3.1.1 Kernel logistic regression

Once again, we are given a training dataset consisting of feature vectors $x_1, \ldots, x_N \in \mathbb{R}^D$ and (binary) class labels $y_1, \ldots, y_N \in \{0, 1\}$. Furthermore, we are given a symmetric similarity measure $\sigma \colon \mathbb{R}^D \times \mathbb{R}^D \to \mathbb{R}$, which in this context is called the **kernel**. A popular choice is once more the Gaussian kernel:

$$\sigma_h(u, v) = e^{-\frac{\|u - v\|^2}{2h^2}}$$

for all $u, v \in \mathbb{R}^D$ with bandwidth $h > 0$. Another possibility is a polynomial kernel:

$$\sigma_{k,b}(u, v) = (\langle u, v \rangle + b)^k$$

where $k \in \mathbb{N}$, $k \geq 1$ and $b \in \mathbb{R}$, $b \geq 0$.

The objective function for **kernel logistic regression** is the following:

$$\ell(\alpha_1, \ldots, \alpha_N) = -\sum_{n=1}^{N} \ln\left(1 + \exp\left((-1)^{y_n+1} \cdot \sum_{m=1}^{N} \alpha_m \sigma(x_m, x_n)\right)\right)$$

A maximum point $\hat{\alpha} = (\hat{\alpha}_1, \ldots, \hat{\alpha}_N)$ of $\ell(\cdot)$ leads to the following classification rule:

$$\hat{f}: \mathbb{R}^D \to \{0,1\}, \ \hat{f}(x_*) = \begin{cases} 1 & \text{if } \sum_{m=1}^{N} \hat{\alpha}_m \sigma(x_m, x_*) > 0 \\ 0 & \text{otherwise} \end{cases}$$

Fig. 6.6 below shows the result of applying the procedure with Gaussian kernel (bandwidth: $h = 0.05$) to the synthetic dataset of spiral distributions. This model is able to capture the shape of the data very well. It is important to note that, in order to apply the learned classifier, we must compute the similarity between the feature vector $x_*$ to be classified and every data point $x_1, \ldots, x_N$ in the training dataset. Procedures of this type are referred to as **instance-based learning**. For instance-based learners, the complexity of the hypothesis increases with the size of the training dataset, which can take up many computational resources. Reducing this complexity may therefore be necessary—for an overview, see [18].

A rationale for how the procedure works can be given as follows. We assume that the kernel has the property that the matrix with entries $\sigma(x_m, x_n)$ is symmetric and positive semidefinite for any finite choice of data points $x_1, x_2, \ldots$. For example, a Gaussian or polynomial kernel satisfies this condition [19, Sect. 2.2].

Suppose we are given a test dataset with $M$ feature vectors $x_{N+1}, \ldots, x_{N+M} \in \mathbb{R}^D$. We may consider the full similarity matrix, including both training and test data:

$$\Sigma = (\sigma(x_m, x_n))_{m,n \in \{1,\ldots,N+M\}}.$$

First, we take the square root of $\Sigma$: we are looking for a square matrix $\Phi$ with $\Sigma = \Phi \cdot \Phi^T$. One way to construct such a matrix is as follows.

Since the kernel is assumed to be symmetric, $\Sigma$ is a symmetric matrix. Therefore, there exists an orthogonal matrix $V$ and a diagonal matrix $\Lambda$ such that the following holds (see Sect. B.2.4 in the appendix):

$$\Sigma = V \cdot \Lambda \cdot V^T$$

The diagonal of $\Lambda$ contains the eigenvalues $\lambda_1, \ldots, \lambda_D$ of $\Sigma$. As $\Sigma$ is also positive semidefinite, the entries of $\Lambda$ are nonnegative. Therefore, we can take their square root, and we write $\sqrt{\Lambda}$ for the resulting matrix. If we define $\Phi = V \cdot \sqrt{\Lambda}$, that matrix has the desired property:

$$\Phi \cdot \Phi^T = V \cdot \sqrt{\Lambda} \cdot \left(V \cdot \sqrt{\Lambda}\right)^T = V \cdot \sqrt{\Lambda} \cdot \sqrt{\Lambda}^T \cdot V^T = V \cdot \Lambda \cdot V^T = \Sigma$$

We note the following:

$$\langle \varphi_m, \varphi_n \rangle = \varphi_m \cdot \varphi_n^T = \Sigma_{mn} = \sigma(x_m, x_n)$$

for all $m, n \in \{1 \ldots, N, \ldots, N+M\}$, where $\varphi_m$ and $\varphi_n$ are the $m$-th and $n$-th row of $\Phi$, respectively. Thus, we can write the decision rule of kernel logistic regression as follows, where $x_*$ is taken from the test dataset:

$$\sum_{m=1}^{N} \hat{\alpha}_m \sigma(x_m, x_*) = \sum_{m=1}^{N} \hat{\alpha}_m \langle \varphi_m, \varphi_* \rangle = \langle w^\perp, \varphi_* \rangle > 0$$

with $w^\perp = \sum_{m=1}^{N} \hat{\alpha}_m \varphi_m$. We conclude that the decision boundary is a hyperplane with normal vector $w^\perp$.

In other words, kernel logistic regression is a linear classifier—but in a (usually higher-dimensional) space defined by the **feature map**

$$\Phi \colon \{x_1, \ldots, x_{M+N}\} \to \mathbb{R}^{N+M}, \ x_n \mapsto \varphi_n = \Phi(x_n)$$

that maps the kernel onto scalar products: $\sigma(x_m, x_n) = \langle \varphi_m, \varphi_n \rangle$.

Our construction of a feature map depends explicitly on the test dataset even though that dataset can be of arbitrary size. For a solid theoretical foundation, it would be nice if this restriction did not exist and the feature map were universal:

$$\Phi \colon \mathbb{R}^D \to \mathcal{H}, \ u \mapsto \Phi(u)$$

where $\mathcal{H}$ is some vector space with scalar product $\langle \cdot, \cdot \rangle$. Such a map would provide a complete representation of the kernel in the target feature space:

$$\sigma(u, v) = \langle \Phi(u), \Phi(v) \rangle$$

for all $u, v \in \mathbb{R}^D$. We do not provide all of the details here, but note that such a construction is possible under fairly general circumstances—see [20, Theorem 6.8]. The target space $\mathcal{H}$ is called a *reproducing kernel Hilbert space*, and it is infinite-dimensional in general.

### 6.3.2 *K*-nearest neighbors classification

Logistic regression produces a linear classifier that allows for a simple geometric interpretation: entities with positive ($\hat{y} = 1$) and negative ($\hat{y} = 0$) prediction are separated by a hyperplane in feature space. If we transform the features in a nonlinear fashion—for example using the kernel trick—the method is able to learn more general classifiers.

So-called $K$-nearest neighbors classification is also best illustrated and understood by imagining feature space as a geometric space. This method makes predictions based on the labeled data points that have the smallest distance to the input.

We are given a training dataset $(x, y)$ where $x = (x_1, \ldots, x_N) \in \mathcal{X}^N$ are data points with class labels $y = (y_1, \ldots, y_N)$. We assume that $\mathcal{X}$ comes equipped with a distance or similarity measure.

Fix the hyperparameter $K \in \mathbb{N}$, $K \geq 1$. The $K$ nearest neighbors of a data point $x_* \in \mathcal{X}$ are those observations $x_{\iota(1)}, \ldots, x_{\iota(K)}$ which have the least distance/greatest similarity to $x_*$.

The **$K$-nearest neighbors classifier** (KNN classifier) commits to a plurality vote among the neighbors of $x_*$: it predicts that $x_*$ is assigned to the mode of $y_{\iota(1)}, \ldots, y_{\iota(K)}$.

We have implicitly assumed that the $K$ nearest neighbors and the mode of their labels are uniquely determined; that there is no need to break ties. There are several sensible ways to handle situations where this is not the case:

- After selecting $K$ observations $x_{\iota(1)}, \ldots, x_{\iota(K)}$ with minimum distance/greatest similarity to $x_*$, we consider all labeled data points within a distance of no more than

$$\delta_{\max} = \max_{k \in \{1, \ldots, K\}} \{\delta(x_*, x_{\iota(k)})\}$$

to $x_*$—even if those may be more than $K$ data points. Here, $\delta(\cdot, \cdot)$ denotes a distance measure. When a similarity measure $\sigma(\cdot, \cdot)$ is used, the plurality vote is conducted among all data points with a similarity measure of at least

$$\sigma_{\min} = \min_{k \in \{1, \ldots, K\}} \{\sigma(x_*, x_{\iota(k)})\}.$$

- If there is a tie among the $K$ nearest neighbors, i.e., if there are two candidates that appear with the same maximum frequency, the $K - 1$ nearest neighbors are considered instead. Repeat this process of reducing the number of neighbors until a plurality vote is reached.

KNN is an instance-based learner. In its most basic form, the KNN classifier computes the distances/similarities of the tested data point to all the data points in the training dataset.

In the case of $K = 1$, KNN just finds the closest/most similar point in the training dataset and assigns its label. However, such a small value for $K$ can easily lead to overfitting. The hyperparameter $K$ can be determined through validation techniques, but in practice rules of thumb such as $K \approx \sqrt{N}$ are also used [21]. In Fig. 6.7, the procedure is demonstrated for two values of $K$ using a synthetic dataset. Note that the ordinary Euclidean metric is used as the distance measure.

### 6.3.3 Bayesian classification algorithms

The Bayes classifier is the optimal decision rule given complete knowledge of the probability distributions underlying the observations.

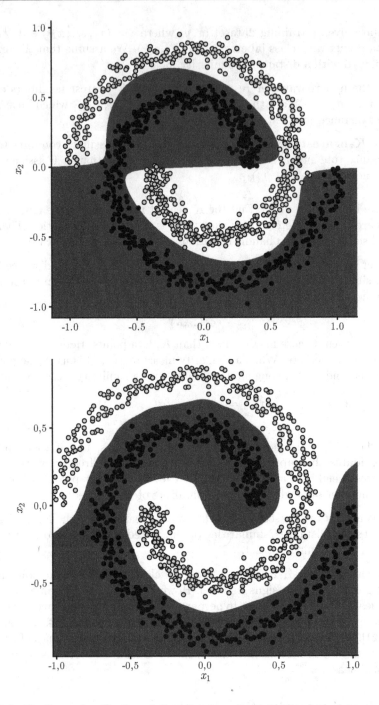

**Fig. 6.6.** *Nonlinear classification via logistic regression: including higher powers (top), and the kernel trick (bottom)*

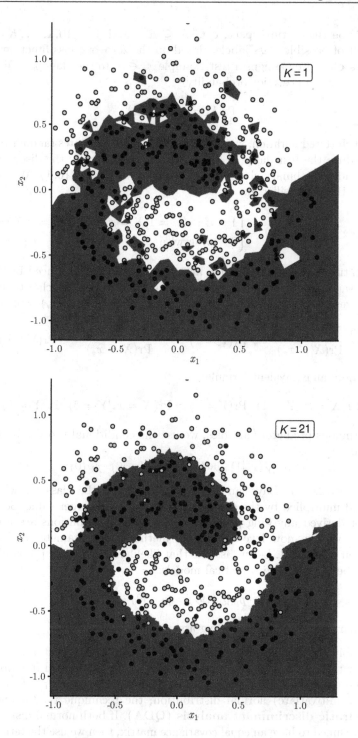

**Fig. 6.7.** *K-nearest neighbor classification*

Let $\mathcal{X}$ be the feature space, e.g., $\mathcal{X} \subseteq \mathbb{R}^D$, and $\mathcal{Y} = \{0, 1, \ldots, K-1\}$ the set of possible class labels. Based on the zero–one loss function, the **Bayes classifier** assigns a test example $x_* \in \mathcal{X}$ to the class $y_* \in \mathcal{Y}$ that maximizes the posterior probability

$$\Pr(Y = y_* | X = x_*)$$

To avoid cluttered notation, we omit the $*$ index that indicates random variables that produce the test data: $Y = Y_*$, $X = X_*$. To simplify the discussion, we only consider the binary classification problem where $\mathcal{Y} = \{0, 1\}$. We can thus write the Bayes classifier like so:

$$f : \mathcal{X} \to \mathcal{Y}, \; f_{\text{Bayes}}(x_*) = \begin{cases} 1 & \text{if } \Pr(Y = 1 | X = x_*) > \Pr(Y = 0 | X = x_*) \\ 0 & \text{otherwise} \end{cases}$$

If the features used for prediction are categorical—i.e., produced by discrete random variables—we can rewrite the condition for a positive class assignment using Bayes' theorem, for any test example $x_*$ with $\Pr(X = x_*) > 0$:

$$\frac{\Pr(X = x_* | Y = 1)}{\Pr(X = x_*)} \cdot \Pr(Y = 1) > \frac{\Pr(X = x_* | Y = 0)}{\Pr(X = x_*)} \cdot \Pr(Y = 0)$$

We can note an equivalent formula:

$$\Pr(X = x_* | Y = 1) \cdot \Pr(Y = 1) > \Pr(X = x_* | Y = 0) \cdot \Pr(Y = 0)$$

For continuous predictor variables, we use the conditional probability density function:

$$p_{X|Y}(x_* | 1) \cdot p_Y(1) > p_{X|Y}(x_* | 0) \cdot p_Y(0)$$

In other words, the Bayesian decision rule assigns the class for which the likelihood multiplied by the prior probability has the larger value. So far, we have not derived anything practical because the distributions are unknown. However, we can now leverage known methods for parameter and density estimation (Sect. 4.4). Let us assume that likelihood and prior probability can be described by suitable statistical models:

$$p(x_* | \theta_1) \cdot p_{\text{prior}}(1 | \alpha) > p(x_* | \theta_0) \cdot p_{\text{prior}}(0 | \alpha)$$

In summary, we make the following assumptions:

- The joint probability mass/density function of the predictor variables $X$ under the condition $Y = 0$ resp. $Y = 1$—i.e. the likelihood function—can be represented by some statistical model $p(\cdot | \theta_0)$ resp. $p(\cdot | \theta_1)$. If this model is a (multivariate) normal distribution, the technique is also known as **quadratic discriminant analysis (QDA)**. If both normal distributions are assumed to have an equal covariance matrix, then we use the term **linear discriminant analysis (LDA)**.

- A priori—before the algorithm sees any data—class labels are distributed according to $p_{\text{prior}}(\cdot\,|\alpha)$.

We recognize the above rationale as a maximum-a-posteriori estimation (see Sect. 4.4.2). The model parameters $\theta_0, \theta_1$ are estimated from the training data. If we were to strictly adhere to the Bayesian paradigm presented thus far, the prior probabilities depended only on the hyperparameters $\alpha$ and would *not* be inferred from the data. However, it is very common to estimate the prior probabilities from the training data as well. This approach is called the **empirical Bayes method**.

**Example.** We present an example that is of little practical importance, but illustrative. We want to predict the sex of a person based on their body height using the CDC dataset [22] as a training dataset. In Sect. 4.4.1, we already established that for a given sex, the distribution of body height $x$ can be modeled well by a normal distribution:

$$p(x|\hat{\mu}_0, \hat{\sigma}_0) = \mathcal{N}(x|\hat{\mu}_0, \hat{\sigma}_0^2), \; p(x|\hat{\mu}_1, \hat{\sigma}_1) = \mathcal{N}(x|\hat{\mu}_1, \hat{\sigma}_1^2)$$

where the parameters estimated from the data are $\hat{\mu}_0 = 178\,\text{cm}$, $\hat{\sigma}_0 = 7.8\,\text{cm}$ for male respondents, and $\hat{\mu}_1 = 163\,\text{cm}$, $\hat{\sigma}_1 = 7.3\,\text{cm}$ for female respondents. These probability density functions will serve us as likelihood functions.

The prior distribution is a Bernoulli distribution, which we can determine by one of the following methods:

- Assign a noninformative prior, i.e., $p_{\text{prior}}(0) = p_{\text{prior}}(1) = \frac{1}{2}$.

- Assign an **empirical prior** inferred from the proportion of respondents of male/female sex, irrespective of body height: $p_{\text{prior}}(0) = 0.45$ and $p_{\text{prior}}(1) = 0.55$.

In this particular case, both methods yield very similar priors. In the following model, we assume a noninformative prior. Under these model assumptions, the Bayesian decision rule for classifying a person of size $x$ as *female* becomes:

$$p(x|\hat{\mu}_1, \hat{\sigma}_1) \cdot p_{\text{prior}}(1) > p(x|\hat{\mu}_0, \hat{\sigma}_0) \cdot p_{\text{prior}}(0) \Leftrightarrow$$

$$\frac{1}{\sqrt{2\pi}\hat{\sigma}_1} \cdot \exp\left(-\frac{(x - \hat{\mu}_1)}{2\hat{\sigma}_1^2}\right) \cdot \frac{1}{2} > \frac{1}{\sqrt{2\pi}\hat{\sigma}_0} \cdot \exp\left(-\frac{(x - \hat{\mu}_0)}{2\hat{\sigma}_0^2}\right) \cdot \frac{1}{2} \Leftrightarrow$$

$$\hat{\sigma}_0|x - \hat{\mu}_1| < \hat{\sigma}_1|x - \hat{\mu}_0|$$

The decision boundary is given by the following value:

$$x = \frac{\hat{\sigma}_0\hat{\mu}_1 + \hat{\sigma}_1\hat{\mu}_0}{\hat{\sigma}_0 + \hat{\sigma}_1} \approx 170\,\text{cm}$$

A person with a body height below this value would be classified as *female*. This decision boundary is the vertical line drawn in Fig. 4.6.

### 6.3.3.1 Naive Bayes classification

Given the Bayesian paradigm, a highly simplifying ("naive") assumption is that—conditioned on class membership—the predictor variables are mutually independent. Under that assumption, the likelihood decomposes into a product over the features.

> The (binary) **naive Bayes classifier**, based on $K$ features and the zero–one loss function, predicts a positive label for a training example $x_* = (x_{*1}, \ldots, x_{*K})$ if the following condition holds:
>
> $$\Pr(Y=1) \cdot \prod_{k=1}^{K} \Pr(X_k = x_{*k} | Y = 1) > \Pr(Y=0) \cdot \prod_{k=1}^{K} \Pr(X_k = x_{*k} | Y = 0)$$

An important use case is the classification of sequences $t = (t_{x_1}, \ldots, t_{x_K})$, where the $t_{x_k}$ are taken from an inventory of $D$ symbols or strings $\{t_1, \ldots, t_D\}$, the **vocabulary**. A more concrete example is the classification of natural language texts. Through the preprocessing step of **tokenization**, the text is broken down into lexical units, the tokens. Often, these tokens are the individual words that make up the text. The vocabulary then represents the collection of these **tokens**.

For example, text case normalization and a simple, punctuation-based tokenization would transform the sentence "The vocabulary represents the collection of tokens." into the following sequence of strings:

(the, vocabulary, represents, the, collection, of, tokens, .)

An additional simplifying assumption is that class membership is independent of the order of the tokens: In this case, we speak of the **bag-of-tokens model** or, more commonly, the **bag-of-words model**. The above sentence would then not be distinguished from the following alphabetical sequence, for example:

(., collection, of, represents, the, the, tokens, vocabulary)

### 6.3.3.2 Multinomial event model

Under the bag-of-tokens/bag-of-words assumption, there are two principal ways of statistically modelling the sequence/text. The first considers each sequence of tokens as a list of categorical variables $x = (x_1, \ldots, x_K)$ with $D$ possible values, where $D$ is the size of the vocabulary: $x_k \in \{1, \ldots, D\}$.

The naive Bayesian decision rule for a test sequence $x_*$ is then of the following form:

$$\Pr(Y=1) \cdot \prod_{k=1}^{K} \Pr(X_k = x_{*k} | Y = 1) > \Pr(Y=0) \cdot \prod_{k=1}^{K} \Pr(X_k = x_{*k} | Y = 0)$$

with $x_{*k} \in \{1, \ldots, D\}$. The bag-of-tokens assumption implies that a token can occur at any position in the sequence with the same probability: $\Pr(X_k = d|Y = y) = \Pr(X_l = d|Y = y)$ for all $k, l \in \{1, \ldots, K\}$, $d \in \{1, \ldots, D\}$, $y \in \{0, 1\}$. We introduce the abbreviations $q = \Pr(Y = 1)$, $p_{d|1} = \Pr(X_k = d|Y = 1)$, and $p_{d|0} = \Pr(X_k = d|Y = 0)$. Consequently, the decision rule becomes:

$$\frac{q}{1-q} \cdot \prod_{k=1}^{K} \frac{p_{x_{*k}|1}}{p_{x_{*k}|0}} > 1$$

An even clearer formula is the following:

$$\frac{q}{1-q} \cdot \prod_{d=1}^{D} \left( \frac{p_{d|1}}{p_{d|0}} \right)^{n_{*d}} > 1$$

where $n_{*d}$ is the absolute frequency with which the token $t_d$ occurs in the sequence $t^{(*)}$ to be classified. This is the **multinomial event model**, which assumes that the occurrence of tokens within a class follows a multinomial distribution. Taking the logarithm on both sides of the inequality yields the following decision rule:

$$\operatorname{logit} q + \sum_{d=1}^{D} n_{*d} \cdot \left( \ln p_{d|1} - \ln p_{d|0} \right) > 0$$

where

$$\operatorname{logit}: \, ]0, 1[ \to \mathbb{R}, \, \operatorname{logit}(q) = \ln \left( \frac{q}{1-q} \right)$$

is the so-called **logit function**.

The probabilities occuring in the above formula are estimated from the training dataset $\left( (t^{(1)}, y_1), \ldots, (t^{(N)}, y_N) \right)$. The prior probability $q$ can be determined via the empirical Bayes method:

$$\hat{q} = \frac{N_+}{N}$$

where $N_+ = \sum_{n=1}^{N} y_n$, the number of sequences with a positive class label.

To estimate the likelihood, we construct a data matrix $\mathbf{N} = (n_{nd})$, the entries of which are the absolute frequency of the $d$-th token in the $n$-th sequence. The relative frequency of occurrence of the $d$-th token in each class gives the following estimates:

$$\hat{p}_{d|1} = \frac{\sum_{n=1}^{N_+} n_{nd}^{(+)}}{\sum_{d=1}^{D} \sum_{n=1}^{N_+} n_{nd}^{(+)}}, \, \hat{p}_{d|0} = \frac{\sum_{n=1}^{N_-} n_{nd}^{(-)}}{\sum_{d=1}^{D} \sum_{n=1}^{N_-} n_{nd}^{(-)}}$$

Here, $n^{(+)}$ and $n^{(-)}$ are those rows of the data matrix $\mathbf{N}$ that are paired with positive and negative class labels, respectively.

In practice, rare tokens may not occur at all within a certain class, implying an ill-defined decision rule because of $\hat{p}_{d|1} = 0$. This challenge can be addressed by **additive smoothing**, which corrects the estimates as follows:

$$\hat{p}_{d|1,s} = \frac{\sum_{n=1}^{N_+} n_{nd}^{(+)} + s}{\sum_{d=1}^{D} \sum_{n=1}^{N_+} n_{nd}^{(+)} + s \cdot D}, \; \hat{p}_{d|0,s} = \frac{\sum_{n=1}^{N_-} n_{nd}^{(-)} + s}{\sum_{d=1}^{D} \sum_{n=1}^{N_-} n_{nd}^{(-)} + s \cdot D}$$

where $s \geq 0$ is the smoothing parameter. The smoothing parameter is a hyperparameter, the optimal value of which can be determined, for example, by cross-validation. Otherwise, $s = 1$ is a commonly used value, which is also called **Laplace smoothing**; $s < 1$ is called **Lidstone smoothing** [23, Sect. 2.6.3].

### 6.3.3.3 Bernoulli event model

Alternatively, we can assign a list of binary variables $x_{n1}, \ldots, x_{nD}$ to each sequence $t^{(n)}$: $x_{nd} = 1$ if the token $t_d$ occurs in the sequence, otherwise $x_{nd} = 0$. That way, we can write the training sequences into a data matrix $\boldsymbol{X} = (x_{nd})$ with binary entries of format $N \times D$. This type of representation is called **one-hot encoding** and leads to the **Bernoulli event model**.

Assuming the Bernoulli event model, the naive Bayesian decision rule becomes the following:

$$\Pr(Y = 1) \cdot \prod_{d=1}^{D} \Pr(X_d = x_{*d}|Y = 1) > \Pr(Y = 0) \cdot \prod_{d=1}^{D} \Pr(X_d = x_{*d}|Y = 0)$$

with $x_{*d} \in \{0, 1\}$.

Using the abbreviations $q = \Pr(Y = 1)$, $p_{d|1} = \Pr(X_d = 1|Y = 1)$, and $p_{d|0} = \Pr(X_d = 1|Y = 0)$, respectively, this decision rule can be written as follows:

$$\text{logit } q + \sum_{d=1}^{D} \ln \left\{ \begin{matrix} \frac{p_{d|1}}{p_{d|0}} & \text{if } x_{*d} = 1 \\ \frac{1-p_{d|1}}{1-p_{d|0}} & \text{if } x_{*d} = 0 \end{matrix} \right\} =$$

$$\text{logit } q + \sum_{d=1}^{D} \ln \left( \frac{1 - p_{d|1} + x_{*d} \cdot (2p_{d|1} - 1)}{1 - p_{d|0} + x_{*d} \cdot (2p_{d|0} - 1)} \right) > 0$$

The likelihood can be estimated from the training dataset by counting tokens/words in addition to making use of additive smoothing:

$$\hat{p}_{d|1,s} = \frac{\sum_{n=1}^{N_+} x_{nd}^{(+)} + s}{N_+ + sD}, \; \hat{p}_{d|0,s} = \frac{\sum_{n=1}^{N_-} x_{nd}^{(-)} + s}{N_- + sD}$$

Here, $x^{(+)}$ and $x^{(-)}$ are those rows of the data matrix $\boldsymbol{X}$ that have positive and negative class assignments, respectively.

## 6.4 Artificial neural networks

**Artificial neural networks** are machine learning systems used for both regression and classification tasks. As the name implies, these methods are inspired by biological nervous systems. In Fig. 6.10, the neural network of the nematode worm *Caenorhabditis elegans* is shown for illustration [24, 25]: each node corresponds to a neuron, each edge to an interneuronal synapse (cf. [26]).

Possibly the first implementation of an artificial neural network, the Mark I Perceptron was developed as early as 1960 by psychologist and computer scientist Frank Rosenblatt [27, 28, 29]. Since about 2010, neural networks have gained immense importance in the field of machine learning and are used for numerous tasks for which large or enormous amounts of training data are available. So-called deep neural networks can have a large or immensely large number of model parameters, in orders of $10^4$–$10^{11}$, to learn highly customized classifiers and regression functions from these training data, yet with an equally high degree of generalizability.

The simplest form of an artificial neural network can be described as follows.

A **feedforward neural network** is a member of the family of functions

$$f \colon \mathbb{R}^{D_0} \to \mathbb{R}^{D_L}, \ f(u) = (f_L \circ f_{L-1} \circ \cdots \circ f_1)(u)$$

where each **layer** $f_l$ is a function of the following form:

$$f_l \colon \mathbb{R}^{D_{l-1}} \to \mathbb{R}^{D_l}, \ f_l(u) = \varphi_l\left(w^{(l)} \cdot u + b^{(l)}\right)$$

for all $l \in \{1, \ldots, L\}$, with so-called **activation functions** $\varphi_l \colon \mathbb{R}^{D_l} \to \mathbb{R}^{D_l}$, matrices of **weights** $w^{(l)}$ of format $D_l \times D_{l-1}$, and the **bias vectors** $b^{(l)} \in \mathbb{R}^{D_l}$.

The functions $f_{l1}, \ldots, f_{lD_l}$ represent the **neurons** of the $l$-th layer, and the value of each function represents the **activation** of the neuron.

We may imagine that a trivial zeroth layer $f_0 \colon u \mapsto u$ is added to the network, the **input layer**. The final layer $f_L$ represents the **output layer**. The intermediate layers with $1 \leq l < L$ are called **hidden layers**. The number of layers determines the **depth** of the network, and the number of neurons in each layer determines its **width**. If the network has more than one hidden layer, i.e., if $L > 2$ applies, we may speak of **deep learning**.

A feedforward neural network can be thought of as an acyclic directed graph. The following figure shows the node–link diagram of a feedforward network with two hidden layers:

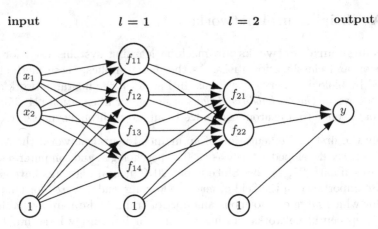

**Fig. 6.8.** *Feedforward neural network*

Each node corresponds to a single neuron. Every neuron of a given layer is linked to each neuron of the following layer via the weights. These links are represented as directed edges and could be called **synapses**. The additional neurons labeled "1" represent the bias vectors. The number of model parameters of a neural network is equal to the number of edges in the associated graph; the network sketched above has 25 parameters.

In principle, any acyclic graph can be used as the architecture of a neural network. In a **residual neural network**, synapses may also skip layers [30].

One possible choice for the activation function is a **sigmoid function**, or **Fermi function**:

$$\text{sig}: \mathbb{R} \to \mathbb{R}, \ \text{sig}(t) = \frac{1}{1 + e^{-t}} = \frac{1}{2} \cdot \left(1 + \tanh\frac{t}{2}\right)$$

The following figure shows the function graph:

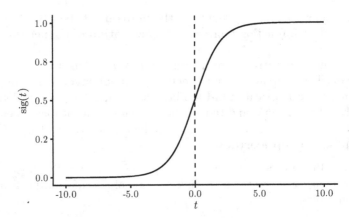

**Fig. 6.9.** *Sigmoid function*

The sigmoid function is applied to each component of a network layer:

$$\varphi_l \colon \mathbb{R}^{D_l} \to \mathbb{R}^{D_l}, \; \varphi_l(u) = (\text{sig}(u_1), \text{sig}(u_2), \ldots, \text{sig}(u_{D_l}))$$

Here is the "biological" interpretation of the sigmoid function: a neuron in the subsequent layer is activated only when the signal transmitted through the synapse has reached the "threshold potential" $t = 0$.

Another option is to apply a **rectifier function**, also called the **rectified linear unit (ReLU)**:

$$\text{rect}_\alpha \colon \mathbb{R} \to \mathbb{R}, \; \text{rect}_\alpha(t) = \begin{cases} t \text{ if } t > 0 \\ \alpha \cdot t \text{ if } t \leq 0 \end{cases}$$

The parameter $\alpha \geq 0$ is chosen to have a much smaller value than one—note that $\alpha = 0$ and $\alpha = 0.01$ are popular values. A rectifier with $\alpha > 0$ is called a **leaky rectifier** [31]. The value for $\alpha$ may also be learned during training, which is called **parametric ReLU (PReLU)** [32].

The rectifier function is not differentiable at $t = 0$. Many optimization methods require calculating the derivative, so we may want to use a smooth approximation:

$$\text{splus}_\alpha(t) = \alpha t + (1 - \alpha) \cdot \ln(1 + e^t)$$

In particular for $\alpha = 0$, this is known as the **softplus function**.

Finally, the **softmax function** is used in the output layer for classification tasks; it is defined as follows:

$$\text{smax} \colon \mathbb{R}^K \to \mathbb{R}^K, \; \text{smax}(u) = \left( \sum_{k=1}^{K} e^{u_k} \right)^{-1} \cdot (e^{u_1}, e^{u_2}, \ldots, e^{u_K})$$

The values of the softmax function can be interpreted as a probability mass distributed over the output neurons, for all $k \in \{1, \ldots, K\}$ and $u \in \mathbb{R}^K$:

$$(\text{smax}(u))_k > 0, \; \sum_{l=1}^{K} (\text{smax}(u))_l = 1$$

### 6.4.1 Regression and classification with neural networks

The model parameters of an artificial neural network are the entries of the weight matrices $w^{(1)}, \ldots, w^{(L)}$ and the components of the bias vectors $b^{(1)}, \ldots, b^{(L)}$. We denote the entries of the weight matrix $w^{(l)}$ and of the bias vector $b^{(l)}$ in the $l$-th layer, $1 \leq l \leq L$, by $w^{(l)}{}_{ji}$ and $b^{(l)}{}_j$, respectively, where $i \in \{1, \ldots, D_{l-1}\}$ and $j \in \{1, \ldots, D_l\}$. For the sake of clarity, in reference to these formulas, we also denote the collection of all of these parameters with the letter $\theta$.

According to the empirical risk minimization paradigm, training the neural network means minimization of the objective function that is the average loss:

$$R(\theta) = \frac{1}{N} \sum_{n=1}^{N} \lambda(y_n, f(x_n; \theta))$$

where $(x_1, y_1), \ldots, (x_N, y_N)$ are the training examples and $(y, \hat{y}) \mapsto \lambda(y, \hat{y})$ is the loss function.

If the neural network is used for regression tasks, $\lambda_2(y, \hat{y}) = (y - \hat{y})^2$ or $\lambda_1(y, \hat{y}) = |y - \hat{y}|$ are popular loss functions. The simplest neural network imaginable consists of exactly one output neuron, has no hidden layers, and the identity map $\varphi: u \mapsto u$ is used as the activation function. Using quadratic loss, the objective function for this (very simple) network is the following:

$$R(w_0, \ldots, w_D) = \frac{1}{N} \sum_{n=1}^{N} (y_n - w_0 - w \cdot x_n)^2$$

$$= \frac{1}{N} \sum_{n=1}^{N} \left( y_n - w_0 - \sum_{d=1}^{D} w_d x_{nd} \right)^2$$

Since there is only one layer, we have written the weight matrix $w^{(1)} = (w^{(1)}_1, \ldots, w^{(1)}_D)$ as a row vector of weights $w = (w_1, \ldots, w_D)$, and $w_0 \in \mathbb{R}$ is the only entry in the distortion vector $b^{(1)}$. Ignoring the constant factor $1/N$, this expression is the residual sum of squares, the objective function for linear regression (see Sect. 6.2.1). Therefore, both methods are equivalent: we have just shown that linear regression is included in the family of neural networks.

For classification tasks, each neuron of the output layer corresponds to one of the total $K$ classes so that the width of the output layer is $D_L = K$. A softmax function ensures that the activation of an output neuron can be interpreted as a probability for the respective class membership. A suitable loss function for classification tasks is **cross-entropy**:

$$\lambda(y, \hat{y}) = \lambda(y_1, \ldots, y_K, \hat{y}_1, \ldots, \hat{y}_K) = - \sum_{k=1}^{K} y_k \ln(\hat{y}_k)$$

Here, $y_k = 1$ holds if the training example belongs to the $k$-th class, otherwise $y_k = 0$: this is again the one-hot encoding. All summands, except the one corresponding to the true class membership, are zero. However, we might also want to train fuzzy labels where class membership is not binary but a matter of degree: $0 \leq y_k \leq 1$ for all $k \in \{1, \ldots, K\}$. For example, **label smoothing** [33] is a method that implies working with fuzzy training labels. Label smoothing can lead to improved generalizability of the trained model [34] and works by granting every training example at least a small but non-zero probability of membership for every class:

$$y_{k,\varepsilon} = (1 - \varepsilon) \cdot y_k + \frac{\varepsilon}{K}$$

where $\varepsilon > 0$ is a small smoothing parameter.

Regardless of whether the training labels are fuzzy/soft or hard, the output of the neural network is generally not a hard class assignment but is a probability distribution of class membership. The final classifier can be obtained by selecting the class with the highest membership probability.

For binary classification, we do not need an output layer with two neurons, as we can make due with a single sigmoid-activated neuron that outputs a positive class membership probability between zero and one. For such a network, cross-entropy takes the following form:

$$\lambda(y, \hat{y}) = -y \ln \hat{y} - (1 - y) \ln(1 - \hat{y})$$

Given this setup, let us consider once again a very simple and "shallow" neural network with no hidden layers. The output of this network, given $u \in \mathbb{R}^D$, takes the following form:

$$
\begin{aligned}
f(u; w_0, \ldots, w_D) &= \text{sig}\,(w \cdot u + w_0) \\
&= \frac{1}{1 + e^{-(w \cdot u + w_0)}}
\end{aligned}
$$

where $w = (w_1, \ldots, w_D)$ is a row vector of weights and $w_0 \in \mathbb{R}$ is the only entry in the bias vector.

The empirical risk, given a training dataset $x_1, \ldots, x_N \in \mathbb{R}^D$, is therefore the following function in the model parameters:

$$
\begin{aligned}
N \cdot R(w_0, \ldots, w_D) &= \sum_{k=1}^{N} \lambda\,(y_n, f\,(x_n; w_0, \ldots, w_D)) \\
&= \sum_{k=1}^{N} \lambda\,(y_n, \text{sig}\,(w \cdot x_n + w_0)) \\
&= -\sum_{k=1}^{N} y_n \ln\,(\text{sig}\,(w \cdot x_n + w_0)) - \\
&\quad \sum_{k=1}^{N} (1 - y_n) \ln\,(\text{sig}\,(-\,(w \cdot x_n + w_0))) \\
&= -\sum_{k=1}^{N} \ln\,(\text{sig}\,((-1)^{y_n}\,(w \cdot x_n + w_0))) \\
&= \sum_{k=1}^{N} \ln\,(1 + \exp\,((-1)^{y_n}\,(w \cdot x_n + w_0))) \\
&= \sum_{n=1}^{N} \ln\left(1 + \exp\left((-1)^{y_n} \cdot \sum_{d=0}^{D} w_d x_{nd}\right)\right)
\end{aligned}
$$

We used the following property of the sigmoid function:

$$1 - \text{sig}(t) = 1 - \frac{1}{1 + e^{-t}} = \frac{e^{-t}}{1 + e^{-t}} = \frac{1}{1 + e^t} = \text{sig}(-t)$$

A comparison with the log-likelihood function $\ell(\cdot)$ of logistic regression (Sect. 6.3.1) implies: $\ell(w_0, \ldots, w_D) = -NR(w_0, \ldots, w_D)$. Thus, maximizing this log-likelihood function corresponds exactly to minimizing the training error of this simple neural network. Finally, the decision rules are also identical, because due to $\text{sig}(t) > 1/2 \Leftrightarrow t > 0$, we have this formula:

$$f(x_*; \hat{w}_0, \ldots, \hat{w}_D) > \frac{1}{2} \Leftrightarrow \sum_{d=0}^{D} \hat{w}_d x_{*d} > 0$$

It is worth summarizing and highlighting those results.

**Linear and logistic regression as special neural networks.**

*Linear regression* is equivalent to a feedforward network with no hidden layers and a single, trivially activated output neuron using quadratic loss.

*Logistic regression* (without the kernel trick) is equivalent to a feedforward network with no hidden layers and with a single sigmoid-activated output neuron using cross-entropy loss.

Fig. 6.11 shows the results of a classification on two similar synthetic datasets using a neural network. The architecture used for this task is shown in Fig. 6.8: two hidden layers with four and two neurons, respectively; sigmoid functions were used as activation functions. The decision boundary is not a straight line: by adding hidden layers, the network is able to learn nonlinear classifiers—unlike logistic regression, the result of which is a linear classifier.

Linear and logistic regression are part of the hypothesis space of neural networks. What other models can be represented as a neural network? It turns out that in some sense, *all* models can be: any continuous function can be approximated by a suitable neural network. Without proof, we state the following theorems that make this statement mathematically precise.

**Universal approximation property of neural networks of arbitrary width [35, Theorem 3.1].** Let $\varphi \colon \mathbb{R} \to \mathbb{R}$ be a continuous function, and $\mathcal{W}(\varphi)$ be the set of feedforward neural networks with a single hidden $\varphi$-activated layer of potentially unlimited width:

$$f \colon \mathbb{R}^D \to \mathbb{R}, \ f(u) = \sum_{k=1}^{K} w^{(2)}{}_k \cdot \varphi \left( \sum_{d=1}^{D} w^{(1)}{}_{kd} \cdot u_d + b_k \right)$$

with parameters $K \in \mathbb{N}$ and $w^{(1)}{}_k, w^{(2)}{}_{kd}, b_k \in \mathbb{R}$.

If $\varphi$ is *not* a polynomial, then for any $\varepsilon > 0$ and any continuous function $g \colon K \to \mathbb{R}$ defined on a compact set $K \subset \mathbb{R}^D$ (e.g., $K = [0,1]^D$) there exists a function $f \in \mathcal{W}(\varphi)$ with the following property:

$$\sup_{u \in K} |f(u) - g(u)| < \varepsilon$$

Thus, any continuous function can be uniformly approximated on compact sets by functions in $\mathcal{W}(\varphi)$. Conversely, if $\mathcal{W}(\varphi)$ possesses this **universal approximation property**, then $\varphi$ cannot be a polynomial.

A similar theorem can be proved for deep networks with bounded width.

**Universal approximation property of neural networks of arbitrary depth and bounded width [36, Theorem 3.2].** Let $\varphi \colon \mathbb{R} \to \mathbb{R}$ be a continuous function that is continuously differentiable in at least one point, with non-vanishing derivative.

Furthermore, let $\mathcal{D}(\varphi)$ be the set of all feedforward neural networks $f \colon \mathbb{R}^D \to \mathbb{R}$ of any depth with $\varphi$-activated hidden layers, where no layer consists of more than $D + 3$ neurons.

If $\varphi \colon \mathbb{R} \to \mathbb{R}$ is not a linear function—that is, it is not of the form $\varphi(u) = m \cdot u + c$—then $\mathcal{D}(\varphi)$ has the universal approximation property: Any continuous function $g \colon \mathbb{R}^D \supset K \to \mathbb{R}$, with $K$ compact, can be uniformly approximated by functions in $\mathcal{D}(\varphi)$.

### 6.4.2 Training neural networks by backpropagation of error

In order to minimize the training error using gradient descent methods, and thus determine the optimal model parameters, with each iteration step, the gradient of a function of the form

$$R(\theta) = \frac{1}{M} \sum_{k=1}^{M} \lambda(y_{n_k}, f(x_{n_k}; \theta))$$

needs to be calculated. Depending on whether a full-batch gradient descent, stochastic gradient descent, or mini-batch gradient descent is performed, we have $M = N$, $M = 1$, or $1 < M < N$ summands, where $N$ is the total number of training examples.

Because derivation is a linear operation, we can also write the gradient as the sum of the gradients of each summand. Hence, in the following formula, we only consider the contribution of a single training example $(x_n, y_n)$, so we are interested in calculating the gradient of the function

$$R_n(\theta) = \lambda(y_n, f(x_n; \theta)).$$

For clarity, instead of the above expression, we write $r(\theta) = \lambda(y, f(x; \theta))$ in the following calculations.

We can compute the gradient of $r(\cdot)$ by an efficient algorithm known as the **backpropagation of error**, or just **backpropagation** or **backprop** for short.

Before describing this algorithm, we first introduce a few abbreviations. We denote the vector of activations in the $l$-th layer by $a^{(l)}$, so it holds that

$$a^{(l)} = (f_l \circ f_{l-1} \circ \cdots \circ f_1)(x)$$

for $1 \leq l \leq L$, and $a^{(0)} = x$. Furthermore, for each layer, we introduce the vector of weighted inputs

$$z^{(l)} = w^{(l)} \cdot a^{(l-1)} + b^{(l)},$$

so that $a^{(l)} = \varphi_l \left( z^{(l)} \right)$ where $\varphi_l$ is the activation function of the $l$-th layer.

Finally, we will denote the loss function with respect to the training example $y$ by $\Lambda$, i.e.: $\Lambda(\cdot) = \lambda(y, \cdot)$.

> **Backpropagation** computes the gradient of the objective function of a feedforward neural network by performing the following steps (cf. [37, Chap. 2]):
>
> 1. **Input.** Activate the input layer $a^{(0)} = x$, and set the current values of the model parameters $w^{(l)}{}_{ji}$ and $b^{(l)}{}_j$.
>
> 2. **Feedforward.** For $l = 1, \ldots, L$, successively compute the weighted inputs and activations: $z^{(l)} = w^{(l)} \cdot a^{(l-1)} + b^{(l)}$ and $a^{(l)} = \varphi_l(z^{(l)})$.
>
> 3. **Output error.** Compute $\Delta^{(L)} := (D\varphi_L(z^{(L)}))^T \cdot \nabla\Lambda(a^{(L)})$.
>
> 4. **Backpropagation of error.** For $l = L - 1, L - 2, \ldots, 1$, successively compute $\Delta^{(l)} = (D\varphi_l(z^{(l)}))^T \cdot (w^{(l+1)})^T \cdot \Delta^{(l+1)}$.
>
> 5. **Output.** Finally, the gradient of the objective function is given by the following partial derivatives:
>
> $$\frac{\partial r}{\partial w^{(l)}{}_{ji}} = \Delta^{(l)}{}_j \cdot a^{(l-1)}{}_i, \quad \frac{\partial r}{\partial b^{(l)}{}_j} = \Delta^{(l)}{}_j$$
>
> for all $l \in \{1, \ldots, L\}$, $i \in \{1, \ldots, D_{l-1}\}$, and $j \in \{1, \ldots, D_l\}$.

At the first iteration step of the gradient descent, the model parameters need to be initialized. In practice, one common method is the following:

- Set every bias $b^{(l)}{}_j$ to zero,

- for each layer—say, the $l$-th layer with $D_{l-1}$ input neurons and $D_l$ output neurons—initialize the weights $w^{(l)}{}_{ji}$ by drawing random, i.i.d. values from a distribution (for example, uniform or normal) with zero mean and variance $\sigma_l^2$.

If $\sigma_l^2 = 2(D_{l-1} + D_l)^{-1}$, this initalization scheme may be called **Glorot initalization**, also known as **Xavier initialization** [38]. If the variance is $\sigma_l^2 = 2(D_{l-1})^{-1}$, it may be called **He initialization** [32]. Glorot initialization is recommended when using sigmoidal activations, while He initialization is geared towards the use of ReLUs.

The backpropagation algorithm is based, firstly, on the observation that the change in the loss function with varying weights and bias in the $l$-th layer can be expressed by the change in the weighted inputs in that layer alone. This fact follows from the chain rule for partial derivatives:

$$
\begin{aligned}
\frac{\partial r}{\partial w^{(l)}{}_{ji}} &= \sum_{k=1}^{D_l} \frac{\partial r}{\partial z^{(l)}{}_k} \cdot \frac{\partial z^{(l)}{}_k}{\partial w^{(l)}{}_{ji}} \\
&= \sum_{k=1}^{D_l} \frac{\partial r}{\partial z^{(l)}{}_k} \cdot \frac{\partial}{\partial w^{(l)}{}_{ji}} \left( \sum_{m=1}^{D_{l-1}} w^{(l)}{}_{km} a^{(l-1)}{}_m + b^{(l)}{}_k \right) \\
&= \frac{\partial r}{\partial z^{(l)}{}_j} \cdot a^{(l-1)}{}_i,
\end{aligned}
$$

$$
\frac{\partial r}{\partial b^{(l)}{}_j} = \sum_{k=1}^{D_l} \frac{\partial r}{\partial z^{(l)}{}_k} \cdot \frac{\partial z^{(l)}{}_k}{\partial b^{(l)}{}_j} = \frac{\partial r}{\partial z^{(l)}{}_j}
$$

The activations $a^{(l-1)}{}_i$ are calculated during the feedforward step. We still need to compute the partial derivatives $\partial r / \partial z^{(l)}{}_j$. We combine them into column vectors, one for each layer:

$$
\Delta^{(l)} = \left( \frac{\partial r}{\partial z^{(l)}{}_1}, \ldots, \frac{\partial r}{\partial z^{(l)}{}_{D_l}} \right)^T
$$

We can compute the transpose of those vectors via the chain rule for Jacobians:

$$
\begin{aligned}
(\Delta^{(L)})^T &= \mathrm{D}\Lambda(a^{(L)}) \cdot \mathrm{D}\varphi_L(z^{(L)}), \\
(\Delta^{(L-1)})^T &= \mathrm{D}\Lambda(a^{(L)}) \cdot \mathrm{D}\varphi_L(z^{(L)}) \cdot w^{(L)} \cdot \mathrm{D}\varphi_{L-1}(z^{(L-1)}),
\end{aligned}
$$

$$
\vdots
$$

$$
\begin{aligned}
(\Delta^{(l)})^T = {}& \mathrm{D}\Lambda(a^{(L)}) \cdot \mathrm{D}\varphi_L(z^{(L)}) \cdot w^{(L)} \cdot \mathrm{D}\varphi_{L-1}(z^{(L-1)}) \cdot \ldots \\
& \cdot w^{(l+2)} \cdot \mathrm{D}\varphi_{l+1}(z^{(l+1)}) \cdot w^{(l+1)} \cdot \mathrm{D}\varphi_l(z^{(l)})
\end{aligned}
$$

$$
\vdots
$$

$$
\begin{aligned}
(\Delta^{(1)})^T = {}& \mathrm{D}\Lambda(a^{(L)}) \cdot \mathrm{D}\varphi_L(z^{(L)}) \cdot w^{(L)} \cdot \mathrm{D}\varphi_{L-1}(z^{(L-1)}) \cdot \ldots \\
& \cdot w^{(3)} \cdot \mathrm{D}\varphi_2(z^{(2)}) \cdot w^{(2)} \cdot \mathrm{D}\varphi_1(z^{(1)})
\end{aligned}
$$

Derivatives of inner functions are successively multiplied from the right side in order to produce the error terms $\Delta^{(l)}$: the error is propagated backwards along the network from the output layer in the direction of the input layer. Multiplication from the left side is the more common representation, which we can derive by taking the transpose:

$$\Delta^{(L)} = (\mathrm{D}\varphi_L(z^{(L)}))^T \cdot \nabla\Lambda(a^{(L)}),$$
$$\Delta^{(L-1)} = (\mathrm{D}\varphi_{L-1}(z^{(L-1)}))^T \cdot (w^{(L)})^T \cdot \Delta^{(L)},$$
$$\vdots$$
$$\Delta^{(l)} = (\mathrm{D}\varphi_l(z^{(l)}))^T \cdot (w^{(l+1)})^T \cdot \Delta^{(l+1)}$$
$$\vdots$$
$$\Delta^{(1)} = (\mathrm{D}\varphi_1(z^{(1)}))^T \cdot (w^{(2)})^T \cdot \Delta^{(2)}$$

### 6.4.2.1 Dropout

**Dropout** is a stochastic modification of backpropagation that is an effective and simple method to avoid overfitting. The idea is to use only part of the network during training, which is randomly selected with each iteration step [39].

Given **dropout** probabilities $0 \le q_l < 1$ for each hidden layer $l \in \{1, \ldots, L-1\}$, the propagation algorithm is modified as follows:

1. With each iteration during the feedforward step, disable a random selection of neurons:

   a) For each index $d \in \{1, \ldots, D_l\}$, select $\rho^{(l)}{}_d \in \{0, 1\}$ independently at random where $q_l$ is the probability for the outcome $\rho^{(l)}{}_d = 0$.

   b) Set $z^{(l)} = w^{(l)} \cdot a^{(l-1)} + b^{(l)}$ and $(a^{(l)})_d = \rho^{(l)}{}_d \cdot (\varphi_l(z^{(l)}))_d$ for all $d \in \{1, \ldots, D_l\}$.

2. Disabled neurons are ignored when calculating backpropagated errors: only weights $w^{(l)}{}_{ji}$ and biases $b^{(l)}{}_j$ with $\rho^{(l-1)}{}_i = \rho^{(l)}{}_j = 1$ or $\rho^{(l)}{}_j = 1$ are updated.

3. After the training is complete, all weights in the $l$-th layer are scaled by a factor of $1 - q_l$.

The dropout probabilities $q_1, \ldots, q_{L-1}$ can be interpreted as new hyperparameters of the model. They only affect the training process: when applying the final, trained model, all neurons remain enabled.

### 6.4.3 Convolutional neural networks

**Convolutional neural networks** (CNNs) are special artificial neural networks that are well-suited for the automated classification of digital images. In this section, we introduce the basic ideas behind their architecture.

A digital image can be represented as a three-dimensional grid of pixels $M$ with width $B$, height $H$, and a depth of $T$ color channels. Each pixel with spatial position $(i, j)$ is assigned a color value $M(i, j, t)$ in color channel $t$ where

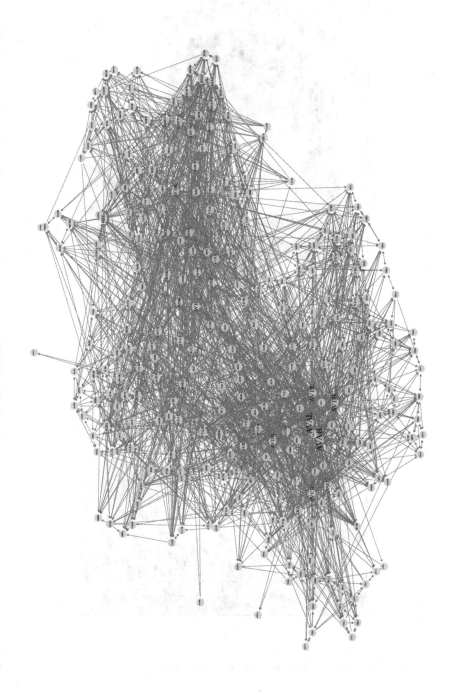

**Fig. 6.10.** *Neurons and interneural synapses of the nematode Caenorhabditis elegans (hermaphrodite)*

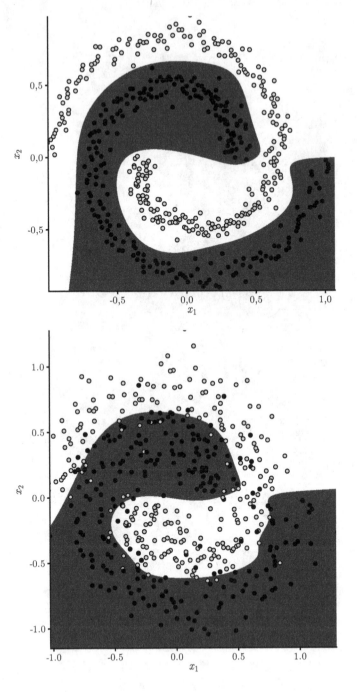

**Fig. 6.11.** *Classification with an artificial neural network*

$j \in \{1, \ldots, B\}$, $i \in \{1, \ldots, H\}$, $t \in \{1, \ldots, T\}$. For example, a high-resolution color image might have a width of $B = 1920$ pixels, a height of $H = 1080$ pixels, and a depth of $T = 3$ color channels. Usually the format of a digital images is specified as width-times-height. However, to stay consistent with the mathematical convention for matrix formats, we will note the number of pixel rows, i.e., the height, first.

Let us call any abstract numerical grid with three dimensions—regardless of whether it represents a human-readable image—a **feature map**, or an **activation map**. In a convolutional neural network, the input image is transformed into new feature maps with each layer. These transformations are implemented through two key operations, **convolution** and **pooling**, which we will describe in more detail below.

A **convolution mask** of bandwidth $2F + 1$, $F \in \mathbb{N}$ is a square feature map of the format $(2F + 1) \times (2F + 1) \times T$. The typical values for convolution masks used in practice are $F = 1$ or $F = 2$. For convolution masks, we want to index the width and height symmetrically around zero: $-F, \ldots, -1, 0, 1, \ldots, F$—this convention makes the definition of the following operation clearer.

The **two-dimensional convolution**, or **2D convolution**, of an image/activation map $M$ of format $H \times B \times T$ with a convolution mask $\kappa$ of the format $(2F + 1) \times (2F + 1) \times T$ yields the following new feature map of format $H \times B \times 1$:

$$(M \star \kappa)(i, j, 1) = \sum_{t=1}^{T} \sum_{f_1=-F}^{F} \sum_{f_2=-F}^{F} M(i + f_1, j + f_2, t) \cdot \kappa(f_1, f_2, t)$$

for all $j \in \{1, \ldots, B\}$ and $i \in \{1, \ldots, H\}$.

Whenever the summation index $(i + f_1, j + f_2)$ leaves the boundaries of the image/activation map $M$, we assume the color values of $M$ to be zero. In other words, we think of the exterior of $M$ as empty or black pixels—this is called **padding**.

2D convolution is a linear map over the vector space of feature maps of a fixed format—for any convolution mask $\kappa$, arbitrary feature maps $M_1, M_2$, and scalars $\lambda \in \mathbb{R}$:

$$(M_1 + M_2) \star \kappa = M_1 \star \kappa + M_2 \star \kappa, \quad (\lambda \cdot M_1) \star \kappa = \lambda \cdot (M_1 \star \kappa)$$

Addition and scalar multiplication are to be understood as pixel-wise.

The above definition of the convolution operation is commonly used in deep learning literature. However, in the fields of signal and **image processing**, the above operation is often referred to as **cross-correlation**, and the following operation for images of format $H \times B$ is commonly known as convolution:

$$(M * k)(i, j) = \sum_{f_1=-F}^{F} \sum_{f_2=-F}^{F} M(i - f_1, j - f_2) \cdot \kappa(f_1, f_2)$$

Cross-correlation is convolution with a mirrored convolution mask, and vice versa. This subtlety plays a minor role when implementing a convolutional neural network, since the entries of the convolution mask are the model parameters and are learned during training. Nevertheless, the different conventions may cause confusion when studying the subject.

**Fig. 6.12.** *2D convolution in image processing*

The above illustration shows the results of applying various convolution filters that are commonly used in image processing. The convolution masks applied to the original photo [40] (from left to right in the figure) are the following:

$$\kappa_1 = \begin{pmatrix} 1 & 2 & 1 \\ 0 & 0 & 0 \\ -1 & -2 & -1 \end{pmatrix}, \kappa_2 = \begin{pmatrix} 0 & 1 & 0 \\ 1 & -4 & 1 \\ 0 & 1 & 0 \end{pmatrix}$$

$$\kappa_3 = \begin{pmatrix} 2 & 1 & 0 \\ 1 & 1 & -1 \\ 0 & -1 & -2 \end{pmatrix}, \kappa_4 = \frac{1}{9} \cdot \begin{pmatrix} 1 & 1 & 1 \\ 1 & 1 & 1 \\ 1 & 1 & 1 \end{pmatrix}$$

The first two convolution masks are the **horizontal Sobel operator** and the **discrete Laplace operator**. Both are used in image processing for **edge detection**. These are followed by the **emboss filter** and the **mean filter**. The mean filter calculates the average color/gray value in the neighborhood of each pixel, which results in a blurring of the image.

For all image filters based on convolution, only the neighborhood of a pixel determines the value of the filtered image at that location. The result is a feature map that reflects *local* image features, such as edges. The basic idea of a convolutional neural network is to let it learn on its own which filters/features are best suited for characterizing and classifying image information.

Reducing image resolution is another important operation within a convolutional neural network. One option is to introduce a **stride** $s > 1$ that skips pixels:

$$(M \star_s \kappa)(i, j)$$

$$= \sum_{f_1=-F}^{F} \sum_{f_2=-F}^{F} M(1 + s \cdot (i-1) + f_1, 1 + s \cdot (j-1) + f_2) \cdot \kappa(f_1, f_2)$$

with $j \in \{1, \ldots, \lfloor B/s \rfloor\}$, $i \in \{1, \ldots, \lfloor H/s \rfloor\}$.

Another option consists of partioning the feature map into patches and selecting only the largest activation in each patch.

> **Maximum pooling**, or **max pooling**, an activation map $M$ of format $B \times H \times T$, with even numbers $B$ and $H$, consists of the following operation:
>
> $$\text{max-pool}(M)(i,j,t) = \max_{k,l \in \{0,1\}} \{M(2i - 1 + k, 2j - 1 + l, t)\}$$
>
> for all $j \in \{1, \ldots, \lfloor B/2 \rfloor\}$, $i \in \{1, \ldots, \lfloor H/2 \rfloor\}$, $t \in \{1, \ldots, T\}$.

The operations on images/feature maps defined above represent the most essential components of a convolutional neural network. The architecture of such a network is an arrangement of the following types of layers:

- A **convolution layer** convolves the input map $M$ of format $H \times B \times T$ with convolution masks $\kappa_1, \ldots, \kappa_K$ of the format $(2F + 1) \times (2F + 1) \times T$, and each convolution comes with a global bias $b_1, \ldots, b_K \in \mathbb{R}$ that is to be added to every entry of the input map. Then, an activation function $\varphi$ is applied (usually pixel-wise). These operations produce $K$ new images or feature maps of the format $H \times B \times 1$:

$$\varphi(M \star \kappa_1 + b_1), \ldots, \varphi(M \star \kappa_K + b_K)$$

These $K$ activation maps are concatenated to yield the output, a feature map of the format $H \times B \times K$. The entries of the convolution mask and the biases represent the model parameters to be learned.

- A **pooling layer** that performs maximum pooling which reduces the size/resolution of the input.

- A **fully-connected layer** first applies a **flattening** operation: the input grid $M$ is reshaped into a (long) vector $\text{flatten}(M)$ of length $D = H \cdot B \cdot T$. This vector is processed by an affine transformation, followed by an activation function—an operation that we already know from ordinary feedforward networks:

$$\varphi(w \cdot \text{flatten}(M) + b)$$

with a matrix of weights $w$ of format $K \times D$ and a bias vector $b$ of length $K$.

For each of these layers, the following table lists the number of model parameters and the format of the output map, given an input map of format $H \times B \times T$:

| layer type | nb. model parameters | output format |
|---|---|---|
| convolution, $K$ masks | $K \cdot (T \cdot (2F + 1)^2 + 1)$ | $H \times B \times K$ |
| max pooling | $0$ | $\lfloor B/2 \rfloor \times \lfloor H/2 \rfloor \times T$ |
| fully-connected, width $K$ | $K \cdot (H \cdot B \cdot T + 1)$ | $1 \times 1 \times K$ |

**Table 6.4.** *Layer types of a convolutional neural network*

A fully-connected layer applies an affine map that can, in principle, be arbitrary:

$$M \mapsto w \cdot \text{flat}(M) + b$$

A convolution layer also applies an affine map but is only allowed to draw from the family of convolutions (plus a shift/bias):

$$M \mapsto M \star \kappa + b$$

Therefore, the number of model parameters to be trained in a convolution layer is usually much smaller than in a fully-connected layer, given the same input.

A typical architecture of a convolutional neural network consists of alternating the execution of convolutional and pooling layers, the output of which is eventually processed by one or more fully-connected layers. Fig. 6.13 illustrates the VGG-16[4] architecture [41], which in 2014 achieved an accuracy of 74.4% on the ImageNet dataset [2, 42]. As of 2022, state-of-the-art architectures achieve an accuracy of more than 88% on ImageNet [43].

---

[4] VGG-16 stands for Visual Geometric Group of the University of Oxford, 16 layers

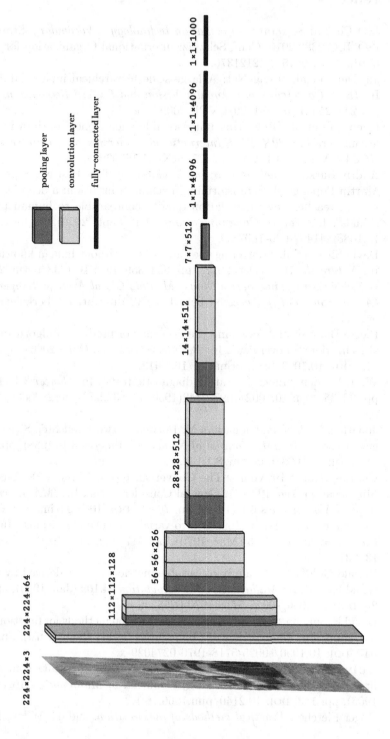

**Fig. 6.13.** *Architecture of the convolutional neural network VGG-16*

# References

[1]   ISO Central Secretary. *Information technology – Vocabulary*. Standard ISO/IEC 2382:2015. Genf, Schweiz: International Organization for Standardization, 2015, p. 2121376.

[2]   Jia Deng et al. "ImageNet: A large-scale hierarchical image database". In: *IEEE Conference on Computer Vision and Pattern Recognition*. 2009, pp. 248–255. DOI: 10.1109/CVPR.2009.5206848.

[3]   Chen Sun et al. "Revisiting unreasonable effectiveness of data in deep learning era". In: *2017 IEEE International Conference on Computer Vision (ICCV)*. Venice: IEEE, Oct. 2017. arXiv:1707.02968.

[4]   Andriy Burkov. *Machine Learning Engineering*. True Positive, Sept. 2020.

[5]   Martin Popel et al. "Transforming machine translation: a deep learning system reaches news translation quality comparable to human professionals". In: *Nature Communications* 11.1 (Sept. 2020), p. 4381. DOI: 10.1038/s41467-020-18073-9.

[6]   David Silver et al. "Mastering the game of Go without human knowledge". In: *Nature* 550.7676 (Oct. 2017), pp. 354–359. DOI: 10.1038/nature24270.

[7]   Berkeley Earth. *Time Series Data – Monthly Global Average Temperature (Annual Summary)*. Accessed Feb. 1, 2020. URL: http://berkeleyearth.org/data/.

[8]   Pierre Baldi et al. "Assessing the accuracy of prediction algorithms for classification: an overview". In: *Bioinformatics* 16.5 (May 2000), pp. 412–424. DOI: 10.1093/bioinformatics/16.5.412.

[9]   W. J. Youden. "Index for rating diagnostic tests". In: *Cancer* 3.1 (1950), pp. 32–35. DOI: 10.1002/1097-0142(1950)3:1<32::aid-cncr2820030106>3.0.co;2-3.

[10]  Jonathan Barzilai and Jonathan M. Borwein. "Two-Point Step Size Gradient Methods". In: *IMA Journal of Numerical Analysis* 8.1 (1988), pp. 141–148. DOI: 10.1093/imanum/8.1.141.

[11]  Charles George Broyden. "The Convergence of a Class of Double-rank Minimization Algorithms 1. General Considerations". In: *IMA Journal of Applied Mathematics* 6.1 (1970), pp. 76–90. DOI: 10.1093/imamat/6.1.76.

[12]  Roger Fletcher. "A new approach to variable metric algorithms". In: *The Computer Journal* 13.3 (Mar. 1970), pp. 317–322. DOI: 10.1093/comjnl/13.3.317.

[13]  Donald Goldfarb. "A family of variable-metric methods derived by variational means". In: *Mathematics of Computation* 24.109 (Jan. 1970), pp. 23–23. DOI: 10.1090/s0025-5718-1970-0258249-6.

[14]  David F. Shanno. "Conditioning of quasi-Newton methods for function minimization". In: *Mathematics of Computation* 24.111 (Sept. 1970), pp. 647–647. DOI: 10.1090/s0025-5718-1970-0274029-x.

[15]  Larry Armijo. "Minimization of functions having Lipschitz continuous first partial derivatives". In: *Pacific Journal of Mathematics* 16.1 (Jan. 1966), pp. 1–3. DOI: 10.2140/pjm.1966.16.1.

[16]  Roger Fletcher. *Practical methods of optimization*. 2nd ed. Wiley, 1987.

[17]   Jan R. Magnus. *Matrix Differential Calculus with Applications in Statistics and Econometrics*. 3rd ed. Wiley, Feb. 2019. DOI: 10.1002/9781119541219.

[18]   D. Randall Wilson and Tony R. Martinez. "Reduction Techniques for Instance-Based Learning Algorithms". In: *Machine Learning* 38 (2000), pp. 257–286. DOI: 10.1023/a:1007626913721.

[19]   Thomas Hofmann, Bernhard Schölkopf, and Alexander J. Smola. "Kernel methods in machine learning". In: *The Annals of Statistics* 36.3 (2008), pp. 1171–1220. DOI: 10.1214/009053607000000677.

[20]   Mehryar Mohri, Afshin Rostamizadeh, and Ameet Talwalkar. *Foundations of Machine Learning*. 2nd ed. MIT Press, 2018.

[21]   microhaus. *Why is $k = \sqrt{N}$ a good solution of the number of neighbors to consider?* Cross Validated. July 2021. URL: https://stats.stackexchange.com/q/535051.

[22]   CDC Population Health Surveillance Branch. *Behavioral Risk Factor Surveillance System (BRFSS) Survey Data 2018*. Accessed Feb. 1, 2020. URL: https://www.cdc.gov/brfss/.

[23]   Sebastian Raschka. *Naive Bayes and Text Classification I – Introduction and Theory*. Feb. 2017. arXiv:1410.5329v4.

[24]   David H. Hall, Zeynep F. Altun, and Laura A. Herndon. *Wormatlas. Neuronal Wiring*. Accessed Dec. 30, 2020. New York, USA. URL: https://www.wormatlas.org/neuronalwiring.html.

[25]   Lav R. Varshney et al. "Structural Properties of the Caenorhabditis elegans Neuronal Network". In: *PLoS Computational Biology* 7.2 (Feb. 2011). Ed. by Olaf Sporns, e1001066. DOI: 10.1371/journal.pcbi.1001066.

[26]   Gang Yan et al. "Network control principles predict neuron function in the Caenorhabditis elegans connectome". In: *Nature* 550.7677 (Oct. 2017), pp. 519–523. DOI: 10.1038/nature24056.

[27]   Frank Rosenblatt. "The perceptron: A probabilistic model for information storage and organization in the brain." In: *Psychological Review* 65.6 (1958), pp. 386–408. DOI: 10.1037/h0042519.

[28]   Frank Rosenblatt. *Principles of Neurodynamics. Perceptrons and the Theory of Brain Mechanisms*. Washington, D.C., USA: Spartan Books, 1962.

[29]   Melanie Lefkowitz. "Professor's perceptron paved the way for AI – 60 years too soon". In: *Cornell Chronicle* (Sept. 2019). URL: https://news.cornell.edu/stories/2019/09/professors-perceptron-paved-way-ai-60-years-too-soon.

[30]   Kaiming He et al. "Deep Residual Learning for Image Recognition". In: *2016 IEEE Conference on Computer Vision and Pattern Recognition (CVPR)*. IEEE, June 2016. DOI: 10.1109/cvpr.2016.90. arXiv:1512.03385.

[31]   Andrew L. Maas, Awni Y. Hannun, and Andrew Y. Ng. "Rectifier nonlinearities improve neural network acoustic models". In: *ICML Workshop on Deep Learning for Audio, Speech and Language Processing*. 2013.

[32]   Kaiming He et al. *Delving Deep into Rectifiers: Surpassing Human-Level Performance on ImageNet Classification*. 2015. DOI: 10.48550/ARXIV.1502.01852.

[33]    Christian Szegedy et al. "Rethinking the Inception Architecture for Computer Vision". In: *2016 IEEE Conference on Computer Vision and Pattern Recognition (CVPR)*. IEEE, June 2016. DOI: 10.1109/cvpr.2016.308. arXiv:1512.00567.

[34]    Rafael Müller, Simon Kornblith, and Geoffrey E Hinton. "When does label smoothing help?" In: *Advances in Neural Information Processing Systems*. Ed. by H. Wallach et al. Vol. 32. Curran Associates, Inc., 2019, pp. 4694–4703. arXiv:1906.02629.

[35]    Allan Pinkus. "Approximation theory of the MLP model in neural networks". In: *Acta Numerica* 8 (Jan. 1999), pp. 143–195. DOI: 10.1017/s0962492900002919.

[36]    Patrick Kidger and Terry Lyons. "Universal Approximation with Deep Narrow Networks". In: *33rd Conference on Learning Theory*. Ed. by Jacob Abernethy and Shivani Agarwal. Vol. 125. Proceedings of Machine Learning Research. PMLR, July 2020, pp. 2306–2327. arXiv:1905.08539.

[37]    Michael A. Nielsen. *Neural networks and deep learning*. Determination Press, 2015. URL: http://neuralnetworksanddeeplearning.com/.

[38]    Xavier Glorot and Yoshua Bengio. "Understanding the difficulty of training deep feedforward neural networks". In: *Proceedings of the Thirteenth International Conference on Artificial Intelligence and Statistics*. Vol. 9. PMLR. 2010, pp. 249–256.

[39]    Nitish Srivastava et al. "Dropout: A Simple Way to Prevent Neural Networks from Overfitting". In: *J. Mach. Learn. Res.* 15.1 (Jan. 2014), pp. 1929–1958.

[40]    Allan G. Weber. *The USC-SIPI Image Database: version 6*. Tech. rep. Los Angeles, USA: Signal and Image Processing Institute, University of Southern California, Feb. 2018. URL: http://sipi.usc.edu/database.

[41]    Karen Simonyan and Andrew Zisserman. "Very Deep Convolutional Networks for Large-Scale Image Recognition". In: *3rd International Conference on Learning Representations, San Diego, USA*. Ed. by Yoshua Bengio and Yann LeCun. May 2015. arXiv:1409.1556.

[42]    Olga Russakovsky et al. "ImageNet Large Scale Visual Recognition Challenge". In: *International Journal of Computer Vision (IJCV)* 115.3 (2015), pp. 211–252. DOI: 10.1007/s11263-015-0816-y. arXiv:1409.0575.

[43]    Papers with Code Community. *ImageNet Benchmark (Image Classification)*. Ed. by Robert Stojnic et al. Accessed Oct. 10, 2022. URL: https://paperswithcode.com/sota/image-classification-on-imagenet.

# 7

## Unsupervised machine learning

Supervised machine learning methods derive a decision rule $f \colon \mathcal{X} \to \mathcal{Y}$ by making generalizations from the patterns observed in an annotated dataset $(\mathcal{X} \times \mathcal{Y})^N$ where $\mathcal{Y} = \mathbb{R}$ or $\mathcal{Y} = \{0, 1, \ldots, K-1\}$ is the space of target values/labels. In **unsupervised learning**, the input data is not annotated: none of the features of the dataset to be analyzed are distinguished a priori as the target variable or class label.

There are two major methods used in unsupervised learning:

- The goal of a **dimensionality reduction** is to reduce the number of features without significantly changing key characteristics of the data, such as the distances between data points. Dimensionality reduction can be understood as an unsupervised analogue to regression.

- **Cluster analysis** aims to group the dataset into sets of similar data records/information objects. Cluster analysis is thus similar to the task of classification—except that the cluster labels are not included with the training data but are learned by the algorithm from the structure of the input data alone.

In the first section of this chapter, we deal with the geometric and topological views on data. This perspective serves as an introductory motivation of dimensionality reduction and for cluster analysis methods, which will be explained in the following sections.

## 7.1 Elements of unsupervised learning

Given a sequence or set of data points $x_1, x_2, \ldots, x_N \in \mathbb{R}^D$, we have already seen that adopting a geometric view is often useful for a better understanding of the algorithms and methods we employ. To give a few examples: The $K$-nearest-neighbor classifier exploits the view of data points as objects spaced at certain distances. Affine subspaces also play an important role: the result

© Springer-Verlag GmbH Germany, part of Springer Nature 2023
M. Plaue, *Data Science*, https://doi.org/10.1007/978-3-662-67882-4_7

of a linear regression is a hyperplane in feature space, and a linear classifier separates classes by such hyperplanes.

### 7.1.1 Intrinsic dimensionality of data

Let us look again at scatter plots of strongly correlated features, for example Fig. 2.3 or Fig. 6.3. In these cases, the data points do not scatter arbitrarily in all directions in the plane, but instead they are close to a regression line or, more generally, regression curve. Thus—except for a small amount of random dispersion—the position of a data point can be described as a point along that curve. In this way, the dimensionality of the dataset was effectively reduced from two (the feature space spanned by the dependent and the independent variable) to one (the position along the regression curve).

The following theorem gives a general indication of whether we may hope to find a given number of data points near a subspace that has low dimensionality.

**Johnson–Lindenstrauss lemma [1].** Let $x_1, \ldots, x_N \in \mathbb{R}^D$ be data points and $\varepsilon \in ]0, 1[$ be a (small) number. Furthermore, let $K \in \mathbb{N}$ be such that:

$$K \geq 4 \cdot \left( \frac{\varepsilon^2}{2} - \frac{\varepsilon^3}{3} \right)^{-1} \cdot \ln N$$

Then, there exists a linear map $f \colon \mathbb{R}^D \to \mathbb{R}^K$ so that for all $m, n \in \{1, \ldots, N\}$ the following bound holds:

$$(1 - \varepsilon) \cdot \|x_m - x_n\|^2 \leq \|f(x_m) - f(x_n)\|^2 \leq (1 + \varepsilon) \cdot \|x_m - x_n\|^2$$

For small values of $\varepsilon$, the pairwise distances between data points are nearly preserved under the map $f$. A less strict bound for the target dimension, but perhaps easier to remember, is $K \geq 20 \cdot \varepsilon^{-2} \cdot \ln N$ if $0 < \varepsilon < 0.9$ (cf. [2, Lemma 15.4]).

We roughly sketch the idea of the proof. First, we randomly choose a $K$-dimensional subspace of $\mathbb{R}^D$. If proj($\cdot$) denotes the orthogonal projection onto that subspace, it can be shown that the map $f(v) := \sqrt{D/K} \cdot \mathrm{proj}(v)$, $v \in \mathbb{R}^D$, satisfies the desired bound with a probability of *at least* $1/N$. In particular, the probability is strictly greater than zero—thus, there must exist a random projection that satisfies the desired bound with certainty.

The estimated target dimension $K$ does not depend on the original dimensionality $D$, and it depends only logarithmically on the number of data points $N$. We illustrate this fact with an example: We are given a dataset of $N = 10,000$ digital color photographs with a resolution of $1920 \times 1080$ pixels. Given three color channels, $D = 1920 \cdot 1080 \cdot 3 = 6{,}220{,}800$ color values must be stored for each uncompressed photo: each image can be represented as a vector in a space with a dimension of about 6 million.

However, the lemma of Johnson and Lindenstrauss implies that the **intrinsic dimensionality** is much lower: there exists a projection $f\colon \mathbb{R}^D \to \mathbb{R}^K$ with $K = 7895$ such that the squared distances between the projected data points do not differ by more than $\varepsilon = 10\%$ from the squared distances in the original data. This potential reduction in dimensionality changes very little with the size of the dataset. For example, even a dataset of $N = 1{,}000{,}000$ photographs has an intrinsic dimensionality of at most $K \approx 12{,}000$.

Note that we assumed nothing about the distribution of the original data points: the contents of the photos could be arbitrary, or consist only of image noise. Furthermore, the theorem is only considered with the *linear* transformation of the data. Therefore, in practice, we can expect the intrinsic dimensionality of any given data to be significantly smaller than what is suggested by the Johnson–Lindenstrauss bound.

The construction of a low-dimensional representation of data is the task of **dimensionality reduction** algorithms, the result of such a reduction is sometimes referred to as an **embedding**. Embeddings commonly used in image analysis [3] or text analysis [4] may have a dimensionality of only a few hundred components.

### 7.1.2 Topological characteristics of data

According to the Johnson–Lindenstrauss lemma, data points of dimensionality $D$ lie near a $K$-dimensional linear subspace. In practice, it often turns out that the intrinsic dimensionality $K$ can be assumed to be much smaller than the extrinsic dimensionality $D$.

This insight is not limited to linear subspaces: it can be extended to more general curves, surfaces, or their higher-dimensional analogues. For example, the following scatter plots show two (synthetic) datasets, both distributed around a closed curve:

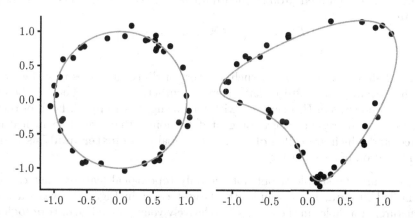

**Fig. 7.1.** *Data points along a curve*

Geometrically, both curves are quite different: the left one is a circle with constant radius, the right curve has variable curvature. However, the qualitative shape is the same: both are continuous, closed curves that we can imagine could be transformed into each other by continuous deformation. **Topology** is the mathematical discipline devoted to the study of qualitative shape, jokingly called "rubber-sheet geometry."

An important topological property is *connectedness*, the possibility to decompose a set into connected parts.

> Let $U$ be a subset of $\mathbb{R}^D$. A set $V \subseteq U$ is called a **path-component** of $U$ if the following holds:
>
> 1. For all $x, y \in V$, there exists a continuous curve that lies entirely in $V$ and connects $x$ to $y$.
>
> 2. There is no set $W$ different from $V$ with $V \subset W \subseteq U$ that also has the above property.

It can be proved that instead of arbitrary continuous curves it is sufficient to consider polygonal chains. The path-components form a partition of $U$, i.e., they are pairwise disjoint and cover $U$. The sets studied in data analysis usually have only finitely many path-connected components $S_1, \ldots, S_K$:

$$U = S_1 \cup \cdots \cup S_K, \ S_k \cap S_l = \emptyset$$

for all $k, l \in \{1, \ldots, K\}$.

A finite set of $N$ data points $U_0 = \{x_1, \ldots, x_N\} \subset \mathbb{R}^D$ has always exactly $N$ path-components: $S_n = \{x_n\}$, $n \in \{1, \ldots, N\}$. Indeed, any path between two different data points necessarily leads out of the set. Thus, the notion does not seem to represent any particular advance in the analysis of the topological structure of the data. Consider instead the following cover of $U_0$ by balls of radius $\varepsilon > 0$, centered around each data point: $U_\varepsilon = B_\varepsilon(x_1) \cup \cdots \cup B_\varepsilon(x_N)$ where

$$B_\varepsilon(x_n) = \{u \in \mathbb{R}^D \,|\, \|u - x_n\| \leq \varepsilon\}$$

for all $n \in \{1, \ldots, N\}$.

Fig. 7.2 shows an example of such a cover for different values of $\varepsilon$ (for $D = 2$, the covering balls are disks) given a synthetic dataset. The number of path-components of $U_\varepsilon$ decreases with increasing disk radius $\varepsilon$. Each of these components represents a grouping of data points that are close in feature space, and which are called **clusters**. The goal of a **cluster analysis** is the identification of such clusters.

Clusters are the simplest, but not the only topological features that can be identified in data. For example, the largest cover in Fig. 7.2 reveals a ring structure, or a "hole" in the data. The relatively young discipline of **topological data analysis** is concerned with the identification of such structures and their higher-dimensional analogues [5, 6, 7, 8].

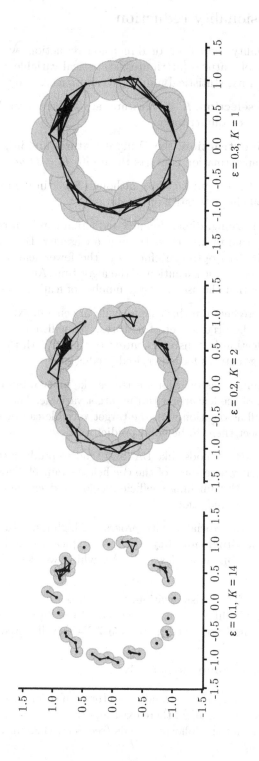

**Fig. 7.2.** *Covering data points by disks with varying radius ε and cluster count K*

## 7.2 Dimensionality reduction

By dimensionality reduction, or dimension reduction, we mean a reduction in the number of features (attributes, statistical variables) that describe each observation (entity, statistical unit). Two approaches can be distinguished:

- In **feature selection**, $K < D$ features are selected from the $D$ input features according to certain criteria.

- In **feature extraction**, the $D$ input features are, in general, modified: a suitable transformation converts them into $K < D$ *new* output features.

In both cases, the output features, although fewer in number, are intended to reflect essential characteristics of the dataset.

In supervised learning, dimensionality reduction can be a processing step that precedes the actual task, such as training a classifier. Dimensionality reduction may be required to optimize efficiency: the fewer features that have to be processed, the faster the execution of the algorithm. Another important factor is the observation that the use of a large number of features can lead to overfitting.

The methods presented in this chapter will, in general, extract features that are different from the original set of features. We will not elaborate on methods for feature selection but instead content ourselves with the following remarks. When selecting features for supervised prediction:

- We can compute association measures like mutual information or Pearson correlation of the features with the target variable. Those features that show only a small association with the target variable can be discarded, since we do not expect them to be good predictors.

- $L_1$-regularized methods (like LASSO) tend to produce sparse solutions, i.e., linear models where some of the coefficients vanish. Those features that are associated with vanishing coefficients can be discarded since they have no impact on the prediction.

Typical challenges associated with processing high-dimensional data are known as the **curse of dimensionality**. One prototypical method to "break the curse" is *principal component analysis (PCA)*, which we describe in the following section.

In the context of data visualization, the goal of a dimension reduction is to be able to visually highlight essential aspects of the high-dimensional input data. The $t$-SNE method presented in Section 7.2.4 is well-suited for this purpose.

### 7.2.1 Principal component analysis

Let $X = (X_1, \ldots, X_D)^T$ be a random vector with covariance matrix $\Sigma[X]$. Since $\Sigma[X]$ is a symmetric matrix, there exist an orthogonal matrix $V$ and a diagonal matrix $\Lambda$ such that the following holds (see Sect. B.2.4 in the appendix):

$$\Sigma[X] = V \cdot \Lambda \cdot V^T$$

The columns of $V$ are the normalized and pairwise orthogonal eigenvectors $v_1, \ldots, v_D \in \mathbb{R}^D$ of $\Sigma[X]$. The diagonal of $\Lambda$ contains the corresponding eigenvalues $\lambda_1, \ldots, \lambda_D$. Since $\Sigma[X]$ is also positive semidefinite, none of these eigenvalues are negative.

If we define the new random vector $Z := V^T \cdot (X - E[X])$, we have $E[Z] = 0$, and moreover:

$$\Sigma[Z] = E[Z \cdot Z^T] = V^T E[(X - E[X]) \cdot (X - E[X])^T]V = V^T \Sigma[X]V = \Lambda$$

The covariance matrix of $Z$ is diagonal: the above transformation turns any random vector $X$ into a new random vector $Z$ that has zero mean and uncorrelated components.

A corresponding result holds for sequences of data points/feature vectors $x_1, \ldots, x_N \in \mathbb{R}^D$. Since the sample covariance matrix $S(x)$ is also symmetric and positive semidefinite, we can repeat the above arguments and start by writing down the same representation:

$$S(x) = V \cdot \Lambda \cdot V^T$$

We assume that the corresponding data matrix $\boldsymbol{X}$ is mean-centered, and therefore write $S(x) = \frac{1}{N}\boldsymbol{X}^T\boldsymbol{X}$ for the covariance matrix. The data matrix of the transformed sample $z_1 = V^T x_1, \ldots, z_N = V^T x_N$ is given by $\boldsymbol{Z} = \boldsymbol{X} \cdot V$, and diagonal:

$$NS(z) = \boldsymbol{Z}^T\boldsymbol{Z} = (\boldsymbol{X}V)^T\boldsymbol{X}V = V^T\boldsymbol{X}^T\boldsymbol{X}V = NV^T S(x)V = N\Lambda$$

Let us compare the data transformation methods that we have learned about so far:

- Mean-centering shifts the data points so that their centroid coincides with the origin of feature space.

- Standardization—i.e., computing $z$-scores, Sect. 5.1—mean-centers the data and scales them so that the variance of every feature is identical to one. (The new features may still be correlated.)

- **Principal component analysis**, as explained above, mean-centers the data and linearly transforms them so that the features have vanishing covariance/correlation.

All of these transformations are affine maps in feature space: a translation of the centroid onto the origin, followed by a linear map. In the case of standardization, that linear map is a scaling transformation (in general, non-uniform). In the case of principal component analysis, it is represented by the matrix $V^T$.

Let us assume that the eigenvalues of the sample covariance matrix have been sorted in descending order: $\lambda_1 \geq \lambda_2 \geq \cdots \geq \lambda_D \geq 0$. Then, the sequence of observations $Z_{\bullet 1}$ has the largest variance $\sigma_1^2 = \lambda_1$, while $Z_{\bullet D}$ has the smallest variance. The basic idea of dimension reduction via principal component analysis is to neglect directions of small variance.

> **Principal component analysis.** Let data points $x_1, \ldots, x_N \in \mathbb{R}^D$ with centroid $\bar{x}$ be given. A **Karhunen–Loève transform** with target dimensionality $1 \leq K \leq D$ is the affine map
>
> $$\mathrm{pca}_K : \mathbb{R}^D \to \mathbb{R}^K, \; \mathrm{pca}_K(u) = (V_K)^T \cdot (u - \bar{x}).$$
>
> The columns of the $D \times K$ matrix $V_K$ are the first $K$ of the normalized eigenvectors of the covariance matrix $S(x)$ that correspond to the $K$ largest eigenvalues.

Conventions for naming the objects involved vary[1]; we will use the following. The eigenvectors $v_1, \ldots, v_K, \ldots, v_D$ of the covariance matrix, sorted in descending order by the size of the associated variance, are called the **principal directions**. The straight line passing through the centroid, spanned by the $k$-th principal direction $v_k$, has the parameter equation

$$\mathbb{R} \to \mathbb{R}^D, \; \lambda \mapsto \bar{x} + \lambda \cdot v_k$$

and represents a **principal axis**. Any two different principal axes are perpendicular. If we project any vector $u \in \mathbb{R}^D$ onto the $k$-th principal axis, the oriented distance of the projection from the centroid is given by $\langle v_k, u - \bar{x} \rangle$: the $k$-th **principal component**, or **principal coordinate**. The first $K$ principal components form the Karhunen–Loève transformed vector:

$$\mathrm{pca}_K : \mathbb{R}^D \to \mathbb{R}^K, \; \mathrm{pca}_K : u \mapsto \begin{pmatrix} \langle v_1, u - \bar{x} \rangle \\ \vdots \\ \langle v_K, u - \bar{x} \rangle \end{pmatrix}$$

In summary, the principal components are the coordinates—with respect to the basis of principal directions—of the projection onto the $K$-dimensional affine subspace spanned by the first $K$ principal axes.

Suppose that the input vector was drawn from a distribution similar to the distribution of the data that the principal component analysis was performed on. Then, the projected vector can be understood as an approximation of the input vector in the sense that it is very likely that it has a small distance to the input vector. That is because, along directions perpendicular to the first principal axes, the variance of the distribution is small. Therefore, it is very likely—we remind ourselves of Chebyshev's inequality, Sect. 3.4.3—that the components in these directions have a small magnitude and can thus be neglected.

---

[1] In particular, the term "principal component" is used very interchangeably for principal axes, directions or coordinates.

Consequently, we can also expect to be able to *reconstruct* the vector approximately from its principal components: we have $u \approx ((\text{pca}_K)^\dagger \circ \text{pca}_K)(u)$ with

$$(\text{pca}_K)^\dagger : \mathbb{R}^K \to \mathbb{R}^D, \ (\text{pca}_K)^\dagger : y \mapsto \bar{x} + \sum_{k=1}^{K} y_k \cdot v_k$$

**Example.** The MNIST dataset [9] consists of digital images of handwritten digits with a resolution of 28 × 28 pixels, see Fig. 8.6.

First, we flatten each image, i.e., we represent it as a vector of gray values with length $D = 28 \cdot 28 = 784$. We can use PCA to reduce the dimensionality of the dataset to $K = 2$: Fig. 7.5 on top shows the result for a small selection of digits, each placed at the position of their respective first two principal coordinates. A certain tendency can be seen with which the same digits are grouped together: this is a first indication that we can identify a handwritten digit from just a few principal components.

Fig. 7.5 on the bottom shows the (approximately) inverse transformation: any linear combination of the first two principal directions can be represented as an image. These artificially generated images show some similarity to the digits zero, one, and nine—but it appears that two principal directions do not span a region of feature space large enough to cover all the digits.

However, twenty principal components can already be used for the reconstruction of acceptable quality. The first twenty eigenvectors of the covariance matrix, brought back to image format, look like this:

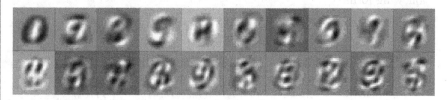

**Fig. 7.3.** *"Eigendigits" of the MNIST dataset*

These images can be seen as a generating set for the MNIST data, and each digit can be represented approximately as a linear combination of these images:

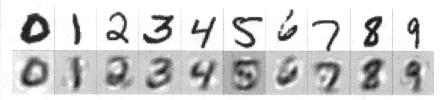

**Fig. 7.4.** *Image reconstruction from principal components*

The figure above shows an example of each digit (top row) and the reconstruction from their first twenty principal components.

## 7.2.2 Autoencoders

Artificial neural networks can also be used for unsupervised learning. **Autoencoders** are based on two essential ideas or assumptions:

- The activations of hidden neurons reflect essential characteristics of the relationship between input and output data.

- We can apply a neural network to any dataset—even when it is unlabeled—by assigning a training example to itself as the output. As a consequence, the network learns an approximation of the identity map $\mathbb{R}^D \to \mathbb{R}^D$, $u \mapsto u$.

An **autoencoder** is an artificial neural network with at least one hidden layer, where the input and output layers have the same width:

$$f \colon \mathbb{R}^D \to \mathbb{R}^D, \ f(u) = (f_L \circ f_{L-1} \circ \cdots \circ f_1)(u), \ L \geq 2$$

One of the hidden layers of the autoencoder—say $f_m \colon \mathbb{R}^{D_{m-1}} \to \mathbb{R}^{D_m}$, $1 \leq m < L$—is singled out as the **latent layer**, and we denote its width by $K := D_m$.

Given a dataset $x_1, \ldots, x_N \in \mathbb{R}^D$, the autoencoder is trained with identical input and output activations—the empirical risk to be minimized has the following form:

$$\hat{R}[f] = \frac{1}{N} \sum_{n=1}^{N} \lambda(x_n, f(x_n))$$

Once the weights and biases have been learned, the maps

$$\text{code} \colon \mathbb{R}^D \to \mathbb{R}^K, \ \text{code}(u) = (\hat{f}_m \circ \hat{f}_{m-1} \circ \cdots \circ \hat{f}_1)(u)$$

and

$$\text{code}^\dagger \colon \mathbb{R}^K \to \mathbb{R}^D, \ \text{code}^\dagger(y) = (\hat{f}_L \circ \hat{f}_{L-1} \circ \cdots \circ \hat{f}_{m+1})(y)$$

represent the **encoding map** and **decoding map**, respectively.

In other words, we represent the data as the activations of the latent layer. The space of possible activations—i.e., the range (domain) of the encoding (decoding) map—is also called **latent space**. The dimensionality of the latent space is given by the width of the latent layer. For the purpose of dimensionality reduction, it is typically chosen to be (significantly) less than the dimensionality of the input data: $K \ll D$.

**Example.** Fig. 7.6 on top shows a scatter plot of an autoencoder's encoding of digits selected from the MNIST dataset; the entire dataset was used for training. The autoencoder itself is an ordinary feedforward network with five hidden layers and the following number of neurons in the hidden layers: 32, 64, 2, 64, 32. The middle layer with two neurons represents the latent layer. Leaky rectifiers were used as activation functions. The training error based on the quadratic loss was minimized via stochastic gradient descent, using the open-source software library Keras [10, 11].

The figure below the scatter plot is a representation of the *decoding* map. When we compare this representation with the results of a principal component analysis (Fig. 7.5 bottom), we can see that the neural network is apparently much better at compressing the entire feature space to a low-dimensional representation with a relatively small loss: we can see the neural network interpolate between a wider variety of digits.

### 7.2.3 Multidimensional scaling

Suppose we are given data points $x_1, \ldots, x_N$ and a symmetric premetric $\delta(\cdot, \cdot)$ that we may apply to those data points to measure their distance/similarity. These can be points in a high-dimensional Euclidean space or other types of data, such as lists of binary features that were compared using the Jaccard distance. We want to map these data points onto a configuration of target points $y_1, \ldots, y_N \in \mathbb{R}^K$ in a Euclidean space of prescribed dimension $K$, usually chosen to be comparatively low. This map should have pairwise distances between points that change as little as possible:

$$\Delta_{mn} = \delta(x_m, x_n) \approx \|y_m - y_n\|$$

We can find such a configuration by minimizing a suitable choice of objective function, also called a **stress function** in this context.

Let $\Delta$ be a distance matrix of format $N \times N$.

**Metric multidimensional scaling** is based on minimizing the following stress function:

$$R_{\mathrm{mMDS}}(y_1, \ldots, y_N) = \sum_{k=1}^{N} \sum_{l=1}^{N} (\Delta_{kl} - \|y_k - y_l\|)^2$$

**Sammon projection** is based on minimizing the following stress function:

$$R_{\mathrm{Samm}}(y_1, \ldots, y_N) = \sum_{k=1}^{N} \sum_{l=1}^{N} \frac{(\Delta_{kl} - \|y_k - y_l\|)^2}{\Delta_{kl}}$$

Here, undefined summands with vanishing denominators are set equal to zero.

Metric multidimensional scaling minimizes a quadratic loss with the goal of preserving pairwise distances between data points.

The Sammon variant differs from metric multidimensional scaling by the addition of weights $1/\Delta_{kl}$: data points located close to each other are given a greater weight than those located further apart. That way, local neighborhood relationships are given more weight to be preserved. This rationale is also behind the $t$-SNE method discussed in the next section.

The following normalized stress functions are also commonly found in the literature, and they deliver the same results as the respective stress functions listed above:

$$S_{\mathrm{mMDS}}(y_1, \ldots, y_N) = \left( \sum_{k=1}^{N} \sum_{l=1}^{N} (\Delta_{kl})^2 \right)^{-1} \cdot \sum_{k=1}^{N} \sum_{l=1}^{N} (\Delta_{kl} - \|y_k - y_l\|)^2$$

$$S_{\mathrm{Samm}}(y_1, \ldots, y_N) = \left( \sum_{k=1}^{N} \sum_{l=1}^{N} \Delta_{kl} \right)^{-1} \cdot \sum_{k=1}^{N} \sum_{l=1}^{N} \frac{(\Delta_{kl} - \|y_k - y_l\|)^2}{\Delta_{kl}}$$

**Example.** Fig. 7.7 shows the result of a Sammon projection applied to a selection of images from the MNIST handwritten digit dataset. The stress was minimized using the BFGS algorithm sketched in Sect. 6.1.4.

### 7.2.4 $t$-distributed stochastic neighbor embedding ($t$-SNE)

Like in the last section, suppose we are given data points $x_1, \ldots, x_N$ and a symmetric premetric $\delta(\cdot, \cdot)$, which we can use to determine the distances between data points. Again, we want to map these data points onto a new set of data points $y_1, \ldots, y_N \in \mathbb{R}^K$ with target dimensionality $K$. This map should preserve distances between *neighboring* data points: in a suitable sense, $\Delta_{mn} = \delta(x_m, x_n) \approx \|y_m - y_n\|$ shall hold for *small* $\Delta_{mn}$.

The basic idea of **stochastic neighbor embedding (SNE)** is the following. First, we imagine a **random walk** on the set of data points, i.e., a succession of random steps where each step connects two data points. We want to impose the following conditions: If the walker is at position $x_m$, at the following step, they move to the position $x_n$ with the *transition probability* $p(n|m)$. This probability does not depend on the walker's past trajectory, and it depends only on the distance $\Delta_{mn}$ between the two data points. We posit that the smaller the distance between $x_m$ and $x_n$, the larger the probability of transitioning between those two points.

For all $m \in \{1, \ldots, N\}$, $\sum_{n=1}^{N} p(n|m) = 1$ holds: this is a family of probability mass functions reflecting the local intrinsic geometry. The idea is to produce a distribution of data points in the target space that is similar to the original distribution, thus exhibiting a similar local geometry.

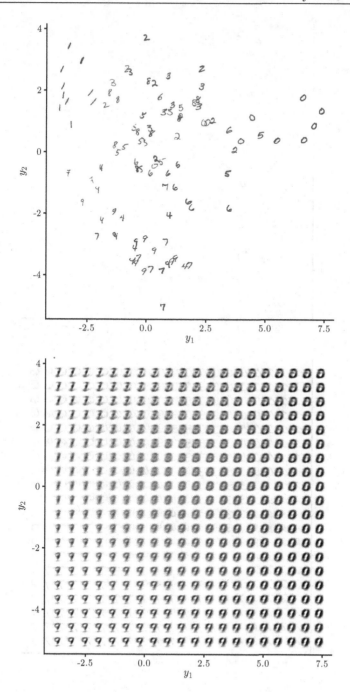

**Fig. 7.5.** *Scatter plot of the first two principal coordinates of a selection of MNIST images (top); reconstruction of linear combinations of the first two principal directions (bottom)*

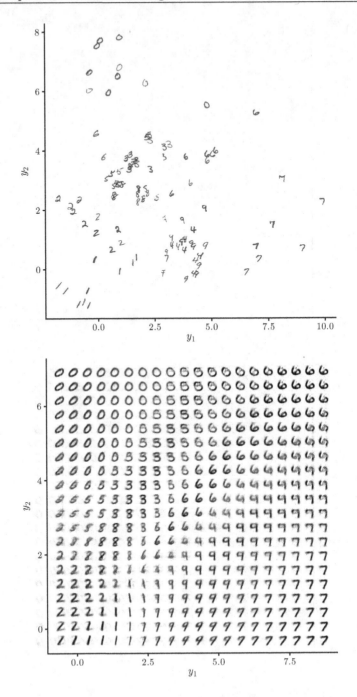

**Fig. 7.6.** *Result of encoding a selection of MNIST images with an autoencoder (top); decoding points in latent space (bottom)*

We write down the following more concrete models, where the bandwidths $\sigma = (\sigma_1, \ldots, \sigma_N)$ are parameters yet to be determined:

$$p_{x,\sigma}(n|m) = e^{-\frac{1}{2}\left(\frac{\Delta_{mn}}{\sigma_m}\right)^2} \cdot \left(-1 + \sum_{k=1}^{N} e^{-\frac{1}{2}\left(\frac{\Delta_{mk}}{\sigma_m}\right)^2}\right)^{-1}$$

and

$$q_y(m, n) = \frac{1}{1 + \|y_m - y_n\|^2} \cdot \left(-N + \sum_{k=1}^{N}\sum_{l=1}^{N} \frac{1}{1 + \|y_k - y_l\|^2}\right)^{-1}$$

if $m, n \in \{1, \ldots, N\}$ and $m \neq n$, otherwise $p_{x,\sigma}(n|m) = q_y(m, n) = 0$.

Thus, for the original distribution, a form of kernel density estimator with a Gaussian kernel is used, while for the target distribution a Cauchy distribution serves as the kernel. The originally proposed SNE method uses a Gaussian kernel to model the target distribution as well [12]. However, using the fat-tailed Cauchy distribution instead avoids the "crowding problem" that prevents natural clusters from being clearly separated among the mapped data points [13, Sect. 3.2]. For positive arguments, the Cauchy distribution $\mathcal{L}(u|0,1) \propto (1+u^2)^{-1}$ agrees with the $t$-distribution with one degree of freedom, hence the name of the method described below.

Let $e^H > 0$ be a fixed parameter, the **perplexity**. With the above definitions, **$t$-distributed stochastic neighborhood embedding**, or **$t$-SNE** for short, is a dimensionality reduction algorithm that consists of the following steps (cf. [13]):

1. The bandwidths $\sigma_1, \ldots, \sigma_N$ are determined such that for all $m \in \{1, \ldots, N\}$, the distribution of transition probabilities matches the prescribed perplexity:

$$-\sum_{n=1}^{N} p_{x,\sigma_m}(n|m) \cdot \ln\left(p_x(n|m)\right) = H$$

2. A symmetric distribution is determined as follows:

$$p_x(n, m) = \frac{1}{2N}\left(p_{x,\sigma_m}(n|m) + p_{x,\sigma_n}(m|n)\right)$$

3. The target data points are determined by minimizing the following objective function, the **cross-entropy** between the distributions $p_x(\cdot, \cdot)$ and $q_y(\cdot, \cdot)$:

$$R(y_1, \ldots, y_N) = -\sum_{k=1}^{N}\sum_{l=1}^{N} p_x(k, l) \ln\left(q_y(k, l)\right)$$

where agree on $0 \cdot \ln 0 = 0$, as usual.

The perplexity can be interpreted as the effective number of neighbors of a data point—a typical range to choose from is given by $5 \leq e^H \leq 50$. The primary application of the $t$-SNE method is data visualization: drawing a low-dimensional scatter plot of high-dimensional data. Thus, $K = 2$ or $K = 3$ usually applies.

The numerical implementation is facilitated by the fact that the gradient of the above objective function can be calculated explicitly. For all $n \in \{1, \ldots, N\}$:

$$\nabla_{y_n} R(y_1, \ldots, y_N) = 4 \sum_{k=1}^{N} \frac{p_x(n, k) - q_y(n, k)}{1 + \|y_n - y_k\|^2} \cdot (y_n - y_k)$$

UMAP[2] is an algorithm that is similar to $t$-SNE but has been developed more recently [14].

**Example.** Fig. 7.7 (bottom) shows the result of applying $t$-SNE with a perplexity of $e^H = 12$ to a selection of images from the MNIST dataset.

## 7.3 Cluster analysis

Suppose that we are given a sequence or set of observations/data points $x_1, \ldots, x_N$ supplied with some notion of similarity or distance. In its most basic form, the goal of a **cluster analysis** is to partition those data into subsets $S_1, \ldots, S_K \subseteq \{x_1, \ldots, x_N\}$, the **clusters**, where each observation is contained in a single cluster. We may label each data point, $f \colon \{x_1, \ldots, x_N\} \to \{1, \ldots, K\}$, in order to indicate **cluster membership**.

There are variants of clustering algorithms where the stated goal is to assign each data point to *at most* one cluster: some data points may be considered outliers and not assigned to any cluster. DBSCAN [15] and HDBSCAN [16] are examples[3] of such algorithms.

There are also **fuzzy clustering** algorithms that may assign a single observation to several clusters [17]. In that case, the assignment comes with a weighting value that reflects the degree of membership; the clusters are interpreted as **fuzzy sets**.

Cluster analysis may or may not learn a decision rule[4] that would allow for determining cluster membership of yet unseen data points, data points which were not included in the original dataset.

---

[2] UMAP stands for Uniform Manifold Approximation and Projection.

[3] (H)DBSCAN stands for (Hierarchical) Density-Based Spatial Clustering of Applications with Noise.

[4] If the clustering algorithm does not provide a decision rule out-of-the-box, there is always the option to learn it from the cluster labels after the fact via a supervised algorithm.

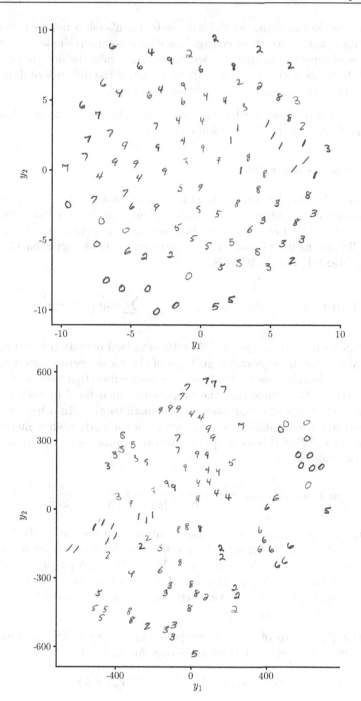

**Fig. 7.7.** *Sammon projection (top) and t-distributed stochastic neighbor embedding (bottom)*

Whatever the details on the definition of cluster membership are, the partitioning of the data is not arbitrary, of course. The goal is to group similar observations together and separate dissimilar observations. As a rule, the data points within a single cluster should have a small distance, while the distance of data points that belong to two different clusters should be large.

In the following sections, we describe two commonly used clustering algorithms: $K$-means clustering and hierarchical clustering.

### 7.3.1 $K$-means algorithm

The $K$-means algorithm can be used to cluster data points $x_1, \ldots, x_N \in \mathbb{R}^D$, where points are assumed to be similar when they have a small (squared) Euclidean distance. Let us assume that our data are generated by a member of the following family of probability density functions, a particular **Gaussian mixture model** $p_{\text{GMM}} \colon \mathbb{R}^D \to \mathbb{R}$:

$$
p_{\text{GMM}}(u|\mu_1, \ldots, \mu_K, h^2) = \frac{1}{K} \cdot \frac{1}{(2\pi)^{\frac{D}{2}}} \sum_{k=1}^{K} \exp\left(-\frac{\|u - \mu_k\|^2}{2h^2}\right)
$$

with mean vectors $\mu_1, \ldots, \mu_K \in \mathbb{R}^D$ and a standard deviation $h > 0$ assumed to be equal for each subpopulation. Each of the mean vectors determines the centroid of an isotropic multivariate normal distribution that produces a cluster of data points. We assume that the subpopulations defined by each Gaussian distribution in the above sum have only a small overlap. In other words, the cluster centroids are sufficiently far apart from each other when measured in multiples of $h$. With this assumption, we can approximate the total density function as follows:

$$
p_{\text{GMM}}(u|\mu_1, \ldots, \mu_K, h^2) \approx \frac{1}{K} \cdot \frac{1}{(2\pi)^{\frac{D}{2}}} \exp\left(-\frac{\|u - \mu_{f(u)}\|^2}{2h^2}\right)
$$

where $\mu_{f(u)} \in \{\mu_1, \ldots, \mu_k\}$ is the cluster centroid that has the smallest distance to $u \in \mathbb{R}^D$. We assume that only this cluster provides a significant contribution to the density at $u$. The decision rule $f \colon \mathbb{R}^D \to \{1, \ldots, K\}$ assigns each point to its cluster label. With the exception of decision boundaries that have equal distance to multiple centroids, this decision rule is determined by the cluster centroids: $f(\,\cdot\,) = f(\,\cdot\,; \mu_1, \ldots, \mu_K)$.

A partition $S_1, \ldots, S_K$ of a set of data points $x_1, \ldots, x_N \in \mathbb{R}^D$ based on such a decision rule can be characterized as follows, for all $k \in \{1, \ldots, K\}$:

$$
S_k = \{x_n | n \in \{1, \ldots, N\}, \ f(x_n) = k\}
$$

Alternatively, we can write down groups $I_1, \ldots, I_K$ of indices:

$$
I_k = \{n \in \{1, \ldots, N\} | f(x_n) = k\} = \{n \in \{1, \ldots, N\} | x_n \in S_k\}
$$

With those simplifying assumptions, the log-likelihood function computed from plugging the data into the Gaussian mixture model takes the following form:

$$\ell(\mu_1, \ldots, \mu_K, h^2) = \sum_{n=1}^{N} \ln\left(p_{\text{GMM}}(x_n | \mu_1, \ldots, \mu_K, h)\right)$$

$$\approx -\frac{1}{2h^2} \cdot \sum_{n=1}^{N} \|x_n - \mu_{f(x_n; \mu_1, \ldots, \mu_K)}\|^2$$

$$- N \cdot \ln(K) - \frac{ND}{2} \ln(2\pi)$$

Maximizing this function is equivalent to the minimization of the sum of squared distances to the respective cluster centroids (the bandwidth $h$ plays no role anymore):

$$R = \sum_{n=1}^{N} \|x_n - \mu_{f(x_n; \mu_1, \ldots, \mu_K)}\|^2 = \sum_{k=1}^{K} \sum_{n \in I_k} \|x_n - \mu_k\|^2$$

In simpler words, we minimize the squared distances of the data points to the centroid of the cluster they belong to.

Each term of the inner sum is nonnegative, and it is minimized by setting each $\mu_k$ to be the empirical centroid of the $k$-th cluster. This argument leads to the following criterion.

Let $x_1, \ldots, x_N \in \mathbb{R}^D$ be data points. The **K-means algorithm** for cluster analysis minimizes the following objective function over the possible assignments of cluster labels $I_1, \ldots, I_K$:

$$R(I_1, \ldots, I_K) = \sum_{k=1}^{K} \sum_{n \in I_k} \|x_n - \mu_k\|^2 \quad \text{with } \mu_k = \frac{1}{|I_k|} \sum_{n \in I_k} x_n$$

**Lloyd's algorithm** can find a local minimum of the objective function:

initialize cluster centroids $\mu_1, \ldots, \mu_K$
**while** *the value of R changes with each iteration* **do**
    **for** *i = 1 to N* **do**
    |  $f(x_i) :=$ index of nearest centroid
    **end**
    **for** *j = 1 to K* **do**
    |  $\mu_j :=$ centroid of the $j$-th cluster
    **end**
    update: $R = \sum_{n=1}^{N} \|x_n - \mu_{f(x_n)}\|^2$
**end**
output:
cluster centroids $\mu_1, \ldots, \mu_K$
cluster memberships $f(x_1), \ldots, f(x_N)$

The centroids can be initialized, for example, by randomly selecting $K$ of the data points to be clustered: $\mu_1 = x_{\iota(1)}, \ldots, \mu_K = x_{\iota(K)}$. Lloyd's algorithm always finds a *local* minimum—however, an unfavorable initialization may lead to a result of low quality that is far away from the global minimum. Therefore, it is recommended to run the algorithm several times with different initial centroids and to compare the results, e.g., using the final goodness-of-fit $R$.

Fig. 7.8 shows the application of the method to a synthetic dataset generated by a Gaussian mixture model with three centroids. The quality of the result depends crucially on the hyperparameter $K$, i.e., the number of clusters that the algorithm is set out to determine.

**$K$-medoids algorithms** are similar to $K$-means [18]. As the name suggests, these algorithms replace cluster centroids with their medoids. This replacement has the advantage that more general distance measures can be used.

### 7.3.1.1 Kernel $K$-means algorithm

In deriving the $K$-means algorithm, we made some rather restrictive assumptions: the clusters have equal size, and the data points in each cluster distribute isotropically around its centroid. Any two clusters can be linearly separated. If these assumptions are not justified, the cluster analysis may yield an inferior result. One such result can be seen in Fig. 7.9 above: the blocks of feature space produced by $K$-means partitioning[5] can not cover the oddly shaped clusters of the "smiley."

The limitations of the ordinary $K$-means method can be countered by using the kernel trick that we had already applied to logistic regression in Sect. 6.3.1. We assume that there exists a feature map $\Phi \colon \mathbb{R}^D \to \mathcal{H}$, where the dimension of the target space $\mathcal{H}$ may be much larger than $D$ or even infinite. The scalar product $\langle \cdot, \cdot \rangle$ in the target space will be induced by a kernel $\sigma \colon \mathbb{R}^D \times \mathbb{R}^D \to \mathbb{R}$, so that for all $u, v \in \mathbb{R}^D$ the following holds:

$$\langle \Phi(u), \Phi(v) \rangle = \sigma(u, v)$$

For example, we can use the popular Gaussian kernel once more: $\sigma_h(u, v) = \exp(-\frac{1}{2} h^{-2} \|u - v\|^2)$.

Furthermore, we write for the cluster centroids under the feature map:

$$\Phi^* \mu_k := \frac{1}{|I_k|} \sum_{n \in I_k} \Phi(x_n)$$

For the squared distance of a transformed data point $u \in \mathbb{R}^D$ to the $k$-th cluster centroid, we get:

---

[5] The kind of partition that ordinary $K$-means is able to produce is called a *Voronoi partition*.

$$r_k^2(u) = \|\Phi(u) - \Phi^*\mu_k\|^2$$
$$= \langle\Phi(u), \Phi(u)\rangle - 2\langle\Phi(u), \Phi^*\mu_k\rangle + \langle\Phi^*\mu_k, \Phi^*\mu_k\rangle$$
$$= \sigma(u, u) - \frac{2}{|I_k|}\sum_{m\in I_k}\sigma(u, x_m) + \frac{1}{|I_k|^2}\sum_{m\in I_k}\sum_{n\in I_k}\sigma(x_m, x_n)$$

Let $x_1, \ldots, x_N \in \mathbb{R}^D$ be data points. The **kernel $K$-Means algorithm** consists of minimizing the following objective function over the possible occupancies of cluster labels $I_1, \ldots, I_K$:

$$R(I_1, \ldots, I_K) = \sum_{k=1}^{K}\sum_{n\in I_k} r_k^2(x_n),$$

where $r_k^2(x_n)$ is calculated as shown in the derivation above.

The Lloyd algorithm can be adapted to the new paradigm as follows:

initialize cluster labels for $x_1, \ldots, x_N$
**while** *the value of $R$ changes with each iteration* **do**
 **for** *$i = 1$ to $N$* **do**
  | new cluster label of $x_i :=$ index $k$ with smallest value for $r_k^2(x_i)$
 **end**
 update: $R = \sum_{k=1}^{K}\sum_{n\in I_k} r_k^2(x_n)$
**end**
output: cluster labels for $x_1, \ldots, x_N$

In Fig. 7.9, the result of a kernel $K$-means cluster analysis is shown below ($K = 4$, Gaussian kernel with $h = 0.5$, cluster labels initialized randomly).

### 7.3.2 Hierarchical cluster analysis

The basic idea behind **agglomerative hierarchical cluster analysis** is the following. We start the iteration with the finest possible partition of the dataset: each cluster contains a single data point. Then, with each subsequent step, the clusters that have minimum distance to each other are merged. The new partition constructed in this way consists of fewer clusters, so it is coarser. The procedure terminates once the partition consists of a single cluster that contains all of the data points.

In this way, we have constructed not one clustering but a whole hierarchy of partitions of the data that we can chose from.

Let $x_1, \ldots, x_N$ be the data points that we wish to group into clusters.

Suppose that we are given a distance measure $\delta(\cdot, \cdot)$ to compare those data points with. In order to determine the distance $D(A, B)$ between two arbitrary clusters of data points $A$ and $B$, we can choose from one of the following alternatives:

$$D_{\min}(A, B) = \min_{u \in A, v \in B} \{\delta(u, v)\},$$

$$D_{\text{avg}}(A, B) = \langle \delta(u, v) \rangle_{u \in A, v \in B},$$

$$D_{\max}(A, B) = \max_{u \in A, v \in B} \{\delta(u, v)\}$$

An **agglomerative hierarchical cluster analysis** generates a hierarchy of partitions of the dataset:

> initialize $t := 0$, $S^{(0)} := \left\{ S_1^{(0)}, \dots, S_N^{(0)} \right\} = \{\{x_1\}, \dots, \{x_N\}\}$
> **while** $|S^{(t)}| > 1$ **do**
> > update $t \leftarrow t + 1$, initialize $S^{(t)} := S^{(t-1)}$
> > # *Determine cluster with smallest distance:*
> > **for** $i = 2$ to $|S^{(t)}|$ **do**
> > > **for** $j = 1$ to $i - 1$ **do**
> > > > compute $\Delta_{ij} := D\left( S_i^{(t)}, S_j^{(t)} \right)$
> > > **end**
> > **end**
> > determine $i, j$ with smallest distance $\Delta_{ij}$
> > # *Merge clusters with smallest distance:*
> > update $S_i^{(t)} \leftarrow S_i^{(t)} \cup S_j^{(t)}$, drop $S_j^{(t)}$
> **end**
> output: partitions $S^{(0)}, S^{(1)}, \dots, S^{(T)}$

Depending on whether $D_{\min}$, $D_{\text{avg}}$, or $D_{\max}$ is used as the distance measure between clusters, we speak of **single-linkage**, **average-linkage**, or **complete-linkage clustering**.

A *divisive* **hierarchical cluster analysis** continuously *refines* partitions, starting from a single cluster that covers the whole dataset. We do not discuss these methods here, see [19, Chap. 6] instead.

In a hierarchy of clusterings, every level of the hierarchy represents a different partition. Hierarchies can also be called rooted trees (look back to Sect. 1.2.3). Every vertex of the tree represents a cluster, starting at the root, which represents the largest possible cluster—the cluster which contains all of the data points. With each level in the direction of the tree's leaves, the partitions become finer and finer, containing smaller and smaller clusters.

We can draw a **dendrogram** to visualize this hierarchy/tree. Fig. 7.10 shows an example that represents the clustering of forty international cities by geographical distance using the average-linkage method. The leaves correspond to the cities to be grouped, and each bifurcation represents the merger of two clusters of cities. The position of the bifurcation can be used to read off the value of the cluster distance function $D_{\mathrm{avg}}(\cdot, \cdot)$ at which the merger occurred.

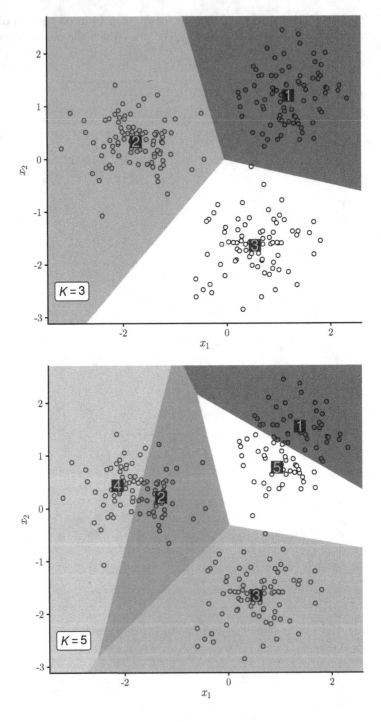

**Fig. 7.8.** *K-means clustering*

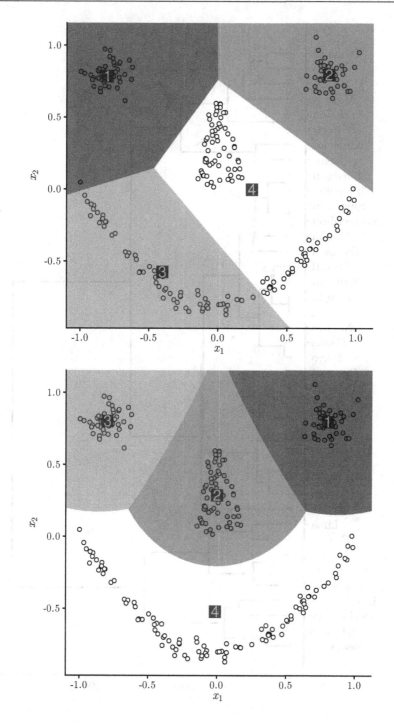

**Fig. 7.9.** *K-means clustering: ordinary (top) and using the kernel trick (bottom)*

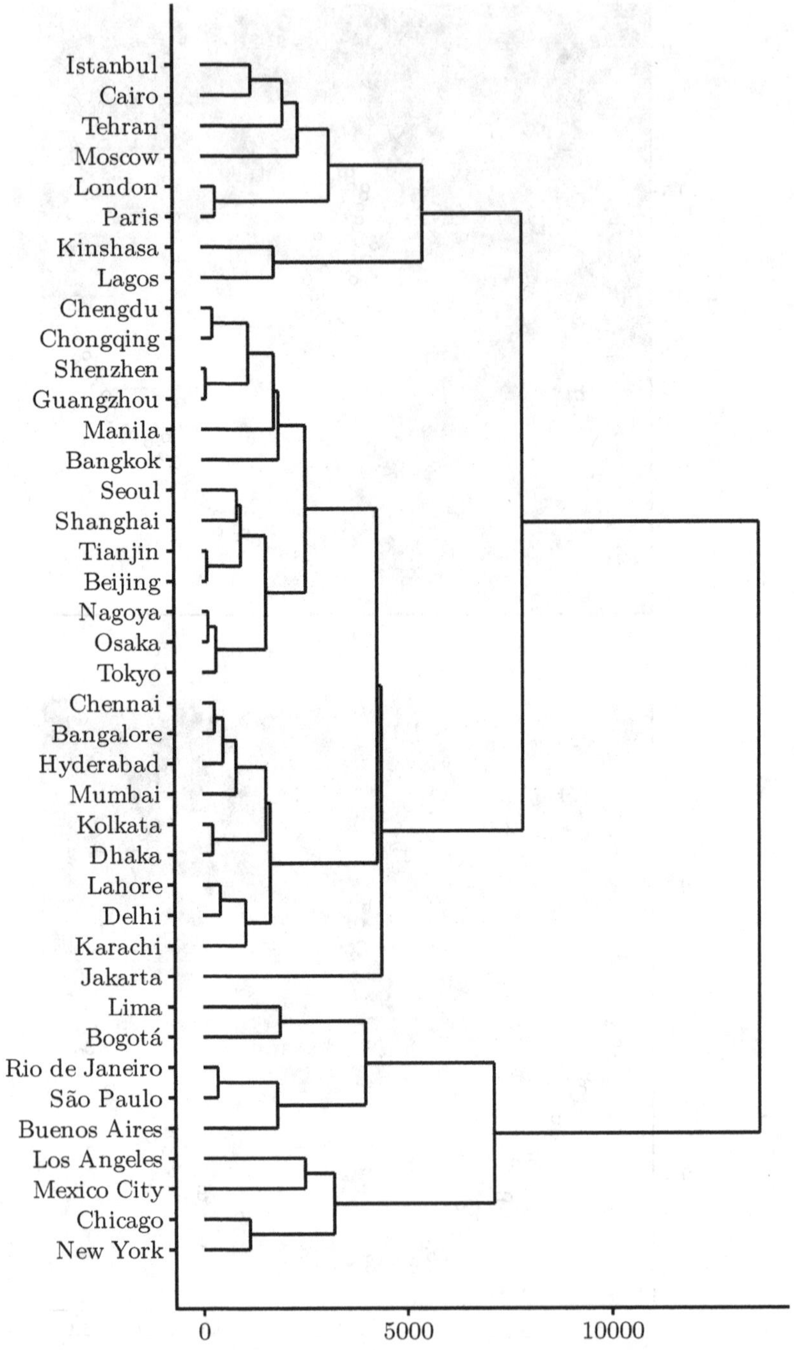

**Fig. 7.10.** *Dendrogram of a hierarchical cluster analysis of international cities*

# References

[1]   Sanjoy Dasgupta and Anupam Gupta. "An elementary proof of a theorem of Johnson and Lindenstrauss". In: *Random Structures & Algorithms* 22.1 (2003), pp. 60–65. DOI: 10.1002/rsa.10073.

[2]   Mehryar Mohri, Afshin Rostamizadeh, and Ameet Talwalkar. *Foundations of Machine Learning*. 2nd ed. MIT Press, 2018.

[3]   Sixue Gong, Vishnu Naresh Boddeti, and Anil K. Jain. "On the Intrinsic Dimensionality of Image Representations". In: *Proceedings of the IEEE/CVF Conference on Computer Vision and Pattern Recognition (CVPR)*. June 2019. arXiv:1803.09672.

[4]   Christian S. Perone, Roberto Silveira, and Thomas S. Paula. *Evaluation of sentence embeddings in downstream and linguistic probing tasks*. June 2018. arXiv:1806.06259v1.

[5]   Gunnar Carlsson. "Topology and data". In: *Bulletin of the American Mathematical Society* 46.2 (Jan. 2009), pp. 255–308. DOI: 10.1090/s0273-0979-09-01249-x.

[6]   Gunnar Carlsson. "Topological pattern recognition for point cloud data". In: *Acta Numerica* 23 (May 2014), pp. 289–368. DOI: 10.1017/s096249291 4000051.

[7]   Gunnar Carlsson and Mikael Vejdemo-Johansson. *Topological Data Analysis with Applications*. Cambridge University Press, Nov. 2021. DOI: 10. 1017/9781108975704.

[8]   Frédéric Chazal and Bertrand Michel. "An Introduction to Topological Data Analysis: Fundamental and Practical Aspects for Data Scientists". In: *Frontiers in Artificial Intelligence* 4 (Sept. 2021). DOI: 10.3389/frai. 2021.667963.

[9]   Yann LeCun, Corinna Cortes, and Christopher J. C. Burges. *The MNIST database of handwritten digits*. 2010. URL: http://yann.lecun.com/exdb/ mnist/.

[10]  François Collet et al. *Keras*. URL: https://keras.io.

[11]  J. J. Allaire and François Chollet. *keras: R Interface to 'Keras'*. R package. 2020. URL: https://CRAN.R-project.org/package=keras.

[12]  Geoffrey E Hinton and Sam Roweis. "Stochastic Neighbor Embedding". In: *Advances in Neural Information Processing Systems*. Ed. by S. Becker, S. Thrun, and K. Obermayer. Vol. 15. MIT Press, 2002.

[13]  Laurens J. P. van der Maaten and Geoffrey E. Hinton. "Visualizing High-Dimensional Data Using t-SNE". In: *Journal of Machine Learning Research* 9 (Nov. 2008), pp. 2579–2605.

[14]  Leland McInnes, John Healy, and James Melville. *UMAP: Uniform Manifold Approximation and Projection for Dimension Reduction*. Sept. 2020. arXiv:1802.03426v3.

[15]  Martin Ester et al. "A density-based algorithm for discovering clusters in large spatial databases with noise". In: *Proceedings of the 2nd ACM International Conference on Knowledge Discovery and Data Mining (KDD)*. 1996, pp. 226–231.

[16]   Ricardo J. G. B. Campello, Davoud Moulavi, and Joerg Sander. "Density-Based Clustering Based on Hierarchical Density Estimates". In: *Advances in Knowledge Discovery and Data Mining*. Springer Berlin Heidelberg, 2013, pp. 160–172. DOI: 10.1007/978-3-642-37456-2_14.

[17]   Enrique H. Ruspini, James C. Bezdek, and James M. Keller. "Fuzzy Clustering: A Historical Perspective". In: *IEEE Computational Intelligence Magazine* 14.1 (Feb. 2019), pp. 45–55. DOI: 10.1109/mci.2018.2881643.

[18]   Erich Schubert and Peter J. Rousseeuw. *Faster k-Medoids Clustering: Improving the PAM, CLARA, and CLARANS Algorithms*. Oct. 2019. arXiv:1810.05691v4.

[19]   Leonard Kaufman and Peter J. Rousseeuw, eds. *Finding Groups in Data: An Introduction to Cluster Analysis*. John Wiley & Sons, Mar. 1990. DOI: 10.1002/9780470316801.

# 8

# Applications of machine learning

Methods of data science and statistics in general, and machine learning methods in particular, are widely used in science and engineering. Here are just a few examples:

- Medical image processing and analysis [1]—for example, for diagnosing diseases such as COVID-19 based on thoracic computed tomography images [2, 3],

- drug discovery [4], protein structure prediction [5],

- processing of astronomical data [6], such as for the morphological classification of galaxies [7, 8] or the discovery of exoplanets [9, 10],

- gesture and speech recognition for human–machine communication [11, 12],

- autonomous vehicle control [13],

- credit fraud detection and prevention [14].

At the time of writing this text in early 2023, state-of-the-art **generative artificial intelligence** that can produce various types of content such as images [15, 16, 17], text [18, 19, 20, 21, 22], and music [23, 24] had been receiving considerable media attention [25, 26, 27, 28].

With the proliferation of increasingly powerful techniques based on ever-growing datasets, legal and ethical issues arise in **data protection and privacy** [29] and the socially and environmentally responsible use of artificial intelligence [30, 31, 32, 33, 34, 35].

In this chapter, we demonstrate how the techniques introduced in Chapters 6 and 7 are already powerful enough to solve various hard problems in computing.

© Springer-Verlag GmbH Germany, part of Springer Nature 2023
M. Plaue, *Data Science*, https://doi.org/10.1007/978-3-662-67882-4_8

## 8.1 Supervised learning in practice

In the following sections, we demonstrate the application of procedures for the automated categorization of data: first, the classification of digital images, and second, text documents.

### 8.1.1 MNIST: handwritten text recognition

The MNIST dataset [36, 37] consists of 70,000 digital grayscale images[1] with a resolution of $28 \times 28$ pixels. Each of the images shows a handwritten digit, see Fig. 8.6 above. Each image is labeled with the digit shown in the image. The problem to solve: implement an algorithm that automatically recognizes the digit that the image depicts. A rule-based solution seems impractical or at least very costly, so we would want to apply machine learning.

For the purpose of exposition, we keep things simple and solve the following binary classification problem instead: recognize the digit one as distinct from all other digits. First, we transform each image by lining up the gray values of the pixels to form a single array. This transformation is also called **flattening** and results in a collection of feature vectors in $\mathbb{R}^D$ that contain the $D = 28 \cdot 28 = 784$ gray values of each image.

We then split the dataset into a training dataset of 60,000 images and a test dataset of 10,000 images. The training dataset can be used to train the algorithms presented in the previous sections. We evaluate the trained classifiers by applying them to the images in the test dataset and compute measures for the goodness of fit.

The following table shows a comparison of logistic regression, quadratic discriminant analysis, the 1-nearest neighbor classifier, and a feedforward network:

|  | precision | recall | $F_1$-score |
|---|---|---|---|
| logistic regression | 93.1% | 97.1% | 95.1% |
| QDA | 89.6% | 95.9% | 92.6% |
| 1-NN | 96.7% | 99.5% | 98.1% |
| feedforward network | 98.7% | 99.0% | 98.9% |
| random (baseline) | 11.4% | 50.0% | 18.5% |

**Table 8.1.** *Performance of different classifiers applied to the MNIST dataset*

Overall, all methods perform relatively well on this dataset. The last line shows the measures that could be expected from an algorithm that were to decide whether the image depicts the digit one or not on a coin toss, i.e., with a probability of 50% each.

To perform the quadratic discriminant analysis, the covariance matrices $\Sigma_0$ and $\Sigma_1$ and the centroids $\mu_0$ and $\mu_1$ were obtained from the training data with class

---

[1] MNIST stands for Modified National Institute of Standards and Technology.

membership $y_n = 0$ and $y_n = 1$, respectively. These parameters determine the likelihood to find a feature vector given that the image shows the digit one or does not show the digit one, assuming a multivariate normal distribution:

$$p(x|y = k) = \mathcal{N}(x|\mu_k, \Sigma_k) = \frac{1}{\sqrt{(2\pi)^D \cdot \det(\Sigma_k)}} \cdot e^{-\frac{1}{2}(x-\mu_k)^T \cdot (\Sigma_k)^{-1} \cdot (x-\mu_k)}$$

with $k \in \{0, 1\}$. However, since the $\Sigma_k$ are singular—i.e., have vanishing determinant and cannot be inverted—we use **regularized covariance matrices** instead: $\Sigma_{k,\varepsilon} = \Sigma_k + \mathrm{diag}(\varepsilon, \varepsilon, \dots, \varepsilon)$ with a small smoothing parameter $\varepsilon > 0$.

The optimal hyperparameter $K = 1$ for the $K$-nearest neighbor classification can be obtained by cross-validation. The following plot shows the range and arithmetic mean of the $F_1$ score over $K$ on the basis of a six-fold cross-validation:

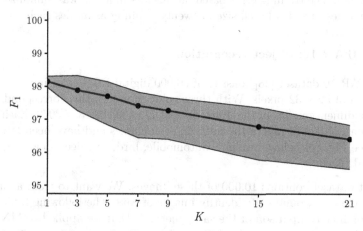

**Fig. 8.1.** *Cross-validation of a K-nearest neighbor model*

The classifier learned by the neural network shows the best goodness-of-fit in terms of the $F_1$ score. Nevertheless, the method misclassified the following images as the digit one:

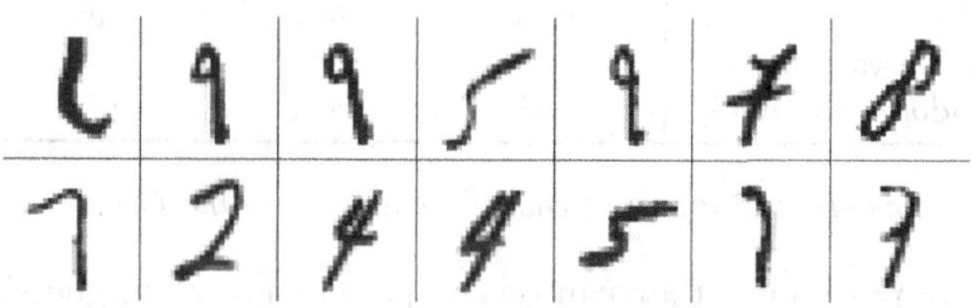

**Fig. 8.2.** *MNIST: false positives for identifying the digit one*

Conversely, the following handwritten variants of the number one were not recognized as such:

**Fig. 8.3.** *MNIST: false negatives for identifying the digit one*

The neural network used for the above classification consists of three hidden layers with 128, 64, and 32 neurons. For the hidden layers, a leaky rectifier has been used as the activation function. The output layer consists of only one neuron activated by a sigmoid function, and cross-entropy was chosen as the loss function. The training error based on this loss function was minimized using a gradient descent at a batch size of twenty training examples each.

### 8.1.2 CIFAR-10: object recognition

The CIFAR-10 dataset [38] consists of 60,000 digital RGB color images[2] with a resolution of $32 \times 32$ pixels. With three color channels, this corresponds to an extrinsic dimensionality of the input data of $D = 32 \cdot 32 \cdot 3 = 3072$. Each image features an object of one of the following ten classes and have been labeled as such—see Fig. 8.6 below: airplane, automobile, bird, cat, deer, dog, frog, horse, ship, and truck.

The test dataset contains 10,000 of these images. We want to train a classifier that is able to automatically identify images of cats. The following table shows a performance comparison of the same methods that we applied to MNIST in the previous section:

|  | precision | recall | $F_1$ score |
|---|---|---|---|
| logistic regression | 22% | 27% | 24% |
| QDA | 16% | 29% | 21% |
| 1-NN | 29% | 24% | 26% |
| feedforward network | 41% | 9% | 15% |
| random (baseline) | 10% | 50% | 17% |

**Table 8.2.** *Performance of different classifiers applied to the CIFAR-10 dataset*

The results are not good—they can be compared to random guessing!

However, using a deep Convolutional Neural Network, we can train a classifier that has an $F_1$ score of 60%, a precision of 70%, and a recall of 53%. Part of the program code in R (cf. [39]) is shown in Fig. 8.7; the program libraries Keras [40, 41] and TensorFlow were used for the implementation [42, 43]. The training can be performed on standard, consumer-level hardware. The average loss was

---

[2] CIFAR-10 stands for Canadian Institute for Advanced Research, 10 object classes.

minimized using Adam[3] [44], an efficient numerical optimization algorithm. A dropout probability of $q = 0.5$ was applied to the final hidden, fully-connected layer.

An additional component of the applied architecture is called **batch normalization** [45]. In this procedure, the activations are standardized with each iteration across the mini-batch, i.e., mean-centered, and brought to a variance of one. This can lead to improved numerical stability.

The following test examples were incorrectly classified by the CNN as images of a cat—the majority are photos of dogs:

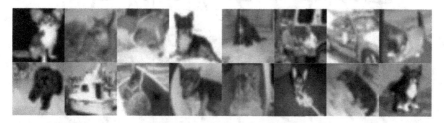

**Fig. 8.4.** *CIFAR-10: false positives for identifying a cat*

Conversely, the following photos of a cat were not recognized as such:

**Fig. 8.5.** *CIFAR-10: false negatives for identifying a cat*

**Explainable AI** (xAI) is a research field that is concerned with the task of applying complex models and algorithms like deep learning while ensuring that humans understand how those systems arrive at a certain result or prediction.

If the output neuron is replaced by a layer of ten softmax-activated neurons, the architecture can also be used to tackle the complete problem of assigning the photos to any of the ten object classes. The accuracy of that classifier applied to the CIFAR-10 test dataset is 80%, and data augmentation may boost it to 90% [39].

Current state-of-the-art algorithms achieve an accuracy of more than 99% [46]. For comparison, humans are able to correctly label about 94% of CIFAR-10 images [47].

---

[3] Adam stands for *Adaptive Moment Estimation*.

**Fig. 8.6.** *MNIST dataset for handwritten digit recognition (top) and CIFAR-10 dataset for object recognition (bottom; original images are in color)*

```
model <- keras_model_sequential() %>%
        layer_conv_2d(filters = 32, kernel_size = c(3,3)
             , input_shape = c(32, 32, 3), padding = 'same') %>%
        layer_activation_leaky_relu() %>%
        layer_batch_normalization(axis = -1) %>%
        layer_conv_2d(filters = 32, kernel_size = c(3,3)
             , padding = 'same') %>%
        layer_activation_leaky_relu() %>%
        layer_batch_normalization(axis = -1) %>%
        layer_max_pooling_2d(pool_size = c(2,2)) %>%
        layer_conv_2d(filters = 64, kernel_size = c(3,3)
             , padding = 'same') %>%
        layer_activation_leaky_relu() %>%
        layer_batch_normalization(axis = -1) %>%
        layer_conv_2d(filters = 64, kernel_size = c(3,3)
             , padding = 'same') %>%
        layer_activation_leaky_relu() %>%
        layer_batch_normalization(axis = -1) %>%
        layer_max_pooling_2d(pool_size = c(2,2)) %>%
        layer_conv_2d(filters = 128, kernel_size = c(3,3)
             , padding = 'same') %>%
        layer_activation_leaky_relu() %>%
        layer_batch_normalization(axis = -1) %>%
        layer_conv_2d(filters = 128, kernel_size = c(3,3)
             , padding = 'same') %>%
        layer_activation_leaky_relu() %>%
        layer_batch_normalization(axis = -1) %>%
        layer_max_pooling_2d(pool_size = c(2,2)) %>%
        layer_flatten() %>%
        layer_dense(units = 512) %>%
        layer_activation_leaky_relu() %>%
        layer_batch_normalization(axis = -1) %>%
        layer_dropout(rate = 0.5) %>%
        layer_dense(units = 1, activation = "sigmoid");

model %>% compile(
    optimizer = 'adam',
    loss = 'binary_crossentropy', metrics = 'accuracy'.
);

set.seed(1234)
history <- model %>%
    fit(
    x = X_train, y = Y_train, epochs = 20, batch_size = 20,
    validation_data = unname(list(x = X_val, y = Y_val))
    );
```

**Fig. 8.7.** *Example code for defining and training a convolutional neural network*

### 8.1.3 Large Movie Review Dataset: sentiment analysis

In the broadest sense, **sentiment analysis** is the systematic identification, extraction, and measurement of information characterized by a subjective feeling. For example, a biometric image or video capture system could be trained to recognize what state of mind a person is in: are they happy, sad, angry, etc.?

In a narrower sense, sentiment analysis is a task of natural language processing (NLP) and consists of analyzing text data with respect to content such as expressions of emotion or opinion. One particular such analysis is the evaluation of the **polarity** of a text, or part of it: does the author express a positive or negative opinion, do they speak positively or negatively about a topic?

A simple way to determine polarity is to use a dictionary of terms that have positive/negative connotations, or usually express a positive/negative sentiment. The text under study would then be matched with the dictionary. For example, the sentence "I love this wonderful movie" contains the words "love" and "wonderful," suggesting positive sentiment. Conversely, if a text contains negatively connotated words like "hate" or "disgusting," the system might classify it as expressing a negative sentiment. The dictionary may also include emoticons like :-) or :-(.

Lexicoder 2015 [48, 49] and VADER[4] [50] are examples of dictionaries for sentiment analysis. Another option is to use supervised learning, as we will demonstrate below.

We use the dataset Large Movie Review Dataset v1.0 [51, 52]. A version simplified in format is available on the data science platform Kaggle [53]. The dataset contains a total of 50,000 English-language reviews of feature films and television series written by users of the Internet Movie Database[5]. Depending on whether the user liked the movie or trashed the movie, the reviews are labeled as positive or negative.

Here are two example texts from the dataset that clearly express the author's sentiment towards the reviewed movie:

| polarity | review |
|---|---|
| *positive* | If you like original gut wrenching laughter you will like this movie. If you are young or old then you will love this movie, hell even my mom liked it. Great Camp!!! |
| *negative* | Hated it with all my being. Worst movie ever. Mentally scarred. Help me. It was that bad. TRUST ME!!! |

**Table 8.3.** *Positive/negative movie reviews*

We want to use a naive Bayes classifier to automatically label a movie review as either positive or negative. First, we use single words as tokens, case-insensitive.

---

[4] VADER stands for Valence Aware Dictionary and sEntiment Reasoner.
[5] https://www.imdb.com/

Although we apply some fairly advanced theory, we should remind ourselves that the underlying rationale is quite intuitive: We perform a **word frequency analysis** and count how often certain words appear in positive/negative reviews. For example, the following words appear almost exclusively in positive reviews:

| token | `flawless` | `superbly` | `perfection` | `wonderfully` | `must-see` |
|---|---|---|---|---|---|
| likelihood (pos.) | 90% | 89% | 89% | 88% | 88% |

**Table 8.4.** *Words that indicate a positive movie review*

On the other hand, an occurrence of the following tokens[6] indicates a negative review:

| token | `stinker` | `mst3k` | `waste` | `unwatchable` | `0` | `unfunny` |
|---|---|---|---|---|---|---|
| likelihood (neg.) | 96% | 96% | 94% | 93% | 92% | 92% |

**Table 8.5.** *Words that indicate a negative movie review*

The words thus identified are then used as features to classify any movie review as either positive or negative. That is, even if the algorithm has not seen a particular review yet: the occurrence of the term "stinker" will indicate a negative opinion for almost any review.

The naive Bayes classifier was trained using 40,000 reviews. A total of $D = 149{,}653$ words can be extracted from those texts. Laplace smoothing was applied to regularize the frequencies of word occurrence. Applied to the test dataset of the remaining 10,000 reviews, machine learning provides a significant improvement over the rule-based methods:

| | precision | recall | $F_1$ score |
|---|---|---|---|
| Lexicoder 2015 | 70.9% | 74.6% | 72.7% |
| VADER | 65.0% | 79.0% | 71.3% |
| multinomial | 87.2% | 81.1% | 84.1% |
| Bernoulli | 88.6% | 81.1% | 84.7% |
| random (baseline) | 50.0% | 50.0% | 50.0% |

**Table 8.6.** *Performance of different methods for sentiment analysis*

A further improvement can be achieved by using not only words but also **N-grams** as features. An $N$-gram is a contiguous sequence of $N$ tokens. For example, the sentence "I love this wonderful movie" consists of the following $N$-grams, $N \leq 4$:

---

[6] Mystery Science Theater 3000, or MST3K, is a television show that spoofs select B movies.

| $N$ | $N$-grams |
|---|---|
| 1 | i, love, this, wonderful, movie |
| 2 | i love, love this, this wonderful, wonderful movie |
| 3 | i love this, love this wonderful, this wonderful movie |
| 4 | i love this wonderful, love this wonderful movie |

**Table 8.7.** *Examples of N-grams*

Both monograms ($N = 1$) and bigrams ($N = 2$) were used as features to arrive at the following performance:

| | precision | recall | $F_1$ score |
|---|---|---|---|
| multinomial, $N$-grams | 88.7% | 87.1% | 87.9% |
| Bernoulli, $N$-grams | 87.0% | 89.5% | 88.2% |

**Table 8.8.** *Performance of naive Bayes classifiers based on N-grams*

Since the use of $N$-grams increases the number of features significantly, feature selection was performed before training: First, $N$-grams occurring in fewer than ten reviews were removed. Additionally, 10% of the $N$-grams with smallest mutual information with the distribution of positive/negative sentiment were removed. As a result of this procedure, a total of $D = 137{,}806$ $N$-grams remained for the final analysis.

Here are two examples of misclassified reviews, where key aspects have been highlighted manually:

| ground truth | prediction | review |
|---|---|---|
| *positive* | *negative* | In Black Mask, Jet Li plays a bio-engineered super-killer turned pacifist, who has to fight against other super-killers. **Bad plot, bad sfx** (60 million dollar budget), **but the fighting scenes were excellent!** Jet Li is the greatest martial-arts star alive! |
| *negative* | *positive* | **The first part** of Grease with John Travolta and Olivia Newton John **is one of the best movie** for teens, **This one is a very bad copy.** The change is only in the sex. In the first one the good one was Sandy, here it's Michael. I prefer to watch the first Grease. |

**Table 8.9.** *Misclassified movie reviews*

The first example is a more balanced review that also lists negative aspects of the film. In the second misclassification, the positive opinion refers to an earlier film in the series—but the actual subject of the review was evaluated negatively. Thus, it could have been advantageous to perform the sentiment analysis not at the document level but at the **aspect level** [54].

## 8.2 Unsupervised learning in practice

For the applications in the following sections, we use data from IMDb, the Internet Movie Database [55], as we did in the previous chapter. The dataset[7] contains a total of 85,855 movies, with attributes such as the title, an English language description, average user rating, number of ratings, genre, etc. The movies have 297,705 people associated with them who worked on their production: actors and actresses, directors, cinematographers, etc.

### 8.2.1 Text mining: topic modelling

First, we would like to examine the films' descriptions. We restrict the analysis to two genres and a certain time period: 3141 family films and 2835 science fiction films from the years 1980–2020. We want to get a quick overview of typical themes that inform the plots of those films. To avoid reading and manually summarizing all of the 6000 plot descriptions, we use methods of *automated text analysis* instead, which is also called **text mining**.

In order to make the unstructured text data accessible to such an analysis, we must preprocess it first. Similar to the sentiment analysis of movie reviews (Sect. 8.1.3), we tokenize the text. In this case, the tokens are not only a means for the algorithm to compute the final result. As we will see later, they are also presented to the analyst as items in a **topic map**. Therefore, we have to be a bit more careful with the feature selection and only extract tokens that have a certain relevance for the content of the text. For example, certain words should be excluded from the analysis that occur very frequently in almost any text corpus, such as the articles "the," "a," or frequently occurring prepositions such as "with" or "from." These **stopwords** are removed during the preprocessing of the data.

Furthermore, we will use only noun phrases: these can be single nouns like "astronaut" but also groups of words like "virtual reality" or "time machine." Noun phrase extraction can be performed with software libraries for natural language processing like quanteda [57] or spaCy [58, 59] Finally, we will only consider phrases composed of no more than two words.

A total of 105,292 noun phrases were extracted from the corpus. These are of variable relevance to the content. The following table shows a small selection of "genre-typical" phrases and their frequency of occurrence in the respective genres in percent. Any proportions missing from 100% are covered by the remaining genres. For example, 73% of all films that contain the phrase "outer space" in the description are science fiction films. We want to focus on such genre-typical phrases and calculate the mutual information of each phrase with the distribution of genres to inform feature selection.

---

[7] As of 2023, the cited dataset is not available at Kaggle anymore. The same author has since then published a similar dataset [56] that can be used instead.

| genre<br>phrase | crime | family | horror | romance | sci-fi | western |
|---|---|---|---|---|---|---|
| undercover cop | 65 | 0 | 4 | 0 | 2 | 0 |
| fairy tale | 2 | 39 | 6 | 16 | 0 | 2 |
| occult | 7 | 2 | 80 | 5 | 2 | 0 |
| young lovers | 8 | 0 | 8 | 43 | 5 | 0 |
| outer space | 2 | 13 | 22 | 4 | 73 | 0 |
| cavalry | 1 | 4 | 0 | 18 | 0 | 70 |

**Table 8.10.** *Noun phrases typically associated with certain film genres*

First, from the 10,522 phrases that occur in descriptions of family films, we remove those that occur either in less than ten descriptions or more often than in 25% of the descriptions. The rationale behind this approach: if a phrase appears in only a few texts, it is not descriptive for the whole corpus or significant parts of the corpus. If the phrase appears too often, it is not well-suited as a feature to distinguish between different sub-types of family films. This procedure reduces the number of tokens to 674. Finally, from these remaining nominal phrases, the top 60 are selected according to their mutual information score.

The features that we want to use to characterize the science fiction films are selected using the same method. Finally, we apply the $t$-SNE algorithm to the similarity matrix between the phrases, which is calculated using the overlap coefficient: if two phrases tend to appear in the same films, that score is going to be high. This results in two topic maps, shown in Fig. 8.8. Moreover, the size of the phrases was scaled with the frequency of their occurrence.

### 8.2.2 Network analysis: community structure

In addition to the description for each film, the IMDb dataset also contains a list of actors and actresses who worked on each film. From these data, we can construct a **collaboration network**: If two actors/actresses appear together in at least one movie, we determine that they are connected in that sense. Mathematically, such a network is an undirected graph: each node represents an actor or actress, and each collaboration an edge. Thus, we can also draw a node–link diagram for such a graph: see Fig. 8.9 for an example of a collaboration network of actors and actresses that played in at least one movie directed by Martin Scorsese. With a few exceptions, all names have been pseudonymized to ensure information privacy.

The position of the nodes in the diagram were determined using the $t$-SNE method, and the visualization was generated using R libraries for network analysis [60, 61, 62]. Examples of software entirely dedicated to network analysis and visualization include Gephi [63] and Cytoscape [64].

Although the chart shows only a very small section of the collaboration network, it may already be quite confusing due to clutter. Especially in the vicinity of strongly networked actors/actresses (so-called **hubs**), it is not easy to visually trace collaborations.

One way to reduce complexity is to detect clusters in the network, the **communities**: Fig. 8.10 shows the dendrogram of a hierarchical cluster analysis using the average-linkage method, based on min–max scaled Jaccard similarity, computed from the number of films in which both performers appeared in.

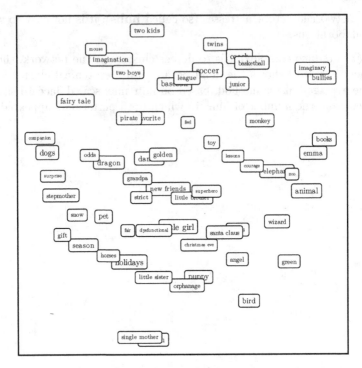

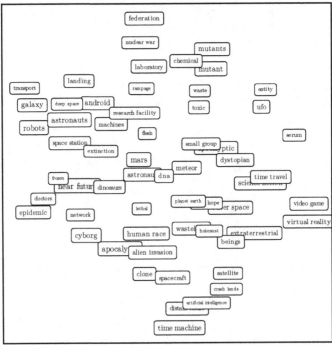

**Fig. 8.8.** *Topic maps for family films and science fiction films (1980-2020)*

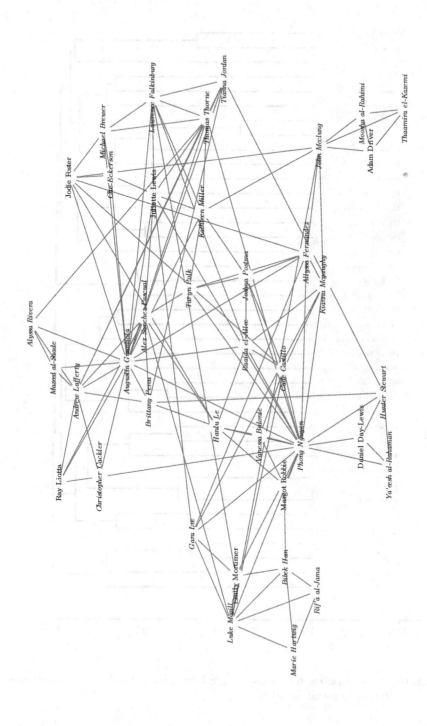

**Fig. 8.9.** Collaboration network of actors/actresses (partially pseudonymized)

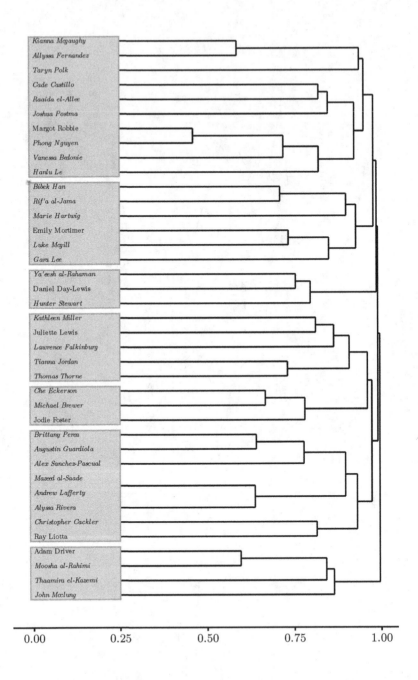

**Fig. 8.10.** *Hierarchical cluster analysis of a collaboration network of actors/actresses (partially pseudonymized)*

# References

[1]   S. Kevin Zhou, Hayit Greenspan, and Dinggang Shen, eds. *Deep Learning for Medical Image Analysis*. Academic Press, Jan. 2017.

[2]   Edward H. Lee et al. "Deep COVID DeteCT: an international experience on COVID-19 lung detection and prognosis using chest CT". In: *npj Digital Medicine* 4.1 (Jan. 2021). DOI: 10.1038/s41746-020-00369-1.

[3]   Nikolas Lessmann et al. "Automated Assessment of COVID-19 Reporting and Data System and Chest CT Severity Scores in Patients Suspected of Having COVID-19 Using Artificial Intelligence". In: *Radiology* 298.1 (Jan. 2021), E18–E28. DOI: 10.1148/radiol.2020202439.

[4]   Austin Robert Clyde. "Artificial Intelligence and High-Performance Computing for Accelerating Structure-Based Drug Discovery". PhD thesis. University of Chicago, Dec. 2022.

[5]   Ewen Callaway. "What's next for AlphaFold and the AI protein-folding revolution". In: *Nature* 604.7905 (Apr. 2022), pp. 234–238. DOI: 10.1038/d41586-022-00997-5.

[6]   Željko Ivezić et al. *Statistics, Data Mining, and Machine Learning in Astronomy*. Revised Edition. Princeton University Press, Dec. 2019.

[7]   Sander Dieleman, Kyle W. Willett, and Joni Dambre. "Rotation-invariant convolutional neural networks for galaxy morphology prediction". In: *Monthly Notices of the Royal Astronomical Society* 450.2 (Apr. 2015), pp. 1441–1459. DOI: 10.1093/mnras/stv632. arXiv:1503.07077.

[8]   Helena Domínguez Sánchez et al. "Improving galaxy morphologies for SDSS with Deep Learning". In: *Monthly Notices of the Royal Astronomical Society* 476.3 (Feb. 2018), pp. 3661–3676. DOI: 10.1093/mnras/sty338. arXiv:1711.05744.

[9]   Carlos Alberto Gomez Gonzalez, Olivier Absil, and Marc van Droogenbroeck. "Supervised detection of exoplanets in high-contrast imaging sequences". In: *Astronomy & Astrophysics* 613 (May 2018), A71. DOI: 10.1051/0004-6361/201731961. arXiv:1712.02841.

[10]  Faustine Cantalloube et al. "Exoplanet imaging data challenge: benchmarking the various image processing methods for exoplanet detection". In: *Adaptive Optics Systems VII*. Ed. by Dirk Schmidt, Laura Schreiber, and Elise Vernet. Vol. 11448. International Society for Optics and Photonics. SPIE, Dec. 2020, pp. 1027–1062. DOI: 10.1117/12.2574803.

[11]  Fan Zhang et al. *MediaPipe Hands: On-device Real-time Hand Tracking*. June 2020. arXiv:2006.10214v1.

[12]  Dong Yu and Li Deng. *Automatic Speech Recognition*. Springer, London, 2015. DOI: 10.1007/978-1-4471-5779-3.

[13]  Sampo Kuutti et al. "A Survey of Deep Learning Applications to Autonomous Vehicle Control". In: *IEEE Transactions on Intelligent Transportation Systems* (2020), pp. 1–22. DOI: 10.1109/tits.2019.2962338. arXiv:1912.10773.

[14]   Fabrizio Carcillo et al. "Combining unsupervised and supervised learning in credit card fraud detection". In: *Information Sciences* (May 2019). DOI: 10.1016/j.ins.2019.05.042.

[15]   Aditya Ramesh et al. *Zero-Shot Text-to-Image Generation.* 2021. DOI: 10.48550/ARXIV.2102.12092.

[16]   Aditya Ramesh et al. *Hierarchical Text-Conditional Image Generation with CLIP Latents.* 2022. DOI: 10.48550/ARXIV.2204.06125.

[17]   Robin Rombach et al. *High-Resolution Image Synthesis with Latent Diffusion Models.* 2021. DOI: 10.48550/ARXIV.2112.10752.

[18]   Tom B. Brown et al. *Language Models are Few-Shot Learners.* May 2020. DOI: 10.48550/ARXIV.2005.14165.

[19]   Leo Gao, John Schulman, and Jacob Hilton. *Scaling Laws for Reward Model Overoptimization.* Sept. 2020. DOI: 10.48550/ARXIV.2210.10760.

[20]   Nisan Stiennon et al. "Learning to summarize with human feedback". In: *Advances in Neural Information Processing Systems.* Ed. by H. Larochelle et al. Vol. 33. Curran Associates, Inc., 2020, pp. 3008–3021. arXiv:2009.01325.

[21]   Aarohi Srivastava et al. *Beyond the Imitation Game: Quantifying and extrapolating the capabilities of language models.* June 2022. DOI: 10.48550/ARXIV.2206.04615.

[22]   Renqian Luo et al. "BioGPT: generative pre-trained transformer for biomedical text generation and mining". In: *Briefings in Bioinformatics* 23.6 (Sept. 2022). DOI: 10.1093/bib/bbac409. arXiv:2210.10341.

[23]   Prafulla Dhariwal et al. *Jukebox: A Generative Model for Music.* Apr. 2020. DOI: 10.48550/ARXIV.2005.00341.

[24]   Andrea Agostinelli et al. *MusicLM: Generating Music From Text.* Jan. 2023. arXiv:2301.11325.

[25]   Cade Metz. "Meet DALL-E, the A.I. That Draws Anything at Your Command". In: *The New York Times* (Aug. 2022). URL: https://www.nytimes.com/2022/04/06/technology/openai-images-dall-e.html.

[26]   Kevin Roose. "The Brilliance and Weirdness of ChatGPT". In: *The New York Times* (Dec. 2022). URL: https://www.nytimes.com/2022/12/05/technology/chatgpt-ai-twitter.html.

[27]   Ajay Agrawal, Joshua Gans, and Avi Goldfarb. "ChatGPT and How AI Disrupts Industries". In: *Harvard Business Review* (Dec. 2022). URL: https://hbr.org/2022/12/chatgpt-and-how-ai-disrupts-industries.

[28]   Ted Chiang. "ChatGPT Is a Blurry JPEG of the Web". In: *The New Yorker* (Feb. 2023). URL: https://www.newyorker.com/tech/annals-of-technology/chatgpt-is-a-blurry-jpeg-of-the-web.

[29]   Elif Kiesow Cortez, ed. *Data Protection Around the World.* T.M.C. Asser Press, 2021. DOI: 10.1007/978-94-6265-407-5.

[30]   Matthias Plaue. "Rise of the Mindless Machines". In: *towards data science* (Nov. 2018). URL: https://towardsdatascience.com/rise-of-the-mindless-machines-c0e578061e65.

[31] Anna Jobin, Marcello Ienca, and Effy Vayena. "The global landscape of AI ethics guidelines". In: *Nature Machine Intelligence* 1.9 (Sept. 2019), pp. 389–399. DOI: 10.1038/s42256-019-0088-2. arXiv:1906.11668.

[32] Miles Brundage et al. *Toward Trustworthy AI Development: Mechanisms for Supporting Verifiable Claims.* Apr. 2020. DOI: 10.48550/ARXIV.2004. 07213.

[33] Markus D. Dubber, Frank Pasquale, and Sunit Das, eds. *The Oxford Handbook of Ethics of AI.* Oxford University Press, July 2020. DOI: 10. 1093/oxfordhb/9780190067397.001.0001.

[34] Emily M. Bender et al. "On the Dangers of Stochastic Parrots". In: *Proceedings of the 2021 ACM Conference on Fairness, Accountability, and Transparency.* ACM, Mar. 2021. DOI: 10.1145/3442188.3445922.

[35] Philipp Hacker, Andreas Engel, and Marco Mauer. *Regulating ChatGPT and other Large Generative AI Models.* Feb. 2023. DOI: 10.48550/ARXIV. 2302.02337.

[36] Yann LeCun, Corinna Cortes, and Christopher J. C. Burges. *The MNIST database of handwritten digits.* 2010. URL: http://yann.lecun.com/exdb/ mnist/.

[37] Jiang Junfeng. *readmnist: Read MNIST Dataset.* R package. 2018. URL: https://CRAN.R-project.org/package=readmnist.

[38] Alex Krizhevsky. *Learning Multiple Layers of Features from Tiny Images.* Tech. rep. 2009.

[39] Moritz Hambach. *Image Augmentation in Keras (CIFAR-10).* Jan. 2018. URL: https://github.com/moritzhambach/Image-Augmentation-in-Keras-CIFAR-10-.

[40] François Collet et al. *Keras.* URL: https://keras.io.

[41] J. J. Allaire and François Chollet. *keras: R Interface to 'Keras'.* R package. 2020. URL: https://CRAN.R-project.org/package=keras.

[42] Martín Abadi et al. *TensorFlow: Large-Scale Machine Learning on Heterogeneous Systems.* URL: https://www.tensorflow.org/.

[43] J. J. Allaire and Yuan Tang. *tensorflow: R Interface to 'TensorFlow'.* R package. 2020. URL: https://CRAN.R-project.org/package=tensorflow.

[44] Diederik P. Kingma and Jimmy Ba. "Adam: A Method for Stochastic Optimization". In: *3rd International Conference on Learning Representations, San Diego, USA.* Ed. by Yoshua Bengio and Yann LeCun. May 2015. arXiv:1412.6980.

[45] Sergey Ioffe and Christian Szegedy. "Batch Normalization: Accelerating Deep Network Training by Reducing Internal Covariate Shift". In: *32nd International Conference on Machine Learning, Lille, France.* Ed. by Francis Bach and David Blei. Vol. 37. Proceedings of Machine Learning Research. PMLR, July 2015, pp. 448–456. arXiv:1502.03167.

[46] Papers with Code Community. *CIFAR-10 Benchmark (Image Classification).* Ed. by Robert Stojnic et al. Accessed Nov. 20, 2022. URL: https://paperswithcode.com/sota/image-classification-on-cifar-10.

[47]  Andrej Karpathy. *Lessons learned from manually classifying CIFAR-10*. Apr. 2011. URL: http://karpathy.github.io/2011/04/27/manually-classifying-cifar10/.

[48]  Lori Young and Stuart Soroka. *Lexicoder Sentiment Dictionary*. 2012.

[49]  Lori Young and Stuart Soroka. "Affective News: The Automated Coding of Sentiment in Political Texts". In: *Political Communication* 29.2 (2012), pp. 205–231. DOI: 10.1080/10584609.2012.671234.

[50]  C. Hutto and Eric Gilbert. "VADER: A Parsimonious Rule-Based Model for Sentiment Analysis of Social Media Text". In: *Proceedings of the International AAAI Conference on Web and Social Media* 8.1 (May 2014), pp. 216–225. DOI: 10.1609/icwsm.v8i1.14550.

[51]  Andrew L. Maas et al. "Learning Word Vectors for Sentiment Analysis". In: *49th Annual Meeting of the Association for Computational Linguistics: Human Language Technologies, Portland, Oregon, USA*. Association for Computational Linguistics, June 2011, pp. 142–150.

[52]  Andrew L. Maas. *Large Movie Review Dataset v1.0*. Accessed Nov. 15, 2020. URL: http://ai.stanford.edu/~amaas/data/sentiment/.

[53]  N. Lakshmipathi. *IMDb dataset of 50k movie reviews. Large Movie Review Dataset*. Accessed Nov. 15, 2020. URL: https://www.kaggle.com/lakshmi25npathi/imdb-dataset-of-50k-movie-reviews.

[54]  Ambreen Nazir et al. "Issues and Challenges of Aspect-based Sentiment Analysis: A Comprehensive Survey". In: *IEEE Transactions on Affective Computing* (2020). DOI: 10.1109/taffc.2020.2970399.

[55]  Stefano Leone. *IMDb movies extensive dataset. 81k+ movies and 175k+ cast members scraped from IMDb*.

[56]  Stefano Leone. *FilmTV movies dataset. 40k+ movies scraped from FilmTV*. Accessed Jan. 22, 2023. URL: https://www.kaggle.com/datasets/stefanoleone992/filmtv-movies-dataset.

[57]  Kenneth Benoit et al. "quanteda: An R package for the quantitative analysis of textual data". In: *Journal of Open Source Software* 3.30 (2018), p. 774. DOI: 10.21105/joss.00774. URL: https://quanteda.io.

[58]  Matthew Honnibal and Ines Montani. *spaCy. Industrial-Strength Natural Language Processing*. Accessed Dec. 9, 2020. URL: https://spacy.io/.

[59]  Kenneth Benoit and Akitaka Matsuo. *spacyr: Wrapper to the 'spaCy' NLP Library*. R package. 2020. URL: https://CRAN.R-project.org/package=spacyr.

[60]  Barret Schloerke et al. *GGally: Extension to 'ggplot2'*. R package. 2020. URL: https://CRAN.R-project.org/package=GGally.

[61]  Carter T. Butts. *network: Classes for Relational Data*. R package. The Statnet Project (http://www.statnet.org). 2020. URL: https://CRAN.R-project.org/package=network.

[62]  Carter T. Butts. "network: a Package for Managing Relational Data in R". In: *Journal of Statistical Software* 24.2 (2008). URL: https://www.jstatsoft.org/v24/i02/paper.

[63] Mathieu Bastian, Sebastien Heymann, and Mathieu Jacomy. *Gephi: An Open Source Software for Exploring and Manipulating Networks.* 2009. URL: https://gephi.org/.

[64] Paul Shannon et al. "Cytoscape: A Software Environment for Integrated Models of Biomolecular Interaction Networks". In: *Genome Research* 13.11 (Nov. 2003), pp. 2498–2504. DOI: 10.1101/gr.1239303. URL: https://cytoscape.org/.

# Appendix

# A

# Exercises with answers

## A.1 Exercises

**Exercise 1. Entity–relationship model.** Draw an entity–relationship diagram in Chen notation that models the domain of a movie dataset that contains the following information about each film:

- title, year, and genre of the film,
- names of the actors and actresses cast in the film,
- names of the directors of the film,
- names and addresses of production companies involved.

The model includes the following entity types: person, company, film.

**Exercise 2. Data quality assessment.** The data table below lists organizations and their attributes:

- **id**: primary key; unique identifier for the organization
- **name**: name of the organization
- **cc**: country of origin/location of headquarters of the organization; two letter country codes according to ISO 3166-1 alpha-2 standard[1]
- **city**: name of the city that the organization's headquarters are located in
- **hospital**: binary Boolean variable, TRUE if the organization is a hospital or other medical facility

Check for invalid, missing, or otherwise corrupted or not optimally cleaned data. Identify as many data defects and data quality issues as possible.

---

[1] https://www.iso.org/iso-3166-country-codes.html

© Springer-Verlag GmbH Germany, part of Springer Nature 2023
M. Plaue, *Data Science*, https://doi.org/10.1007/978-3-662-67882-4

| id | name | cc | city | hospital |
|---|---|---|---|---|
| 5 | Apple Inc. | US | | TRUE |
| 1 | Samsung Electronics Co., Ltd. | KR | ⟨NA⟩ | TRUE |
| 10 | Boston Children's Hospital | US | Boston | FALSE |
| 7 | Mayo Clinic | US | Rochester | FALSE |
| 8 | Uber Technologies | US | ⟨NA⟩ | TRUE |
| 6 | Tesla | ZZ | Austin | TRUE |
| 9 | Oxford Univeristy | UK | Oxford | TRUE |
| 10 | Oxford University | GB | Oxford | TRUE |
| 4 | Seoul National Univeristy | KR | NULL | TRUE |
| 2 | Broadcom Limited | US | San Jose | TRUE |
| 3 | Stanford Univeristy | US | | TRUE |

**Table A.1.** *Dirty organization data*

**Exercise 3. Exploratory data analysis.** The UC Irvine Machine Learning Repository provides the results of a survey of faculty members from two Spanish universities on teaching uses of Wikipedia [1, 2]:

https://doi.org/10.24432/C50031

Download the data and process them with a tool for data analysis and visualization of your choosing, e.g., RStudio [3] or Colab [4].

1. Provide an overview of the respondents' demographics that includes age, years of experience, domain/field of expertise, and job position.

2. The answers to the survey questions are given on a **Likert scale** that ranges from one (strongly disagree/never) to five (strongly agree/very often). Which of the questions do a majority of survey participants select extreme responses for?

3. Members from which domains are the most/least likely to consult Wikipedia for issues related to their field of expertise? Compare with the practice of citing Wikipedia in academic papers.

Use charts to visualize your findings.

*Hint:* As of April 2023, the codebook supplied by the UCI website is incomplete and does not explain the meaning of the variable assignment DOMAIN = 6. Comparison with the original publication [2], however, shows that this value almost certainly corresponds to "Social Sciences."

**Exercise 4. Base rate fallacy.** In February 2023, a member of the UK parliament posted on Twitter [5]: "devastating stats from New Zealand, how can anyone deny the [COVID] vaccine harms now," referencing the following statistics [6]:

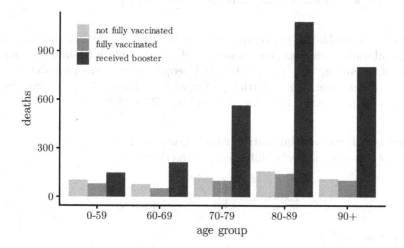

**Fig. A.1.** *COVID-19 deaths by vaccination status*

We can compute the percentages by vaccination status from those data. Depending on age group, among all fatalities, between 69% and 89% died of COVID even though they had been fully vaccinated against the disease:

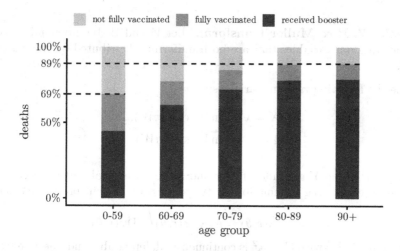

**Fig. A.2.** *Vaccination status of people that died of COVID-19*

Do those data actually support the hypothesis that COVID vaccines are generally harmful? If not, provide an (educational) explanation for the presumably high proportion of vaccinated individuals among the fatalities.

*Hint:* As of February 2023, at least 80% of New Zealand's population has been fully vaccinated/boosted against COVID-19 [7]. In the age groups 80+, at least 95% of the population are fully vaccinated/boosted [8].

**Exercise 5. Birthday problem.** Assume that when selecting a person from the population at random, they have one of 365 possible birthdays with equal probability. Among $N$ randomly selected people, what is the probability that at least two of them share a birthday? For which values of $N$ is the probability that two people share a birthday at least 50%, 95%, 100%?

**Exercise 6. Exponential distribution.** Consider the following parameterized family of exponential probability density functions:

$$\text{Exp}(u|\lambda) = \begin{cases} \lambda \cdot e^{-\lambda u} & \text{if } u > 0 \\ 0 & \text{if } u \leq 0 \end{cases}$$

with parameter $\lambda > 0$.

Suppose that $X$ is a random variable that is exponentially distributed. Compute the mean, the median, and the variance of $X$ (as a function of the parameter $\lambda$).

Given numeric data $x_1, \ldots, x_N$, derive the formula to determine the parameter $\lambda$, based on the maximum likelihood method.

**Exercise 7. Box–Muller transform.** Let $V$ and $W$ be independent continuous random variables that are both uniformly distributed on the interval $[0, 1]$.

Define the following random variables:

$$X = \sqrt{-2 \ln V} \cdot \cos(2\pi W),$$
$$Y = \sqrt{-2 \ln V} \cdot \sin(2\pi W)$$

Prove that $X$ and $Y$ are independent and standard normally distributed. You may use the following formula for the transformation of joint probability densities:

$$p_{\varphi^{-1}(X,Y)}(r, \theta) = p_{X,Y}(\varphi(r, \theta)) \cdot |D\varphi(r, \theta)|$$

where $\varphi \colon U \to \mathbb{R}^2$ with $U \subseteq \mathbb{R}^2$ is continuously differentiable and injective (with the possible exception of isolated points).

*Hint:* As the naming of the arguments already suggests, the polar coordinate map

$$\varphi \colon [0, \infty[ \times [0, 2\pi[ \to \mathbb{R}^2, \ (r, \theta) \mapsto \begin{pmatrix} r \cdot \cos(\theta) \\ r \cdot \sin(\theta) \end{pmatrix}$$

is of particular interest here.

**Exercise 8. Simpson's paradox.** At a university's six largest departments, $p_x = 30\%$ of $N_x = 1835$ female applicants are admitted, compared to $p_y = 45\%$ of $N_y = 2691$ male applicants. Use the one-sided $Z$-test to show that the difference in admission rate is statistically significant at a 99% confidence level, which would support the hypothesis that female applicants are more likely to be rejected.

These are the numbers broken down by university department:

| dept. | $N_x$ | $p_x$ | $N_y$ | $p_y$ |
|---|---|---|---|---|
| A | 108 | 82% | 825 | 62% |
| B | 25 | 68% | 560 | 63% |
| C | 593 | 34% | 325 | 37% |
| D | 375 | 35% | 417 | 33% |
| E | 393 | 24% | 191 | 28% |
| F | 341 | 7% | 373 | 6% |
| all | 1835 | 30% | 2691 | 45% |

**Table A.2.** *University admission rates by department and gender*

Show that an analysis of the data for each department does *not* support the hypothesis that female applicants are more likely to be rejected. Try to find an explanation for this paradoxical result.

*Hint:* For each department, compute the admission rate $p$, irrespective of gender. Which departments are the easiest to be admitted to, and persons of which gender typically apply for them?

**Exercise 9. Expected root mean squared error of random guessing.** Suppose that we are given a test dataset for a regression task where the target variable takes values $y_1, \ldots, y_M$ between zero and one. A machine guesses every outcome by randomly choosing a number from the unit interval with uniform probability, yielding the predictions $\hat{y}_1, \ldots, \hat{y}_M$. Prove the following inequality:

$$E\left[\left(\frac{1}{M} \sum_{m=1}^{M} (Y_m - \hat{Y}_m)^2\right)^{\frac{1}{2}}\right] < 0.58$$

where $Y_1, \ldots, Y_M$ and $\hat{Y}_1, \ldots, \hat{Y}_M$ are i.i.d. random variables that produce the target values and predictions, respectively.

*Hint:* Since the random variable $Y$ takes values within the unit interval, $E[(1 - Y) \cdot Y] \geq 0$.

**Exercise 10. Data transformations.** Consider the tasks of training a $K$-nearest neighbor classifier (using Euclidean distance), applying a linear regression, applying a logistic regression, and performing a $K$-means cluster analysis.

Determine for each task whether mean-centering or standardizing the features beforehand would have an impact on the results, assuming that the mean and standard deviation is estimated once and the same transformation applied to all datasets involved. What about performing a full Karhunen–Loève transform that preserves the dimensionality of the dataset? What about arbitrary affine transformations that are invertible?

*Hints:* Which of those transformations preserve distances between data points? How does the objective function change?

**Exercise 11. Parkinson's disease detection.** For this exercise and the following exercises, use of the Python libraries pandas [9, 10] and scikit-learn [11] is recommended.

Speech disorders can be an early sign of motor impairment in Parkinson's disease [12]. The Oxford Parkinson's Disease Detection Dataset [13, 14] can be used to demonstrate that biomedical voice measurements are indicative of whether the patient suffers from Parkinson's. The dataset is available at the UC Irvine Machine Learning Repository:

$$\text{https://doi.org/10.24432/C59C74}$$

Write a computer program that will:

- load the dataset; clean/normalize the data if necessary,
- standardize the features (i.e., compute the $z$-scores),
- split the data into training and testing data,
- train an $L_2$-regularized logistic regression classifier to predict whether the voice recording comes from a patient who suffers from Parkinson's disease, with the regularization parameter determined from cross-validation aimed at optimizing the $F_1$ score,
- train a $K$-nearest neighbor classifier for the same purpose with $K$ determined from cross-validation,
- train a neural network with a single hidden layer, the width of which is determined from cross-validation,
- test all three classifiers and print the optimal hyperparameter, precision, recall, $F_1$ score, and accuracy.

Compare the performance of the classifiers with *class majority assignment* as the baseline, i.e., with the performance of a machine that would assign every data record the class that appears most frequently in the dataset.

**Exercise 12. Customer segmentation.** The UC Irvine Machine Learning Repository includes a dataset [15] of the annual spending of clients of a wholesale distributor on diverse product categories:

https://doi.org/10.24432/C5030X

Write a computer program that will:

- load the dataset and drop the *Region* column,

- filter the data records for clients that are restaurants, hotels, and cafés (i.e., drop clients from retail),

- normalize the data by computing the *proportionate* spending per product category (i.e., each row in the data table sums to one),

- standardize the data,

- use $K$-means to cluster the distributor's customers based on those data into $K = 4$ groups,

- print the cluster centroids.

Offer a short description of each group of clients based on their annual spending across product categories.

**Exercise 13. Concrete compressive strength prediction.** The following dataset [16, 17] is available at the UC Irvine Machine Learning Repository which can be used to demonstrate how the compressive strength of concrete can be predicted from the composition and age of the material:

https://doi.org/10.24432/C5PK67

We can apply a Yeo–Johnson transform to each feature in a way that the distribution of the variable more closely resembles a normal distribution. Write a computer program that will:

- load the dataset and perform the necessary preprocessing,

- perform a ridge regression and train a neural network, with hyperparameters tuned by cross-validation,

- using the raw data and the Yeo–Johnson transformed data,

- test all four regression algorithms and print the optimal hyperparameter, coefficient of determination $r^2$, and the root mean squared error.

Compare the performance of the regression models with the performance of a machine that would assign every data record the mean of the target variable.

## A.2 Answers

**Answer 1. Entity–relationship model.**

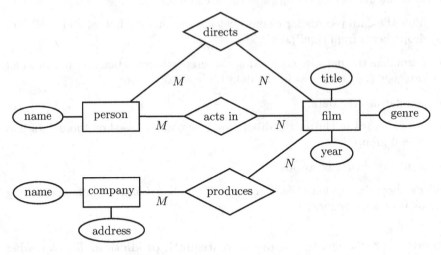

**Fig. A.3.** *Entity–relationship diagram of movie productions*

**Answer 2. Data quality assessment.**

- typographic errors: almost all instances of "University" are misspelled as "Univeristy"

- duplicates, incomplete disambiguation: the organization Oxford University is represented by two data records

- **id**: Boston Children's Hospital and Oxford University both reference the same primary key

- **name**: incomplete/inconsistent harmonization: some company names include the organization type ("Inc.", "Ltd."), some do not; types are abbreviated but not always ("Broadcom Limited")

- **cc**: "UK" is not a valid country code; "ZZ" may indicate missing data

- **city**: missing data (5 out of 11 data records); inconsistent format for missing data: NULL, ⟨NA⟩, empty string

- **hospital**: all the Boolean values have been flipped

**Answer 3. Exploratory data analysis.**

1. 46% of the respondents have a PhD, and 42% are female. The distribution of age and years of experience can be visualized with histograms; bar charts can be used to chart the distribution of domain and position: see Fig. A.6.

For each respondent, if more than one job position was stated, the higher position was selected.

2. More than 50% strongly agree with at least one of the following statements:

   - PEU1: "Wikipedia is user-friendly"

   - SA3: "It is important that students become familiar with online collaborative environments"

   Less than 50% engage in more than one of the following activities:

   - Use2: using Wikipedia as a platform to develop educational activities with students

   - Exp4: contributing to Wikipedia

   Charts for illustration are shown in Fig. A.7.

3. For each domain, we can summarize the responses to the question Exp1: "I consult Wikipedia for issues related to my field of expertise." The following stacked bar chart shows that in Engineering & Architecture, at least 50% of faculty members use Wikipedia often or very often. On the other hand, the majority of faculty members from Law & Politics rarely or never consult Wikipedia to inform on issues related to their field.

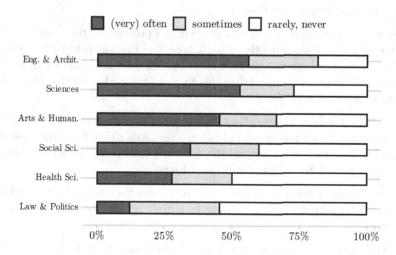

**Fig. A.4.** *Frequency of faculty members consulting Wikipedia by field of expertise*

As the following figure shows, a similar association can be observed for the practice of citing Wikipedia—which, however, appears to be much less common across all fields. This suggests that Wikipedia is being used by academic authors as a starting point for research more often than what could be inferred from citation counts.

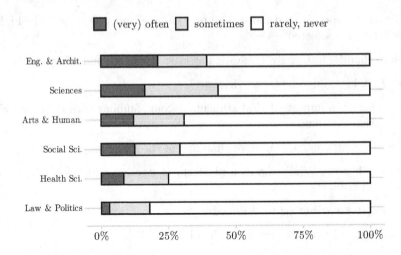

**Fig. A.5.** *Frequency of faculty members citing Wikipedia in academic papers*

**Answer 4. Base rate fallacy.** Cf. [18, 19]. It is true that if COVID vaccines were to provide full protection from the disease, then no vaccinated individuals would get infected or sick. Consequently, we would find no such individuals among the fatalities of the disease.

However, in reality, vaccines do not provide full protection; they are not 100% effective [20, 21]. Consequently, a high base vaccination rate in the general population implies high vaccination rates among COVID fatalities as well. In order to understand and internalize this fact, it may be instructive to consider the following hypotheticals:

- If COVID vaccines had no effect on mortality at all, we would find that the proportion of vaccinated individuals among those dying from the disease is equal to the proportion among the general population, i.e., 95% for higher age groups and 80% for individuals of lower age.

- If every individual among the population was vaccinated, we would find a vaccination rate of 100% among fatalities, no matter how effective the vaccine.

The actual data show that the proportion of vaccinated individuals among those dying from COVID-19 is considerably lower than the proportion of vaccinated people among the general population. Thus, the data do not support the hypothesis that COVID vaccines are generally harmful.

Note that adverse, possibly severe reactions after vaccination do happen. However, they are rare [22].

**Answer 5. Birthday problem.** Cf. [23, Sect. 4.7]. First, let us consider a situation when $N > 365$, i.e., there are more people than possible birthdays. In

**Fig. A.6.** *Demographics of the wiki4HE survey*

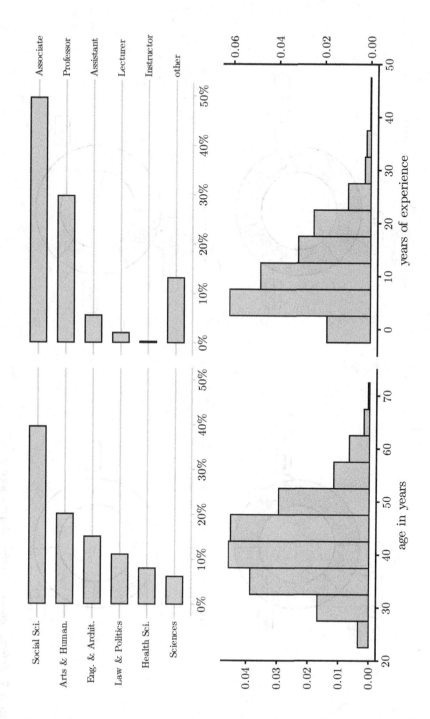

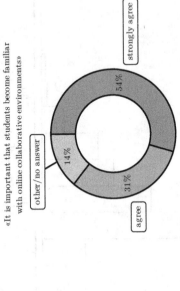

«It is important that students become familiar
with online collaborative environments»

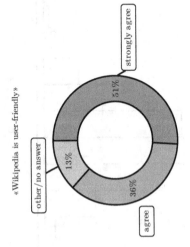

«Wikipedia is user-friendly»

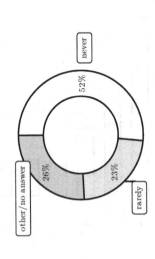

«I contribute to Wikipedia
(editions, revisions, articles improvement…) »

«I use Wikipedia as a platform to develop
educational activities with students»

**Fig. A.7.** *Selected answers to the wiki4HE survey*

that case, there must be at least two persons who share a birthday because there is no way to distribute the birthdays among the people without[2] repetition.

Now, say that we have selected one person, and afterwards we then select another. The two persons have different birthdays if the second person doesn't happen to have the same birthday as the first. This event has the probability $364/365 = 1 - \frac{1}{365}$. A third person has one of the remaining 363 birthdays with probability $363/365 = 1 - \frac{2}{365}$. And so on and so forth. Independent probabilities multiply, so if $N \leq 365$, then the probability that all of the $N$ people have a different birthday is given by the following product:

$$\left(1 - \frac{1}{365}\right) \cdot \left(1 - \frac{2}{365}\right) \cdots \left(1 - \frac{N-1}{365}\right) = \prod_{n=1}^{N-1} \left(1 - \frac{n}{365}\right)$$

The probability $p_N$ that at least two people share the same birthday is given by one minus this number because that is the complementary event. The following figure is a plot of that probability as a function of $N$:

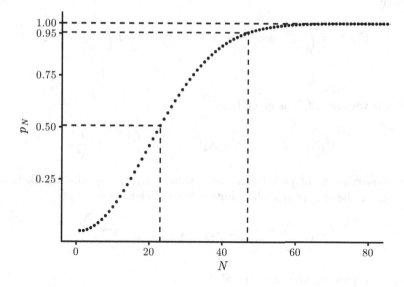

**Fig. A.8.** *Probability that two people share a birthday*

As noted before, $p_N = 100\%$ if $N \geq 366$. Furthermore, $p_N \geq 50\%$ if $N \geq 23$ and $p_N \geq 95\%$ if $N \geq 47$. Thus, in a room with more than 50 people, it is very likely that at least two of them share a birthday.

**Answer 6. Exponential distribution.** The cumulative distribution function of an exponentially distributed random variable $X$ is given as follows, with $u > 0$:

---

[2] This is an application of the so-called *pigeonhole principle*.

$$F_X(u) = \int_{-\infty}^{u} \text{Exp}(\xi|\lambda)\,\mathrm{d}\xi = \int_0^u \lambda e^{-\lambda\xi}\,\mathrm{d}\xi$$

$$= -e^{-\lambda\xi}\Big|_{\xi=0}^{u} = 1 - e^{-\lambda u}$$

The median of $X$ is the number $m[X]$ where $F_X(m[X]) = 1/2$, that is:

$$1 - e^{-\lambda \cdot m[X]} = \frac{1}{2} \Leftrightarrow m[X] = \frac{\ln(2)}{\lambda}$$

The mean is given by the following integral which can be solved by integration by parts:

$$E[X] = \int_0^u \xi \cdot \lambda e^{-\lambda\xi}\,\mathrm{d}\xi = \lambda\xi e^{-\lambda\xi}\Big|_{\xi=0}^{\infty} + \int_0^\infty e^{-\lambda\xi}\,\mathrm{d}\xi$$

$$= 0 - 0 + \frac{-1}{\lambda}e^{-\lambda\xi}\Big|_{\xi=0}^{\infty} = \frac{1}{\lambda}$$

Similarly:

$$E[X^2] = \int_0^u \xi^2 \cdot \lambda e^{-\lambda\xi}\,\mathrm{d}\xi = \lambda\xi^2 e^{-\lambda\xi}\Big|_{\xi=0}^{\infty} + 2\int_0^\infty \xi e^{-\lambda\xi}\,\mathrm{d}\xi$$

$$= \frac{2}{\lambda^2}$$

Thus, the variance of $X$ is given by:

$$\sigma^2[X] = E[X^2] - (E[X])^2 = \frac{2}{\lambda^2} - \left(\frac{1}{\lambda}\right)^2 = \frac{1}{\lambda^2}$$

Given observations of positive numeric values $x_1, \ldots, x_N$, the log-likelihood function for the exponential distribution is the following:

$$\ell(\lambda) = \sum_{n=1}^{N} \ln(\lambda e^{-\lambda x_n}) = N\ln(\lambda) - \lambda\sum_{n=1}^{N} x_n$$

We can compute its first and second derivative:

$$\frac{\mathrm{d}\ell}{\mathrm{d}\lambda}(\lambda) = \frac{N}{\lambda} - \sum_{n=1}^{N} x_n,$$

$$\frac{\mathrm{d}^2\ell}{\mathrm{d}\lambda^2}(\lambda) = -\frac{N}{\lambda^2} < 0$$

The second derivative is negative everywhere, so the log-likelihood is maximized at the zero of the first derivative:

$$\hat{\lambda} = N \cdot \left(\sum_{n=1}^{N} x_n\right)^{-1} = (\bar{x}_{\text{arithm}})^{-1}$$

**Answer 7. Box–Muller transform.** Cf. [24]. We define the following new random variables which represent the random vector $(X, Y)^T$ in polar coordinates:

$$R = \sqrt{-2\ln V}, \; \Theta = 2\pi W$$

The joint probability distribution functions are related as follows:

$$p_{X,Y}(r \cdot \cos\theta, r \cdot \sin\theta) = \frac{1}{r} \cdot p_{R,\Theta}(r, \theta)$$

where the factor "$1/r$" is produced by the Jacobian determinant.

Since $V$ and $W$ are independent, $R$ and $\Theta$ are independent as well:

$$p_{R,\Theta}(r, \theta) = p_R(r) \cdot p_\Theta(\theta)$$

Clearly, $\Theta$ provides a uniformly distributed polar angle. Thus, on the interval of interest, the density is constant: $p_\Theta(\theta) = \frac{1}{2\pi}$. As for the radial variable, we notice that

$$f(v) = \sqrt{-2\ln v} \Rightarrow f^{-1}(r) = e^{-\frac{r^2}{2}}$$

and apply the usual transformation formula:

$$p_R(r) = p_{f(V)}(r) = p_V(f^{-1}(r)) \cdot \left| \frac{\mathrm{d}}{\mathrm{d}r} f^{-1}(r) \right| = r \cdot e^{-\frac{r^2}{2}}$$

Plugging these results into the formula for the joint probability density yields the following result:

$$p_{X,Y}(r \cdot \cos\theta, r \cdot \sin\theta) = \frac{1}{r} \cdot p_R(r) \cdot p_\Theta(\theta)$$
$$= e^{-\frac{r^2}{2}} \cdot \frac{1}{2\pi}$$

Expressing this result in Cartesian coordinates $(x, y)$ with $r^2 = x^2 + y^2$ shows that it is the product of two standard normal distributions:

$$p_{X,Y}(x, y) = e^{-\frac{x^2+y^2}{2}} \cdot \frac{1}{2\pi}$$
$$= \frac{1}{\sqrt{2\pi}} e^{-\frac{x^2}{2}} \cdot \frac{1}{\sqrt{2\pi}} e^{-\frac{y^2}{2}}$$

**Answer 8. Simpson's paradox.** Cf. [25, 26] and [27, Chap. 2, Sect. 4]. The data are Bernoulli distributed with an estimated probability of success/admission $p_x$ and $p_y$ for female and male applicants, respectively. We can calculate the variance from those probabilities: $s^2(x) = p_x(1 - p_x)$ and $s^2(y) = p_y(1 - p_y)$. Thus, the test score for the one-sided $Z$-test computes to:

$$z^*(0.99) \cdot \sqrt{\frac{p_x(1 - p_x)}{N_x} + \frac{p_y(1 - p_y)}{N_y}} =$$

$$2.33 \cdot \sqrt{\frac{0.3 \cdot (1 - 0.3)}{1835} + \frac{0.45 \cdot (1 - 0.45)}{2691}} = 3.3\%$$

This $Z$-test gives a smaller value than the observed difference of 15%. Thus, at a confidence level of 99%, the data are not consistent with the null hypothesis "female applicants are at least as likely to be admitted as male applicants."

For each department, we can compute the two-sided test score ($z(0.99) = 2.58$) and compare that score with the absolute value of the observed difference in admission rate, $|p_y - p_x|$. From the available data, we can also compute the rate of admission for each department, irrespective of gender:

$$p = \frac{N_x \cdot p_x + N_y \cdot p_y}{N_x + N_y}$$

The results are summarized in a table:

| dept. | $p$ | $N_x$ | $N_y$ | $|p_y - p_x|$ | $z$ |
|-------|-----|-------|-------|---------------|-----|
| A | 64% | 108 | 825 | -20% | 10% |
| B | 63% | 25 | 560 | -5% | 25% |
| C | 35% | 593 | 325 | +3% | 9% |
| D | 34% | 375 | 417 | -2% | 9% |
| E | 25% | 393 | 191 | +4% | 10% |
| F | 6% | 341 | 373 | -1% | 5% |
| all | 39% | 1835 | 2691 | +15% | 4% |

**Table A.3.** *Analysis of university admission rates*

Only department A exhibits a significant difference in admission rates. Contrary to the comparison across all departments, it is *in favor* of female applicants.

Departments A and B are the departments where applicants are the most likely to succeed in being admitted, by a large margin. 51% of male applicants choose these departments to apply for but only 7% of all female applicants do so. Therefore, the data are consistent with the hypothesis that female applicants are more likely to apply for more competitive studies, which implies that they are more likely to be rejected.

**Answer 9. Expected root mean squared error of random guessing.** For convenience, we write $Y = Y_m$ and $\hat{Y} = \hat{Y}_m$.

The predictions are chosen from a continuous uniform distribution on the unit interval, and we know the mean and the variance of this distribution:

$$E[\hat{Y}] = \frac{1}{2}, \ \sigma^2[\hat{Y}] = \frac{1}{12}$$

Those values imply $E[\hat{Y}^2] = (E[\hat{Y}])^2 + \sigma^2[\hat{Y}] = 1/3$. Furthermore:

$$E[(Y - \hat{Y})^2] = E[Y^2] - 2E[Y] \cdot E[\hat{Y}] + E[\hat{Y}^2]$$
$$= E[Y^2] - E[Y] + \frac{1}{3} = -E[Y \cdot (1 - Y)] + \frac{1}{3}$$
$$\leq \frac{1}{3}$$

We can apply Jensen's inequality (the square root is a concave function) and the fact that the random variables are i.i.d.:

$$E\left[\left(\frac{1}{M} \sum_{m=1}^{M} (Y_m - \hat{Y}_m)^2\right)^{\frac{1}{2}}\right] \leq \left(E\left[\frac{1}{N} \sum_{m=1}^{M} (Y_m - \hat{Y}_m)^2\right]\right)^{\frac{1}{2}}$$
$$= \sqrt{E[(Y - \hat{Y})^2]} \leq \sqrt{\frac{1}{3}} < 0.58$$

**Answer 10. Data transformations.** An invertible affine transformation applied to every data point has no impact on the results of either linear regression or logistic regression.

The argument in more detail: the linear term in either objective function that depends on the model parameters $w_0, \ldots, w_D$ can be written as $\sum_{d=0}^{D} x_d w_d$ where $x = (x_0, \ldots, x_D)^T$ is one of the data points augmented by a one in the zeroth component. Suppose that we transform the data point(s): $z = A \cdot x$ with $A$ being an invertible matrix of format $(D + 1) \times (D + 1)$. A short calculation shows that applying the transformation to the data is the same operation as applying its transpose to the model parameters:

$$\sum_{d=0}^{D} z_d w_d = \sum_{d=0}^{D} \left(\sum_{k=0}^{D} A_{dk} x_k\right) w_d = \sum_{d=0}^{D} x_d \left(\sum_{k=0}^{D} A_{kd} w_k\right)$$

In terms of the objective function $R(\text{parameters}; \text{data})$, we may therefore write: $R(w; z) = R(w; A \cdot x) = R(A^T \cdot w; x)$. Applying the chain rule yields the following result:

$$\text{grad } R(w; z) = A^T \cdot \text{grad } R(A^T \cdot w; x)$$

Thus, $\hat{v}$ is a critical point of the objective function with respect to the transformed data if and only if $\hat{w} = A^T \cdot \hat{v}$ is a critical point with respect to the original data. Both situations are equivalent because they yield the same predictions for the target/latent variable when applied to test data $x_*$ or $z_*$, respectively:

$$f(x_*; \hat{w}) = \hat{w}^T \cdot x_* = \left(A^T \cdot \hat{v}\right)^T \cdot A^{-1} \cdot z_*$$
$$= \hat{v}^T \cdot A \cdot A^{-1} \cdot z_* = \hat{v}^T \cdot z_*$$
$$= f(z_*; \hat{v})$$

Mean-centering preserves pairwise distances between data points. Applied to the already mean-centered data, the full Karhunen–Loève transform is an orthogonal map, and therefore it preserves distances as well. This means that neither mean-centering nor the full Karhunen–Loève transform have an impact on determining the nearest neighbors of data points. They also don't change the value of the objective function for $K$-means cluster analysis.

Computing the standard scores, on the other hand, implies dividing by the variance which is a scaling operation that is, in general, non-uniform. Non-uniform scaling does not preserve distances. Thus, if the distance measure is not transformed as well, standardizing the data has an impact[3] on determining the $K$ nearest neighbors, or clusters with the $K$-means algorithm.

**Answer 11. Parkinson's disease detection.** This and the following solutions have been written in Python. The code can be executed, for example, in Colab [4].

First, we can download the data directly from the UC Irvine Machine Learning Repository:

```
import pandas as pd

url = "https://archive.ics.uci.edu/ml/
    machine-learning-databases/parkinsons/parkinsons.data"
df = pd.read_csv(url)
df.set_index("name", inplace = True, drop = True)
```

Next, we apply some prepocessing to the data:

```
# drop rows with missing values (if present)
df.dropna(inplace = True)

# set proper data types
df = df.astype(float)

# set named labels
labels = ['healthy', 'PD']
df["status"] = df["status"].replace({0: labels[0], 1:
    labels[1]})
```

The following code will randomly select 60% of the data (117 records) to be used for training. The remaining data records (78) are held back for testing:

---

[3] In practice, this effect of standardization is often welcomed, as standardization eliminates the need for a choice of units and leads to an equal contribution from each feature.

```
from sklearn.model_selection import train_test_split

X = df.drop(["status"], axis = 1)
y = df["status"]
X_train, X_test, y_train, y_test = train_test_split(X, y,
→test_size = 0.4, random_state = 42)
```

With a few lines of code, we can standardize the data. Since the algorithm should not be informed by the test data in any way during the training step, we estimate the mean and standard deviation from the training data only:

```
from sklearn.preprocessing import StandardScaler

scaler = StandardScaler()
Z_train = scaler.fit_transform(X_train)
Z_test = scaler.transform(X_test)
```

We can summarize the models for the experiments to be run in a single dictionary. For each model, we define a range of hyperparameters to be explored via grid search. The distance measure for the K-nearest neighbor classifier is not specified in the exercise; we can decide to use the Euclidean metric. The hidden layer of the neural network is activated with a rectified linear unit:

```
import numpy as np
from sklearn.linear_model import LogisticRegression
from sklearn.neighbors import KNeighborsClassifier
from sklearn.neural_network import MLPClassifier
from sklearn.model_selection import GridSearchCV
from sklearn.metrics import f1_score, make_scorer,
→classification_report

experiments = [
            {"name": "logistic regression w/ L2",
             "model": LogisticRegression(penalty = "12",
→random_state = 42, solver = "lbfgs", max_iter = 1000),
             "param_grid": {"C": np.arange(0.05, 1.5, 0.05)}},
            {"name": "KNN",
             "model": KNeighborsClassifier(metric = "euclidean"),
             "param_grid": {"n_neighbors": np.arange(1, 20)}},
            {"name": "neural network",
             "model": MLPClassifier(solver = "lbfgs", alpha = 0.
→0, random_state = 42, max_iter = 1000, activation = "relu"),
             "param_grid": {"hidden_layer_sizes": np.arange(5,
→20)}}
        ]
```

Finally, for every model, we train and determine the hyperparameters via 5-fold cross-validation, and afterwards we then print a classification report that includes the measures for goodness of fit asked for by the exercise:

```
scorer = make_scorer(f1_score, pos_label = "PD")
result = pd.DataFrame(columns = ["name", "accuracy",␣
    ↪"precision", "recall", "f1 score"])
for exp in experiments:
    # train and validate the model
    model = GridSearchCV(exp["model"], exp["param_grid"], cv =␣
    ↪5, scoring = scorer)
    model.fit(Z_train, y_train)

    # test the model
    y_pred = model.predict(Z_test)

    print(exp["name"], "-- opt. hyperparam.:", model.
    ↪best_params_)
    print("\n", classification_report(y_test, y_pred,␣
    ↪target_names = model.classes_), "\n")
```

The optimal regularization parameter for the logistic regression is given by $\beta = 1.43$; the optimal number of neighbors for KNN is given by $K = 1$; and the optimal width of the hidden layer of the neural network was determined to be $D_1 = 9$.

75% of the data records in the test dataset describe audio recordings of patients suffering from Parkinson's disease. Therefore, any machine learning algorithm we train should achieve an accuracy of at least 75% in order to beat majority class assignment. We can summarize the information from the reports as follows:

| | precision | recall | $F_1$-score | accuracy |
|---|---|---|---|---|
| logistic regression ($L_2$) | 89% | 95% | 92% | 87% |
| KNN | 98% | 97% | 97% | 96% |
| neural network | 93% | 95% | 94% | 91% |
| majority vote (baseline) | 74% | 100% | 85% | 74% |

**Table A.4.** *Performance of classifiers applied to the Oxford Parkinson's Disease Detection Dataset*

**Answer 12. Customer segmentation.** We download the data from the UC Irvine Machine Learning Repository and then apply the necessary steps of normalization and standardization:

```
import pandas as pd
```

```
url = "https://archive.ics.uci.edu/ml/
    ↪machine-learning-databases/00292/
    ↪Wholesale%20customers%20data.csv"
X = pd.read_csv(url)

# Channel 1: hotel/restaurant/cafe
# Channel 2: retail
X.drop(X[X["Channel"] == 2].index, inplace = True)
X.drop(columns = ["Region", "Channel"], inplace = True)

# store proportions of spendings instead
X = X.div(X.sum(axis = 1), axis = 0)

# standardize data
Z = (X - X.mean()) / X.std()
```

We run the *K*-means algorithm on the prepared data:

```
from sklearn.cluster import KMeans
K = 4

kmeans = KMeans(n_clusters = K, random_state=42, n_init
    ↪="auto")
kmeans.fit(Z)
labels = pd.DataFrame({"cluster": kmeans.predict(Z)}, index =
    ↪X.index)
Z = Z.join(labels)
```

Finally, we summarize the cluster centroids into one table:

```
Z.groupby("cluster").mean().round(2)
```

The above command produces the following output when printed:

| cluster | Fresh | Milk | Grocery | Frozen | Detergents_Paper | Delicassen |
|---|---|---|---|---|---|---|
| 0 | -1.31 | 0.94 | 1.63 | -0.67 | 1.73 | -0.40 |
| 1 | 0.85 | -0.49 | -0.45 | -0.43 | -0.33 | -0.34 |
| 2 | -0.30 | -0.34 | -0.34 | 1.43 | -0.25 | -0.18 |
| 3 | -0.93 | 1.13 | 0.44 | -0.30 | -0.10 | 1.58 |

**Table A.5.** *Cluster centroids of customer segments produced by K-means*

Compared with customers from other groups, those in group #1 spend mostly on fresh products, whereas those in group #2 prefer frozen products. Customers from the remaining groups spend above average on milk and grocery products.

> However, customer group #0 spends significantly more on detergents/paper products, whereas #3 predominantly buys delicatessen products.

**Answer 13. Concrete compressive strength prediction.** The loading and preprocessing of the data are similar to what is required for the exercise on Parkinson's detection. For the solution of this exercise, we select 70% of the data (721 records) for training and 309 data records for testing.

Afterwards, we apply the Yeo–Johnson transform. To make sure that all variables operate on the same scale, we standardize both raw data and transformed data.

```
from sklearn.preprocessing import StandardScaler
from sklearn.preprocessing import PowerTransformer

linscaler = StandardScaler()
Z_train = linscaler.fit_transform(X_train)
Z_test = linscaler.transform(X_test)

powerscaler = PowerTransformer(method = "yeo-johnson",
  ↪standardize = True)
W_train = powerscaler.fit_transform(X_train)
W_test = powerscaler.transform(X_test)
```

The code for training and testing is also very similar; we use 8-fold cross validation to tune the hyperparameters:

```
import numpy as np
from sklearn.linear_model import Ridge
from sklearn.neural_network import MLPRegressor
from sklearn.model_selection import GridSearchCV
from sklearn.metrics import r2_score, mean_squared_error

experiments = [
    {"name": "ridge regression",
     "model": Ridge(max_iter = 1000, solver = "svd",
  ↪random_state = 42),
     "param_grid": {"alpha": np.arange(0.0, 5.0, 0.01)},
     "powertrans": False},
    {"name": "Yeo-Johnson + ridge regression",
     "model": Ridge(max_iter = 1000, solver = "svd",
  ↪random_state = 42),
     "param_grid": {"alpha": np.arange(0.0, 5.0, 0.01)},
     "powertrans": True},
    {"name": "neural network",
     "model": MLPRegressor(solver = "lbfgs", alpha = 0.0,
  ↪random_state = 42, max_iter = 2000, activation = "relu"),
```

```
      "param_grid": {"hidden_layer_sizes": np.arange(3, 7)},
      "powertrans": False},
     {"name": "Yeo-Johnson + neural network",
      "model": MLPRegressor(solver = "lbfgs", alpha = 0.0,␣
 ↪random_state = 42, max_iter = 2000, activation = "relu"),
      "param_grid": {"hidden_layer_sizes": np.arange(3, 7)},
      "powertrans": True}
]

for exp in experiments:
  # train and validate the model
  model = GridSearchCV(exp["model"], exp["param_grid"], cv =␣
 ↪8, scoring = "r2")
  if exp["powertrans"]:
    model.fit(W_train, y_train)
    y_pred = model.predict(W_test)
  else:
    model.fit(Z_train, y_train)
    y_pred = model.predict(Z_test)

  # test the model
  print(exp["name"], "-- opt. hyperparam.:", model.
 ↪best_params_)
  print("\nR^2 score:", round(r2_score(y_test, y_pred), 3))
  print("RMSE:", round(mean_squared_error(y_test, y_pred,␣
 ↪squared = False), 3), "\n")
```

The optimal regularization parameter for the ridge regression is given by $\beta = 2.17$ when using the raw data and $\beta = 3.67$ based on the transformed data. The optimal number of neurons in the hidden layer of the neural network is given by $D_1 = 3$ on the raw data and $D_1 = 6$ on the transformed data.

The mean squared error for the predictor that assigns every outcome the mean of the target variable is equal to the variance of the target variable, which implies that the coefficient of determination vanishes. The standard deviation of the compressive strength values (estimated from the test data) is given by 16.5 MPa.

|  | $r^2$ | RMSE |
|---|---|---|
| ridge regression | 60% | 10.5 MPa |
| Yeo–Johnson + ridge regression | 80% | 7.4 MPa |
| neural network | 82% | 7.1 MPa |
| Yeo–Johnson + neural network | 86% | 6.3 MPa |
| mean (baseline) | 0% | 16.5 MPa |

**Table A.6.** *Performance of regression models applied to the Concrete Compressive Strength Dataset*

# References

[1] Eduard Aibar et al. *wiki4HE*. UCI Machine Learning Repository. 2015. DOI: 10.24432/C50031.

[2] Antoni Meseguer-Artola et al. "Factors that influence the teaching use of Wikipedia in higher education". In: *Journal of the Association for Information Science and Technology* 67.5 (Feb. 2015), pp. 1224–1232. DOI: 10.1002/asi.23488.

[3] RStudio Team. *RStudio: Integrated Development Environment for R*. Posit, PBC. Boston, MA, 2022. URL: http://www.rstudio.com/.

[4] Google Research. *Colaboratory. Frequently Asked Questions*. Accessed April 1, 2023. URL: https://research.google.com/colaboratory/faq.html.

[5] Andrew Bridgen [@ABridgen]. *Devastating stats from New Zealand, how can anyone deny the vaccine harms now? UK stats May to Dec 2022 should be released tomorrow*. Twitter. Feb. 20, 2023. URL: https://twitter.com/ABridgen/status/1627687430835916802.

[6] Statistics New Zealand. *COVID-19 data portal*. Accessed Feb. 22, 2023. URL: https://www.stats.govt.nz/experimental/covid-19-data-portal/.

[7] Edouard Mathieu et al. "A global database of COVID-19 vaccinations". In: *Nature Human Behaviour* 5.7 (May 2021), pp. 947–953. DOI: 10.1038/s41562-021-01122-8.

[8] New Zealand Ministry of Health. *COVID-19: Vaccine data*. Accessed Feb. 22, 2023. URL: https://www.health.govt.nz/covid-19-novel-coronavirus/covid-19-data-and-statistics/covid-19-vaccine-data.

[9] The pandas development team. *pandas-dev/pandas: Pandas 1.4.4*. Aug. 2022. DOI: 10.5281/zenodo.3509134.

[10] Wes McKinney. "Data Structures for Statistical Computing in Python". In: *Proceedings of the 9th Python in Science Conference*. Ed. by Stéfan van der Walt and Jarrod Millman. 2010, pp. 56–61. DOI: 10.25080/Majora-92bf1922-00a.

[11] F. Pedregosa et al. "Scikit-learn: Machine Learning in Python". In: *Journal of Machine Learning Research* 12 (2011), pp. 2825–2830.

[12] Andrew Ma, Kenneth K Lau, and Dominic Thyagarajan. "Voice changes in Parkinson's disease: What are they telling us?" In: *Journal of Clinical Neuroscience* 72 (Feb. 2020), pp. 1–7. DOI: 10.1016/j.jocn.2019.12.029.

[13] Max A. Little. *Oxford Parkinson's Disease Detection Dataset*. UCI Machine Learning Repository. 2008. DOI: 10.24432/C59C74.

[14] Max A. Little et al. "Suitability of Dysphonia Measurements for Telemonitoring of Parkinson's Disease". In: *IEEE Transactions on Biomedical Engineering* 56.4 (Apr. 2009), pp. 1015–1022. DOI: 10.1109/tbme.2008.2005954.

[15] Margarida Cardoso. *Wholesale Customers Dataset*. UCI Machine Learning Repository. 2014. DOI: 10.24432/C5030X.

[16] I-Cheng Yeh. *Concrete Compressive Strength*. UCI Machine Learning Repository. 2007. DOI: 10.24432/C5PK67.

[17] I.-C. Yeh. "Modeling of strength of high-performance concrete using artificial neural networks". In: *Cement and Concrete Research* 28.12 (Dec. 1998), pp. 1797–1808. DOI: 10.1016/s0008-8846(98)00165-3.

[18] Martin Bauer [@martinmbauer]. *This is a @ABridgen, a member of parliament in the UK with a science degree, unable to grasp the concept of a base rate. He is making a very strong case for more mandatory math education here.* Twitter. Feb. 21, 2023. URL: https://twitter.com/martinmbauer/status/1627820248144441346.

[19] Ulrich Hoffrage et al. "Communicating Statistical Information". In: *Science* 290.5500 (Dec. 2000), pp. 2261–2262. DOI: 10.1126/science.290.5500.2261.

[20] Daniel R. Feikin et al. "Duration of effectiveness of vaccines against SARS-CoV-2 infection and COVID-19 disease: results of a systematic review and meta-regression". In: *The Lancet* 399.10328 (Mar. 2022), pp. 924–944. DOI: 10.1016/s0140-6736(22)00152-0.

[21] Jonathan J. Lau et al. "Real-world COVID-19 vaccine effectiveness against the Omicron BA.2 variant in a SARS-CoV-2 infection-naive population". In: *Nature Medicine* 29.2 (Jan. 2023), pp. 348–357. DOI: 10.1038/s41591-023-02219-5.

[22] U.S. National Center for Immunization and Respiratory Diseases (NCIRD), Division of Viral Diseases. *Selected Adverse Events Reported after COVID-19 Vaccination.* Accessed Feb. 23, 2023. URL: https://www.cdc.gov/coronavirus/2019-ncov/vaccines/safety/adverse-events.html.

[23] I. J. Good. *Probability and the Weighing of Evidence.* Charles Griffin, 1950.

[24] George E. P. Box and Mervin E. Muller. "A Note on the Generation of Random Normal Deviates". In: *The Annals of Mathematical Statistics* 29.2 (June 1958), pp. 610–611. DOI: 10.1214/aoms/1177706645.

[25] Colin R. Blyth. "On Simpson's Paradox and the Sure-Thing Principle". In: *Journal of the American Statistical Association* 67.338 (June 1972), pp. 364–366. DOI: 10.1080/01621459.1972.10482387.

[26] P. J. Bickel, E. A. Hammel, and J. W. O'Connell. "Sex Bias in Graduate Admissions: Data from Berkeley". In: *Science* 187.4175 (Feb. 1975), pp. 398–404. DOI: 10.1126/science.187.4175.398.

[27] David Freedman, Robert Pisani, and Roger Purves. *Statistics.* 4th ed. W. W. Norton & Company, Feb. 2007.

# B

## Mathematical preliminaries

This appendix provides a quick reference for the mathematical concepts from linear algebra and multivariate calculus that were used in this book. It is not intended to serve as a textbook (see, for example, [1, 2, 3, 4, 5]), nor does it replace a proper course on the subject matter. Basic arithmetic and univariate calculus are assumed to be familiar subjects to the reader.

## B.1 Basic concepts

### B.1.1 Numbers and sets

**Real numbers** can be used to quantify the result of a measurement or computation to arbitrary precision. A real number may be positive, zero, or negative. Examples of real numbers are: $0.35$, $-\frac{1}{2}$, $\sqrt{2}$, $\pi = 3.14159\ldots$.

Real numbers can be *ordered*: either two real numbers are equal or one of them is smaller than the other. We write $x < y$ if the number $x$ is smaller than $y$, for example: $1.1 < 1.2$. We write $x \leq y$ to mean "$x$ is less than or equal to $y$."

The **absolute value** $|x|$ of a real number $x$ is its nonnegative value without regard to its sign, for example: $|-2| = 2$, or $|2| = 2$. The **floor function** rounds a real number down to the closest integer, while the **ceiling function** rounds up: $\lfloor 2.3 \rfloor = 2$, $\lceil 2.3 \rceil = 3$.

A **natural number** is a nonnegative integer. Natural numbers can be used for counting or enumerating objects.

A **set** $M$ is a "gathering together of definite, distinct objects of our intuition or of our thought (which are called the 'elements' of $M$) into a whole" (translated[1] from [6]). An example of a set that contains three natural numbers is the

---

[1] *Unter einer ‚Menge' verstehen wir jede Zusammenfassung M von bestimmten wohlunterschiedenen Objecten unsrer Anschauung oder unseres Denkens (welche die ‚Elemente' von M genannt werden) zu einem Ganzen.*

© Springer-Verlag GmbH Germany, part of Springer Nature 2023
M. Plaue, *Data Science*, https://doi.org/10.1007/978-3-662-67882-4

following: $M = \{2, 3, 5\}$. The order of elements plays no role in the definition of a set: $\{2, 3, 5\} = \{5, 3, 2\}$.

We write $|M|$ for the number of elements contained in $M$. However, a set may also contain an infinite number of elements. The set of *all* natural numbers is an example of such a set: $\mathbb{N} = \{0, 1, 2, 3, 4, \ldots\}$. The set of all real numbers is another example, denoted by $\mathbb{R}$.

A set $M$ is called of **countably infinite** size if there exists a one-to-one correspondence between elements of $\mathbb{N}$ and elements of $M$. Such a correspondence is also called a *bijective map*, see further below. In other words, although $M$ has an infinite number of elements, we may enumerate all of them without repetition. The set of integers is an example of a countably infinite set: $\mathbb{Z} = \{0, -1, 1, -2, 2, \ldots\}$.

Sets can be declared via properties of their elements. The set of prime numbers, for example, can be written as:

$$P = \{n|n \text{ is a natural number that has exactly two distinct divisors}\}$$

Given a set $M$, if $x$ is an element of $M$, we write $x \in M$ as a shorthand notation, and $x \notin M$ otherwise. The set with no elements is called the **empty set** and is denoted by $\emptyset$.

Basic operations on sets are the **intersection**, **union**, and **difference** of two sets $M$ and $N$:

$$M \cap N = \{x|x \in M \text{ and } x \in N\},$$
$$M \cup N = \{x|x \in M \text{ or } x \in N\},$$
$$M \setminus N = \{x|x \in M \text{ and not } x \in N\}$$

Note that, in mathematics, the words "and," "or," and "not" have well-defined meanings as part of a propositional expression that might not always coincide with everyday usage. In particular, "or" is *not* exclusive: an element that belongs to *both* sets is also an element of their union.

These operations can be illustrated with the following diagrams:

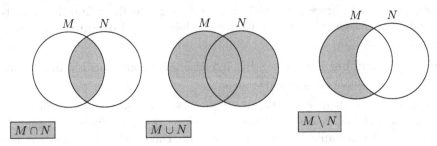

**Fig. B.1.** *Set operations*

Two sets $M$ and $N$ are called **disjoint** if they have no element in common: $M \cap N = \emptyset$.

If every element of $M$ is also an element of $N$, we call $M$ a **subset** of $N$. We write $M \subseteq N$ as the shorthand notation. Any set is also a subset of itself—if we want to exclude that situation, we write $M \subset N$ and call $M$ a *proper* subset of $N$.

Important subsets of the set of real numbers are the **intervals**, i.e., all numbers that lie between two given numbers $a$ and $b$:

$$]a, b[ = \{x | x \in \mathbb{R}, \, x > a, \, x < b\}$$

The above formula represents the definition of an *open* interval—if the endpoints are included, it is a *closed* interval:

$$[a, b] = \{x | x \in \mathbb{R}, \, x \geq a, \, x \leq b\}$$

Furthermore, we refer to *half-open* intervals that exclude only one of the endpoints, and *unbounded* intervals like the following:

$$[a, \infty[ = \{x | x \in \mathbb{R}, \, x \geq a\}$$

## B.1.2 Maps and functions

Let $X$ and $Y$ be sets. A **map** $f$ from $X$ to $Y$ assigns each element $x$ of $X$ to exactly one element $y$ of $Y$. The set $X$ is called the **domain** of $f$, and $Y$ the **codomain**. Given a map $f$ and some fixed element in its domain $x$, we write $f(x)$ to denote the unique element that $x$ is assigned to. We write the following as a shorthand for those statements:

$$f \colon X \to Y, \, x \mapsto f(x)$$

We may also write $f(\cdot)$ instead of $f$ to make it clear that the object in question is a map that takes an argument to produce a result.

A map can also be called a **function**, a term that is often used when the map produces real numbers. An example of a function is

$$\sqrt{\cdot} \colon [0, \infty[ \to \mathbb{R}, \, x \mapsto \sqrt{x}$$

which maps every nonnegative real number onto its square root. The term *transformation* is generally reserved for maps where domain and codomain coincide.

The **image** of some subset of the domain, $U \subseteq X$, under $f$ is the following subset of the codomain:

$$f(U) = \{y | y \in Y \text{ and there exists some } x \in U \text{ with } y = f(x)\}$$

The image of the whole domain under $f$—i.e., $f(X)$—is also simply called the image of $f$, or the **range** of $f$. The **preimage** of some subset of the codomain, $V \subseteq Y$, is defined as follows:

$$f^{-1}(V) = \{x | x \in X \text{ and there exists some } y \in V \text{ with } y = f(x)\}$$

By the definition of a map, we must always have $f^{-1}(Y) = X$, because every element of the domain needs to be assigned a value from the codomain. A map does not necessarily map onto its whole codomain, though. For example, the range of the exponential function

$$\exp: \mathbb{R} \to \mathbb{R}, \, x \mapsto e^x,$$

where e = 2.71828... is Euler's number, is given by the set of all positive real numbers.

We say that a map $f: X \to Y$ is **surjective** if $Y = f(X)$. It is called **injective** if $u \neq v$ implies $f(u) \neq f(v)$ for all $u, v \in X$. A map is **bijective** if and only if it is both surjective and injective. A simple but important example for a bijective map is the **identity map** on some set $X$, which assigns every element to itself:

$$\mathrm{id}_X: X \to X, \, x \mapsto x$$

Given two maps $f: X \to Y$ and $g: W \to Z$ with $f(Y) \subseteq W$, we may define the **composition** of those maps as follows:

$$g \circ f: X \to Z, \, x \mapsto g(f(x))$$

Given some map $f: X \to Y$, a map $g: Y \to X$ is called an **inverse** of $f$ if $g \circ f = \mathrm{id}_X$ and $f \circ g = \mathrm{id}_Y$ hold. It turns out that, if an inverse exists, it must be unique, and we may denote it by $f^{-1}$. Furthermore, a map has an inverse if and only if it is bijective.

If we restrict the codomain of the exponential function to the positive numbers, it has an inverse that is given by the natural logarithm:

$$\ln: \, ]0, \infty[ \to \mathbb{R}, \, x \mapsto \ln(x) = \log_e(x)$$

The following figure shows the graphs of those functions.

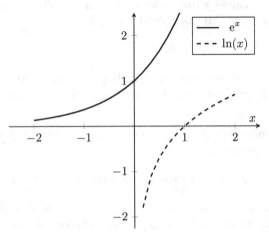

**Fig. B.2.** *Graphs of exponential and logarithmic functions*

We will make extensive use of both the exponential function and logarithms throughout this book. Therefore, we should recall some basic properties:

$$e^x \cdot e^y = e^{x+y}, \ (e^x)^y = e^{x \cdot y}, \ \ln(e^x) = x$$

for any $x, y \in \mathbb{R}$, and

$$\ln(x) + \ln(y) = \ln(x \cdot y), \ y \cdot \ln(x) = \ln(x^y), \ e^{\ln(x)} = x$$

for any $x, y \in \mathbb{R}$ with $x, y > 0$.

### B.1.3 Families, sequences and tuples

A **sequence** $s$ is an ordered, not necessarily finite list of not necessarily distinct objects. The famous Fibonacci sequence in which each number is the sum of the two preceding ones may serve as an example:

$$s = (s_1, s_2, \dots) = (0, 1, 1, 2, 3, 5, 8, 13, 21, 34, 55, \dots)$$

A **tuple** $t$ is a finite sequence, for example $t = (2, 5, 2)$. For sequences and tuples, the order of elements matters: $(2, 5, 2) \neq (2, 2, 5)$. A tuple with exactly two elements may be called an **(ordered) pair**.

The **Cartesian product** of a collection $M_1, \dots, M_D$ of sets is the set of all tuples with $D$ elements that we may build from elements of those sets:

$$M_1 \times \cdots \times M_D = \{(x_1, \dots, x_D) | x_1 \in M_1, \dots, x_D \in M_D\}$$

We can also define *Cartesian powers*. For example, the set of all tuples of real numbers of length $D$ may be written as follows:

$$\underbrace{\mathbb{R} \times \cdots \times \mathbb{R}}_{D \text{ times}} = \mathbb{R}^D = \{(x_1, \dots, x_D) | x_1 \in \mathbb{R}, \dots, x_D \in \mathbb{R}\}$$

We may identify nested tuples with simple tuples by appropriate omitting of parentheses, for example:

$$(x, (y, z)) \cong ((x, y), z) \cong (x, y, z)$$

By a similar abuse of notation, we may also identify a *singleton* with the unique element that it contains: $(x) \cong x$.

Finally, an **(indexed) family** is a map $A \colon I \to X$ where we call $I$ the *index set*, and that we rather write as $(A_i)_{i \in I}$, or just $(A_i)$ for short, instead of $i \mapsto A(i)$.

For example, an infinite sequence $s$ can be seen as a family that is indexed by the set of natural numbers: $(s_n)_{n \in \mathbb{N}}$—or $(s_n)_{n \in \{1, 2, \dots\}}$ if we prefer to start counting from one.

For an indexed family of sets—i.e., $A_i$ is a set for every $i \in I$—we may define the intersection/union of all of the sets in the family:

$$\bigcap_{i\in I} A_i = \{x | \text{for every } i \in I, x \in A_i\}$$

$$\bigcup_{i\in I} A_i = \{x | \text{there exists some } i \in I \text{ with } x \in A_i\}$$

### B.1.4 Minimum/maximum and infimum/supremum

Suppose that $M$ is a set of real numbers. The **minimum** of $M$ is the smallest of the numbers contained in $M$, and we denote this number by $\min M$. Similarly, the **maximum** is denoted by $\max M$. For example, $\min\{2,3,5\} = 2$ and $\max\{2,3,5\} = 5$. Similarly, we can determine the smallest and largest elements in a sequence of real numbers $(x_n)$, written as $\min_n\{x_n\}$ and $\max_n\{x_n\}$.

The minimum or maximum may not exist. For example, the open interval $]2,5[$ has neither minimum nor maximum.

We define a lower bound of $M$ to be any $a \in \mathbb{R}$ such that $a \le x$ for all $x \in M$. The **infimum** of $M$ is a greatest lower bound, i.e., a lower bound $b \in \mathbb{R}$ such that $a \le b$ for any lower bound $a$ of $M$. It turns out that if an infimum exists, it is uniquely determined. If the set has a minimum, it coincides with the infimum. The **supremum** is defined as the least upper bound. For example: $\inf ]2,5[ = 2$ and $\sup ]2,5[ = 5$.

We can apply those definitions to the range of a function $f\colon X \to \mathbb{R}$. In this case, we write $\inf_{\xi\in X}\{f(\xi)\}$ and $\sup_{\xi\in X}\{f(\xi)\}$. For example, $\inf_{\xi\in\mathbb{R}}\{e^\xi\} = 0$. The supremum of the exponential function does not exist, but it is custom to write $\sup_{\xi\in\mathbb{R}}\{e^\xi\} = \infty$ if the function has no upper bound.

## B.2 Linear algebra

### B.2.1 Vectors and points

The set of all tuples of real numbers with $D$ elements is denoted by $\mathbb{R}^D$. An element $x$ of $\mathbb{R}^D$ may be written as $x = (x_1, \ldots, x_D)$, where each $x_d$, $d \in \{1, \ldots, D\}$ denotes a real number.

An element of $\mathbb{R}^D$ can be interpreted as both a **vector** or a **point**. When we say "vector," we mean the object to represent a direction and a scale. Vectors may change direction and scale via the operations of vector addition and scalar multiplication, as explained below. On the other hand, a "point" denotes a specific location in $D$-dimensional space.

For $D = 2$, we may illustrate vectors and points in the coordinate plane—here is a selection of three points and their *position vectors* drawn as arrows that connect them with the origin:

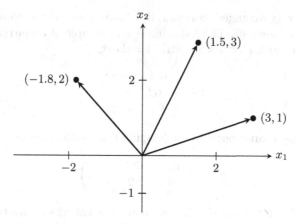

**Fig. B.3.** *Points and vectors*

We define the *dot product* between two vectors $x$ and $y$ with the same number of entries as follows:

$$x \bullet y = x_1 \cdot y_1 + x_2 \cdot y_2 + \cdots + x_D \cdot y_D = \sum_{d=1}^{D} x_d \cdot y_d$$

Here, we have also introduced summation notation "$\sum$." Products can be denoted with "$\prod$" instead. Two vectors $v, w$ of real numbers are **orthogonal** if and only if $v \bullet w = 0$.

The **Euclidean distance** between points is the following nonnegative number, which is the length of the line segment that connects those points:

$$\delta(x, y) = \sqrt{\sum_{d=1}^{D} (x_d - y_d)^2}$$

The **Euclidean norm, length,** or **magnitude** of a vector is closely related:

$$\|x\| = \sqrt{\sum_{d=1}^{D} (x_d)^2}$$

Any vector $v$ that is not the zero vector (i.e., at least one entry of $v$ is not zero) can be **normalized** (i.e., scaled to unit length) by dividing by its norm: $v \mapsto v/\|v\|$.

### B.2.2 Matrices

A **column vector** is a vector that we arrange vertically, for example:

$$x = \begin{pmatrix} 0.3 \\ -0.5 \\ 1.1 \end{pmatrix}$$

A **row vector** is arranged horizontally. More generally, a **matrix** $A$ is a rectangular arrangement. The following is an example of a matrix with three rows and two columns, or $3 \times 2$-matrix for short:

$$A = \begin{pmatrix} 0.3 & -0.1 \\ -0.5 & 2.2 \\ 1.1 & -2.2 \end{pmatrix}$$

A matrix can be **transposed** by exchanging rows with columns:

$$A^T = \begin{pmatrix} 0.3 & -0.5 & 1.1 \\ -0.1 & 2.2 & -2.2 \end{pmatrix}$$

If $A = A^T$ holds for a matrix $A$, that matrix is called **symmetric**. It must necessarily be **square**, i.e., have as many columns as it has rows.

A computer program may store vectors and matrices as *arrays*. Vectors or matrices of the same format can be added by adding them entry by entry. For example:

$$\begin{pmatrix} 0.3 \\ -0.5 \end{pmatrix} + \begin{pmatrix} 0.3 \\ 0.5 \end{pmatrix} = \begin{pmatrix} 0.3 + 0.3 \\ -0.5 + 0.5 \end{pmatrix} = \begin{pmatrix} 0.6 \\ 0.0 \end{pmatrix}$$

Similarly, we may multiply a matrix/vector by a single real number—in this context called a *scalar*—by multiplying each entry of the matrix/vector with that number.

We may also compute the product of two vectors or matrices $A$ and $B$. However, regular **matrix multiplication** is *not* multiplication entry by entry. Rather, the product $A \cdot B$ is computed according to the following rule: the entry located in the $m$-th row and the $n$-th column of $A \cdot B$ is given by the dot product of the $m$-th row of $A$ with the $n$-th column of $B$. For example:

$$\begin{pmatrix} 0.3 & -0.1 \\ -0.5 & 2.2 \\ 1.1 & -2.2 \end{pmatrix} \cdot \begin{pmatrix} 0.6 \\ 0.0 \end{pmatrix} = \begin{pmatrix} 0.3 \cdot 0.6 + (-0.1) \cdot 0.0 \\ (-0.5) \cdot 0.6 + 2.2 \cdot 0.0 \\ 1.1 \cdot 0.6 + (-2.2) \cdot 0.0 \end{pmatrix} = \begin{pmatrix} 0.18 \\ -0.30 \\ 0.66 \end{pmatrix}$$

In order for this product to be defined, the number of columns of $A$ must be equal to the number of rows of $B$. Matrix multiplication relates to transposition via the following rule: $(A \cdot B)^T = B^T \cdot A^T$.

The **determinant** of a square matrix of format $2 \times 2$ may be computed as follows:

$$\det \begin{pmatrix} a & b \\ c & d \end{pmatrix} = a \cdot d - b \cdot c$$

This number is the signed area of the parallelogram spanned by the column vectors. It turns out to be equal to the signed area spanned by the row vectors, as illustrated in the following figure:

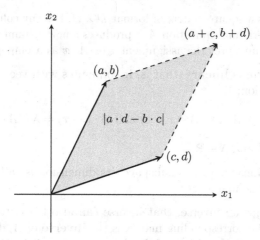

**Fig. B.4.** *Area of a parallelogram; determinant of a matrix*

The determinant of a $3 \times 3$-matrix is the signed volume of the parallelepiped spanned by the column/row vectors. More generally, the determinant of a square matrix of format $D \times D$ is determined by the signed volume of the $D$-dimensional *parallelotope* that is spanned by the column/row vectors.

### B.2.3 Subspaces and linear maps

**Linear combinations** of vectors $v_1, \ldots, v_K$ are terms of the form $\lambda_1 v_1 + \cdots + \lambda_K v_K$ with scalars $\lambda_1, \ldots, \lambda_K$.

A finite set or tuple of vectors is called **linearly independent** if none of the vectors can be written as a linear combination of the others.

Given vectors $v_1, \ldots, v_K$, we can study the **subspace** that they **span**:

$$\mathrm{span}(v_1, \ldots, v_K) = \left\{ \sum_{k=0}^{K} \lambda_k v_k \,\middle|\, \lambda_1, \ldots, \lambda_K \in \mathbb{R} \right\}$$

The **dimension** of a subspace is the minimum number of linearly independent vectors needed to span that space.

For example, let us consider the subspace in three-dimensional space that is spanned by the following column vectors:

$$v_1 = \begin{pmatrix} 0 \\ 1 \\ 0 \end{pmatrix}, \; v_2 = \begin{pmatrix} 1 \\ 0 \\ 0 \end{pmatrix}, \; v_3 = \begin{pmatrix} 2 \\ 1 \\ 0 \end{pmatrix}$$

We notice that $v_3 = v_1 + 2 \cdot v_2$, so those vectors are linearly dependent. The linearly independent vectors $v_1$ and $v_2$ suffice to span the subspace, which is a two-dimensional plane. More generally, subspaces of $D$-dimensional space that have dimension $D - 1$ are called **hyperplanes**.

Suppose that $A$ is a square matrix of format $D \times D$. For any column vector $x$ of length $D$, the matrix multiplication $A \cdot x$ produces a new column vector of length $D$. We may thus interpret the assignment $x \mapsto A \cdot x$ as a map $f_A \colon \mathbb{R}^D \to \mathbb{R}^D$.

A map of this form is **linear**—that is, it commutes with vector addition and scalar multiplication:

$$A \cdot x + A \cdot y = A \cdot (x + y) \text{ and } A \cdot (\lambda \cdot x) = \lambda \cdot (A \cdot x)$$

for any $x, y \in \mathbb{R}^D$ and $\lambda \in \mathbb{R}$.

The image of a linear map is a subspace, its dimension is called the **rank** of that linear map/the matrix $A$.

If a linear map has an inverse, that inverse can also be written as a matrix multiplication. The corresponding matrix is the **inverse** of $A$, denoted by $A^{-1}$. An inverse of $A$ exists if and only if the determinant of $A$ does not vanish. The transpose of an invertible matrix is invertible, and $\left(A^T\right)^{-1} = \left(A^{-1}\right)^T$ holds.

An **orthogonal matrix** $V$ is a matrix that we can simply invert by transposition, i.e., it is invertible and $V^{-1} = V^T$ holds. An orthogonal matrix may also be characterized by the fact that its columns are pairwise orthogonal and have unit length. Orthogonal maps represent rotations, reflections, and compositions thereof. As such, they preserve Euclidean distance: $\delta(V \cdot x, V \cdot y) = \delta(x, y)$ for all vectors $x, y$.

### B.2.4 Eigenvectors and eigenvalues

Let $A$ be some square matrix, representing a linear map. Assume that $v$ is a vector that is not the zero vector and that there is some $\lambda \in \mathbb{R}$ such that $A \cdot v = \lambda \cdot v$. In that case, $v$ is called an **eigenvector** of $A$ to **eigenvalue** $\lambda$.

In general, the linear map may change the direction of its input, so that the output $A \cdot x$ is *not* a multiple of $x$. The eigenvectors of $A$ denote those directions where it is.

For example, the following matrix reflects every vector with respect to the $x_2$-axis:

$$R = \begin{pmatrix} -1 & 0 \\ 0 & 1 \end{pmatrix}$$

As can be easily checked, $v_1 = (1, 0)^T$ is an eigenvector of $R$ with eigenvalue $\lambda_1 = -1$, and $v_2 = (0, 1)^T$ is an eigenvector with eigenvalue $\lambda_2 = 1$:

$$R \cdot v_1 = -v_1, \; R \cdot v_2 = v_2$$

These directions correspond to the axis of reflection and the direction orthogonal to that axis:

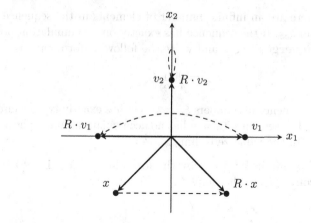

**Fig. B.5.** *Reflection in the plane*

For **diagonal matrices** like $R$, i.e., when all entries outside the main diagonal are zero, determining the eigenvalues is easy: they are the entries which are exactly on the diagonal.

Suppose that $A$ is a square matrix. We may say that $A$:

- is **positive definite** if all of its eigenvalues are positive,

- is **positive semidefinite** if all of its eigenvalues are nonnegative,

- is **indefinite** if it has both positive and negative eigenvalues.

*Negative (semi-)definite* matrices satisfy analogous conditions.

One very useful fact that has many applications is the following (see, for example, [3, Sect. 6.4] or [5, Sect. 1.14]): If $S$ is a symmetric matrix of format $D \times D$, there exists an orthogonal matrix $V$ and a diagonal matrix $\Lambda$ such that the following holds:

$$S = V \cdot \Lambda \cdot V^T$$

Furthermore, the diagonal entries $\lambda_1, \ldots, \lambda_D$ of $\Lambda$ are the eigenvalues, while the columns $v_1, \ldots, v_D$ of $V$ are the corresponding normalized eigenvectors of $S$.

# B.3 Multivariate calculus

## B.3.1 Limits

In this section, by "sequence" we mean an infinite sequence with values in $\mathbb{R}^D$ for some fixed $D \geq 1$. Such a sequence can be written as a list of numbers $(D = 1)$ or vectors $(D > 1)$ that does not end: $(\xi_n) = (\xi_1, \xi_2, \ldots)$ with every $\xi_n$ being an element of $\mathbb{R}^D$.

Given some sequence $(\xi_n)$ with values in $\mathbb{R}^D$, an *accumulation point* of that sequence is any $\xi_\infty \in \mathbb{R}^D$, so that no matter how small we choose some positive distance $\varepsilon > 0$ there are an infinite number of elements in the sequence within that distance from $\xi_\infty$. If the sequence has exactly one accumulation point $\xi_\infty$, we say that it **converges** to $\xi_\infty$ and write the following formula:

$$\xi_\infty = \lim_{n \to \infty} \xi_n$$

For example, the sequence of numbers $\xi_n = (-1)^n$ has exactly two accumulation points: $-1$ and $1$. The sequence $\xi_n = n$ has no accumulation point. The sequence $\xi_n = \frac{1}{n}$ converges to the value zero: $\lim_{n \to \infty} \frac{1}{n} = 0$.

The sequence of points in $\mathbb{R}^2$ given by the formula $\xi_n = (\frac{1}{n}, 1 - \frac{1}{n})$ starts at $\xi_1 = (1, 0)$ and converges to the point $\xi_\infty = (0, 1)$:

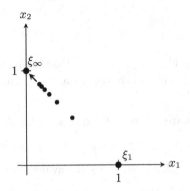

**Fig. B.6.** *Convergent sequence in the plane*

### B.3.2 Continuous functions

Let $f$ be some vector-valued function with domain $X \subseteq \mathbb{R}^D$:

$$f : X \to \mathbb{R}^K, \ (x_1, \ldots, x_D) \mapsto \begin{pmatrix} f_1(x_1, \ldots, x_D) \\ \vdots \\ f_K(x_1, \ldots, x_D) \end{pmatrix}$$

The number $D$ denotes the dimension of the domain, and $K$ is the dimension of the codomain. The **graph** of $f$ is the set of points $(x, f(x)) \in \mathbb{R}^D \times \mathbb{R}^K$, where $x$ varies over the domain.

Now, suppose that $K = 1$, i.e., the codomain is the real number line. In other words, $f$ is a *scalar-valued* function that just returns a single value. If the domain has low dimensionality, we may illustrate the function by drawing its graph:

- **Univariate functions.** The simplest case is $D = 1$, which includes the elementary functions studied in basic univariate calculus, like logarithms, exponential functions, trigonometric functions (like sine or cosine), polynomials, etc. These functions can be graphed in the usual way in the 2-dimensional coordinate plane.

- **Bivariate functions, surfaces.** For $D = 2$, we can also draw the graph of $f$ but in 3-dimensional space. If the function is sufficiently regular (i.e., continuous, differentiable—as explained in the following), that graph is now a *surface*. See Fig. B.9 for an example.

If the function is *vector-valued* ($K > 1$), we may still be able to interpret it geometrically and draw a visual representation.

- **Curves.** For $D = 1$ and $K = 2$, the *image* of $f$ (if sufficiently regular) traces a curve in the plane. For $D = 1$ and $K = 3$, the function traces a curve in 3-dimensional space.

- **Vector fields.** For $D = 2$ and $K = 2$, we may interpret the function as a vector field in the plane: every point in the domain is assigned a vector. We may draw that vector as an arrow (possibly rescaled in length to avoid clutter) for a suitable selection of points, e.g., on a regular grid.

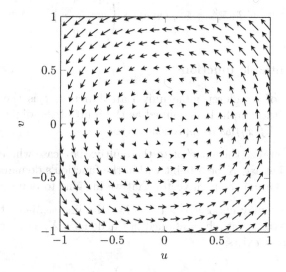

**Fig. B.7.** *Plot of the vector field* $(u, v) \mapsto (-v, u)$

The functions, graphs, curves, and vector fields that we study and like to model data and observations with should be *sufficiently regular*. For example, we would frequently impose the condition that the function does not exhibit large changes in value given small changes in its argument.

In order to formalize that idea, let us assume that we are given some function $f : X \to \mathbb{R}^K$ with $X \subseteq \mathbb{R}^D$, and fix some point $\xi_\infty \in \mathbb{R}^D$. Suppose that there exists at least one sequence $(\xi_n)$ that takes values in $X$ and converges to $\xi_\infty$, and that for all those sequences the corresponding $(f(\xi_n))$ converge to the same value $y_\infty$. Then, the following limit is well-defined:

$$y_\infty = \lim_{\xi \to \xi_\infty} f(\xi)$$

Now, fix a point $x \in X$. We say that the function $f$ is **continuous** at $x$ if:

$$f(x) = \lim_{\xi \to x} f(\xi)$$

More informally, we can imagine walking around the domain towards some value $x$, taking the steps $\xi_1, \xi_2, \ldots$ along the way. We may say that $f$ is continuous at $x$ if the corresponding function values $f(\xi_1), f(\xi_2), \ldots$ always approach $f(x)$ just the same.

It turns out that a function is continuous if and only if every one of its scalar component functions $f_1, \ldots, f_K$ is continuous. Sums, products, and compositions of continuous scalar functions are continuous. Multivariate monomials—i.e., $(x_1)^{i_1} \cdots (x_D)^{i_D}$ with $i_1, \ldots, i_D \in \mathbb{N}$—are continuous functions. Thus, an example of a continuous function would be the following:

$$f \colon \mathbb{R}^2 \to \mathbb{R}^3, \ (u, v) \mapsto \begin{pmatrix} u^2 + v^2 \\ u + 5 \cdot v \\ u \cdot v \end{pmatrix}$$

### B.3.3 Differentiable functions

The derivative of a differentiable univariate function $f$ is the slope of the tangent line to the graph. It is defined as the limit of the difference quotient $\alpha^{-1} \cdot (f(x + \alpha) - f(x))$ as $\alpha$ approaches zero.

We want to generalize this concept to the multivariate case where $f \colon X \to \mathbb{R}^K$ with $X \subseteq \mathbb{R}^D$. Let $x \in X$ be a point within the function's domain, and $h \in \mathbb{R}^D$ with $\|h\| = 1$ be a unit column vector that we imagine to point away from $x$.

Suppose that we may choose $\alpha > 0$ so that the line segment that connects $x - \alpha h$ and $x + \alpha h$ is entirely contained in the domain $X$, and secondly so that the following limit exists:

$$\mathrm{D}_h f(x) = \lim_{\alpha \to 0} \frac{f(x + \alpha \cdot h) - f(x)}{\alpha}$$

In that case, the above vector is the **derivative of $f$ in the direction of $h$** at $x$. If all directional derivatives exist and are continuous, the function is called **continuously differentiable** at $x$. The derivatives in the direction of the coordinate axes $e_1, \ldots, e_D$ are called the **partial derivatives** of $f$ and are also denoted as follows:

$$\partial_d f(x) = \frac{\partial f}{\partial x_d}(x) = \mathrm{D}_{e_d} f(x)$$

Furthermore, it turns out that if the function is continuously differentiable, we can write every directional derivative as $\mathrm{D}_h f(x) = \mathrm{D}f(x) \cdot h$ where $\mathrm{D}f(x)$ is the **Jacobian matrix** that contains all of the partial derivatives:

$$\mathrm{D}f(x) = (\partial_1 f(x), \partial_2 f(x), \ldots, \partial_D f(x)) = \begin{pmatrix} \partial_1 f_1(x) & \partial_2 f_1(x) & \ldots & \partial_D f_1(x) \\ \partial_1 f_2(x) & \partial_2 f_2(x) & \ldots & \partial_D f_2(x) \\ \vdots & & & \vdots \\ \partial_1 f_K(x) & \partial_2 f_K(x) & \ldots & \partial_D f_K(x) \end{pmatrix}$$

We may also denote the Jacobian as a *total derivative*, $\frac{\mathrm{d}}{\mathrm{d}x} f(x)$.

Given known derivatives of univariate elementary functions, partial derivatives are fairly easy to compute by keeping all but one variable fixed, treating the others as constants. For example:

$$f(u, v) = \begin{pmatrix} u^2 + v^2 \\ u + 5 \cdot v \\ u \cdot v \end{pmatrix} \Rightarrow \mathrm{D}f(u, v) = \begin{pmatrix} 2u & 2v \\ 1 & 5 \\ v & u \end{pmatrix}$$

The *chain rule* in multivariate calculus is given by the following formula:

$$\mathrm{D}(g \circ f)(x) = \mathrm{D}g(f(x)) \cdot \mathrm{D}f(x)$$

In the following discussion, we will assume that $f$ is continuously differentiable and scalar-valued, i.e., $f \colon X \to \mathbb{R}$ with $X \subseteq \mathbb{R}^D$. The **gradient** of $f$ is given by the transposed Jacobian, which is the column vector that contains all partial derivatives:

$$\mathrm{grad}\, f(x) = \nabla f(x) = (\mathrm{D}f(x))^T = \begin{pmatrix} \frac{\partial f}{\partial x_1}(x) \\ \vdots \\ \frac{\partial f}{\partial x_D}(x) \end{pmatrix}$$

A **level set** of $f$ is the preimage of some fixed constant value $c \in \mathbb{R}$, i.e., the set $f^{-1}(\{c\})$. The gradient of a scalar function at some point is either zero, or perpendicular to the level set at that point. If it is nonzero, the gradient points in the direction in which the values of the function exhibit the largest increase, and its magnitude is equal to the rate of that increase. For $D = 2$, a level set is either empty, a single point, or a curve that is called a **contour line**: see Fig. B.9.

Let us further assume that $f$ is twice continuously differentiable, i.e., the vector field $\mathrm{grad}\, f \colon X \to \mathbb{R}^D$ is continuously differentiable. In that case, we may write down the **Hessian matrix** (also referred to as just Hessian) of *second* partial derivatives:

$$\mathrm{Hess}\, f(x) = \mathrm{D}\, \mathrm{grad}\, f(x) = \begin{pmatrix} \partial_1 \partial_1 f(x) & \partial_2 \partial_1 f(x) & \ldots & \partial_D \partial_1 f(x) \\ \partial_1 \partial_2 f(x) & \partial_2 \partial_2 f(x) & \ldots & \partial_D \partial_2 f(x) \\ \vdots & & & \vdots \\ \partial_1 \partial_D f(x) & \partial_2 \partial_D f(x) & \ldots & \partial_D \partial_D f(x) \end{pmatrix}$$

Under the conditions given here, it turns out that the Hessian must be a symmetric matrix: it holds that $\partial_m \partial_n f(x) = \partial_n \partial_m f(x)$ for all $m, n \in \{1, \ldots, D\}$.

We may use the gradient and the Hessian to locate minimum and maximum points of $f$. A **local minimum point** of $f$ is any point $x_{min} \in X$ such that there exists some $\varepsilon > 0$ so that for any point $x \in X$ within distance $\varepsilon$ of $x_{min}$ it holds that $f(x) \geq f(x_{min})$. A **local maximum** is defined analogously.

A *boundary point* of $X$ is a point so that at any positive distance from that point we can find points in $X$ as well as points *not* in $X$. It turns out that every local minimum or maximum point that is *not* a boundary point of $X$ must be a **stationary point**, i.e., the gradient vanishes at that point. Suppose that $x_c$ is a stationary point of $f$: $\operatorname{grad} f(x_c) = 0$. Then, the following holds:

- $x_c$ is a local maximum of $f$ if $\operatorname{Hess} f(x_c)$ is negative definite,

- $x_c$ is a local minimum of $f$ if $\operatorname{Hess} f(x_c)$ is positive definite,

- $x_c$ is neither a local minimum nor a maximum of $f$ if $\operatorname{Hess} f(x_c)$ is indefinite.

For example, the gradient of the function $f \colon \mathbb{R}^2 \to \mathbb{R}$ with $f(u, v) = u \cdot e^{-(u^2 + v^2)}$ is the following vector field:

$$\operatorname{grad} f(u, v) = e^{-(u^2 + v^2)} \cdot \begin{pmatrix} 1 - 2u^2 \\ -2uv \end{pmatrix}$$

The gradient vanishes at the points $(-1/\sqrt{2}, 0)$ and $(1/\sqrt{2}, 0)$. Calculating second derivatives and plugging those stationary points into the Hessian matrix yields the following results:

$$\operatorname{Hess} f(\mp 1/\sqrt{2}, 0) = \sqrt{2} e^{-\frac{1}{2}} \cdot \begin{pmatrix} \pm 2 & 0 \\ 0 & \pm 1 \end{pmatrix}$$

These are diagonal matrices, so we may read off their eigenvalues from the diagonal: the Hessian at $(-1/\sqrt{2}, 0)$ is a positive definite matrix, at $(1/\sqrt{2}, 0)$ it is negative definite. Thus, the function has a local minimum at $(-1/\sqrt{2}, 0)$ and a local maximum at $(1/\sqrt{2}, 0)$.

We conclude this section with a list of some derivatives known from basic calculus that are used in this book ($r \in \mathbb{R}$ is an arbitrary constant):

| $f(x)$ | $x^r$ | $e^x$ | $\ln(x)$ | $\sin(x)$ | $\cos(x)$ | $\arcsin(x)$ | $\arctan(x)$ |
|---|---|---|---|---|---|---|---|
| $f'(x)$ | $r \cdot x^{r-1}$ | $e^x$ | $1/x$ | $\cos(x)$ | $-\sin(x)$ | $1/\sqrt{1-x^2}$ | $1/1+x^2$ |

**Table B.1.** *Derivatives of elementary functions*

### B.3.4 Integrals

In basic univariate calculus, we may compute the signed area under the graph of a continuous function $f \colon [a, b] \to \mathbb{R}$ via an integral. The *fundamental theorem of calculus* implies that we may compute the integral as follows:

$$\int_a^b f(\xi) \, d\xi = F(b) - F(a)$$

where $F\colon [a, b] \to \mathbb{R}$ is an antiderivative of $f$. That is, $F$ is continuous, and the derivative of $F$ exists on the open interval and is equal to $f$.

Suppose now that we are given instead a *bivariate*, continuous function $f\colon X \to \mathbb{R}$ with $X \subseteq \mathbb{R}^2$, and that we may write $X$ as a *normal domain* of integration:

$$X = \{(u, v) \in \mathbb{R}^2 | a \le u \le b, \, \alpha(u) \le v \le \beta(u)\}$$

where $\alpha\colon [a, b] \to \mathbb{R}$ and $\beta\colon [a, b] \to \mathbb{R}$ are continuous functions with $\alpha(u) \le \beta(u)$ for all $u \in [a, b]$.

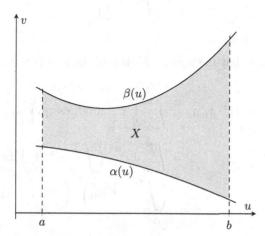

**Fig. B.8.** *Normal domain of integration*

We want to compute the signed *volume* under the graph of $f$. It turns out that this volume is given by the following iterated integral:

$$\iint\limits_{X} f(\xi) \, \mathrm{d}\xi = \int_a^b \left( \int_{\alpha(u)}^{\beta(u)} f(u, v) \, \mathrm{d}v \right) \mathrm{d}u$$

If the domain of integration is a rectangle, $X = [a, b] \times [c, d]$, the order of integration can be exchanged—this is known as **Fubini's theorem**:

$$\iint\limits_{X} f(\xi) \, \mathrm{d}\xi = \int_a^b \left( \int_c^d f(u, v) \, \mathrm{d}v \right) \mathrm{d}u = \int_c^d \left( \int_a^b f(u, v) \, \mathrm{d}u \right) \mathrm{d}v$$

*Integration by substitution* in univariate calculus is given by the following formula:

$$\int_{\varphi(a)}^{\varphi(b)} f(\eta) \, \mathrm{d}\eta = \int_a^b f(\varphi(\xi)) \cdot \varphi'(\xi) \, \mathrm{d}\xi$$

where $\varphi\colon [a, b] \to \mathbb{R}$ is continuously differentiable with range inside the domain of $f$.

The corresponding formula in bivariate calculus is the following:

$$\iint_{\varphi(X)} f(\eta)\, d\eta = \iint_X f(\varphi(\xi)) \cdot |\det(D\varphi(\xi))|\, d\xi$$

where $\varphi\colon X \to \mathbb{R}^2$ is continuously differentiable and injective.

We will apply the above formula to compute the **Gaussian integral** that plays a prominent role in probability theory and statistics:

$$G = \int_{-\infty}^{\infty} e^{-\xi^2} d\xi$$

First of all, we note that the square of our goal integral $G$ is a double integral[2] over the plane:

$$\iint_{\mathbb{R}^2} e^{-(u^2+v^2)}\, du\, dv = \int_{-\infty}^{\infty} \left( \int_{-\infty}^{\infty} e^{-u^2} e^{-v^2} dv \right) du$$

$$= \int_{-\infty}^{\infty} e^{-u^2} \cdot \left( \int_{-\infty}^{\infty} e^{-v^2} dv \right) du$$

$$= \left( \int_{-\infty}^{\infty} e^{-u^2} du \right) \cdot \left( \int_{-\infty}^{\infty} e^{-v^2} dv \right)$$

$$= G^2$$

As an integral over a positive function, $G$ is a positive number, so the square root of the double integral is equal to $G$.

We compute this double integral via the following change of variables[3] (introducing **polar coordinates**):

$$\varphi\colon [0,\infty[\, \times\, [0, 2\pi[\, \to \mathbb{R}^2, (r, \theta) \mapsto \begin{pmatrix} r \cdot \cos(\theta) \\ r \cdot \sin(\theta) \end{pmatrix}$$

The Jacobian matrix is given as follows:

$$D\varphi(r, \theta) = \begin{pmatrix} \cos(\theta) & -r \cdot \sin(\theta) \\ \sin(\theta) & r \cdot \cos(\theta) \end{pmatrix}$$

The Jacobian determinant:

$$\det(D\varphi(r, \theta)) = r \cdot (\cos(\theta))^2 - (-r \cdot (\sin(\theta))^2) = r$$

where we have used the trigonometric identity $(\cos(\theta))^2 + (\sin(\theta))^2 = 1$.

---

[2] This is an *improper integral*, and mathematical rigour would call for discussing definition and existence of limits. However, we will omit these details.

[3] That transformation is not injective at the origin, but single points may be excluded without changing the value of the integral.

Thus, via the substitution $z(r) = r^2$ in the end:

$$\iint\limits_{\mathbb{R}^2} e^{-(u^2+v^2)} \, du \, dv = \int_0^{2\pi} \left( \int_0^\infty e^{-r^2} r \, dr \right) d\varphi$$

$$= 2\pi \cdot \int_0^\infty e^{-r^2} r \, dr$$

$$= 2\pi \cdot \frac{1}{2} \int_0^\infty e^{-z} \, dz$$

$$= \pi$$

Therefore, the final result for the Gaussian integral is given as follows:

$$\int_{-\infty}^\infty e^{-\xi^2} \, d\xi = \sqrt{\pi}$$

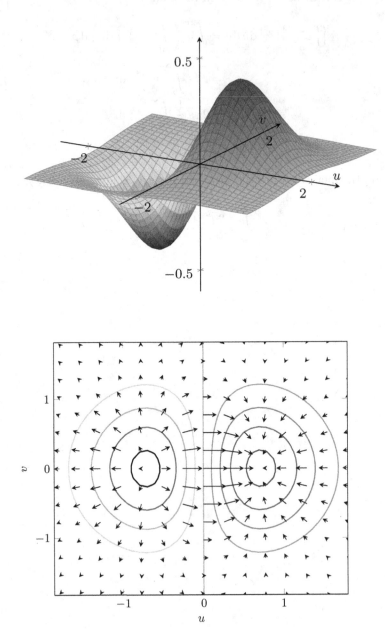

**Fig. B.9.** *Graph of the bivariate function* $(u, v) \mapsto u \cdot e^{-(u^2+v^2)}$ *(top), contour lines and gradient (bottom)*

# References

[1] Michael Spivak. *Calculus on Manifolds. A Modern Approach to Classical Theorems of Advanced Calculus.* Philadelphia, PA: Westview Press, Jan. 1971.

[2] Michael Spivak. *Calculus.* 3rd ed. Cambridge, England: Cambridge University Press, June 2006.

[3] Gilbert Strang. *Introduction to Linear Algebra.* 5th ed. Wellesley, MA: Wellesley-Cambridge Press, Aug. 2016.

[4] Gilbert Strang. *Calculus.* 3rd ed. Wellesley, MA: Wellesley-Cambridge Press, Sept. 2017.

[5] Jan R. Magnus. *Matrix Differential Calculus with Applications in Statistics and Econometrics.* 3rd ed. Wiley, Feb. 2019. DOI: 10.1002/9781119541219.

[6] Georg Cantor. "Beiträge zur Begründung der transfiniten Mengenlehre". In: *Mathematische Annalen* 46.4 (Nov. 1895), pp. 481–512. DOI: 10.1007/bf02124929.

# Supplementary literature

The mathematical prerequisites necessary for understanding this book are taught in academic courses on linear algebra and (multivariate) calculus. *Matrix Differential Calculus with Applications in Statistics and Econometrics* by Jan R. Magnus is a self-contained text that covers all that groundwork [1]. The books by Strang [2, 3] and Spivak [4, 5] come recommended. The mathematically inclined reader might enjoy *Foundations of Modern Analysis* [6] by Dieudonné, and advanced books on linear algebra have been written by Kostrikin and Manin [7] and Halmos [8].

For further reading, I would like to refer to the following texts.

**Data organization.** The *Data Management Body of Knowledge* [9] by the international Data Management Association provides a comprehensive overview that extends far beyond the scope of this book. Entity–relationship modeling is covered by Bernhard Thalheim [10], whereas graph-based data models in general—and the graph database neo4j in particular—are the subject of a book by Ian Robinson et al. [11]. For those who would like to study the mathematical theory of graphs in more detail, the classic book by Harary is recommended [12].

**Probability theory and statistics.** With *Probability: Theory and Examples*, Rick Durrett provides a mathematically well-founded overview of probability theory, covering important topics like central limit theorems [13]. Also recommended to mathematically inclined readers is the classic text by Fisz [14] and the somewhat more recent volume by Casella and Berger [15]. Roman Vershynin, in his book *High-Dimensional Probability*, highlights the essential mathematical concepts and results that are the basis for the analysis of high-dimensional data [16]. Gunnar Carlsson and Mikael Vejdemo-Johansson have recently published a book on topological data analysis [17]. Finally, the *Encyclopedia of Distances* is entirely devoted to the topic of distance measures [18].

**Machine learning.** Recommended classics on machine learning and pattern recognition are *Pattern Classification*, by Duda, Hart, and Stork [19], and

*The Elements of Statistical Learning*, by Hastie, Tibshirani, and Friedman [20], together with its companion book that aims at making the methods accessible to a wider audience [21]. Theodoridis and Koutroumbas also share a wealth of knowledge [22]. The volume by Ian Goodfellow et al. contains a comprehensive account of artificial neural networks [23]. François Chollet, initiator of the Keras deep learning program library, is the author of practical books on the subject [24, 25]. A compact and clear presentation of the core aspects of statistical learning theory is provided in a review article by Ulrike von Luxburg and Bernhard Schölkopf [26]. A recommended book on the subject is the text by Mehryar Mohri et al. [27].

**Other topics.** The book by Tufte may serve as a classic reference on data visualization [28], and *Storytelling with Data* by Knaflic [29] comes recommended. A standard work on digital image processing is the textbook by Bernd Jähne [30]. Jacob Eisenstein offers a modern introduction to natural language processing [31]. Those who wish to know more about sentiment analysis can consult the book by one of the field's top experts, Bing Liu [32]. Perhaps the leading introductory textbook in the scientific study of networks is the book by Mark Newman [33].

Furthermore, *Foundations of Data Science* by Avrim Blum, John Hopcroft, and Ravindran Kannan should not go unmentioned, which provides insight into a wide variety of topics [34].

Finally, I'd like to call attention to Grant Sanderson's YouTube channel 3Blue1Brown, which—among other interesting mathematical topics—features entertaining and insightful introductions to neural networks [35] and Bayes' theorem [36].

# References

[1] Jan R. Magnus. *Matrix Differential Calculus with Applications in Statistics and Econometrics*. 3rd ed. Wiley, Feb. 2019. DOI: 10.1002/9781119541219.

[2] Gilbert Strang. *Introduction to Linear Algebra*. 5th ed. Wellesley, MA: Wellesley-Cambridge Press, Aug. 2016.

[3] Gilbert Strang. *Calculus*. 3rd ed. Wellesley, MA: Wellesley-Cambridge Press, Sept. 2017.

[4] Michael Spivak. *Calculus*. 3rd ed. Cambridge, England: Cambridge University Press, June 2006.

[5] Michael Spivak. *Calculus on Manifolds. A Modern Approach to Classical Theorems of Advanced Calculus*. Philadelphia, PA: Westview Press, Jan. 1971.

[6] Jean Dieudonné. *Foundations of Modern Analysis*. Academic Press, 1960.

[7] Alexei I. Kostrikin and Yuri I. Manin. *Linear Algebra and Geometry*. Gordon and Breach Science Publishers, 1997.

[8] Paul R. Halmos. *Finite-Dimensional Vector Spaces*. Springer New York, 1974. DOI: 10.1007/978-1-4612-6387-6.

[9] DAMA International. *DAMA-DMBOK: Data Management Body of Knowledge*. Bradley Beach, NJ: Technics Publications, July 2021.

[10] Bernhard Thalheim. *Entity–Relationship Modeling*. Springer Berlin Heidelberg, 2000. DOI: 10.1007/978-3-662-04058-4.

[11] Ian Robinson, Jim Webber, and Emil Eifrem. *Graph Databases*. 2nd ed. Sebastopol, USA: O'Reilly, 2015.

[12] Frank Harary. *Graph Theory*. Reading, USA: Addison Wesley, 1969.

[13] Rick Durrett. *Probability: Theory and Examples*. 5th ed. Cambridge University Press, May 2019.

[14] Marek Fisz. *Probability Theory and Mathematical Statistics*. 3rd ed. John Wiley & Sons, Dec. 1963.

[15] George Casella and Roger L. Berger. *Statistical Inference*. Pacific Grove, USA: Duxbury Thomson Learning, 2001.

[16] Roman Vershynin. *High-Dimensional Probability: An Introduction with Applications in Data Science*. Cambridge University Press, Sept. 2018. DOI: 10.1017/9781108231596.

[17] Gunnar Carlsson and Mikael Vejdemo-Johansson. *Topological Data Analysis with Applications*. Cambridge University Press, Nov. 2021. DOI: 10.1017/9781108975704.

[18] Michel Marie Deza and Elena Deza. *Encyclopedia of Distances*. Springer, Berlin, Heidelberg, 2009. DOI: 10.1007/978-3-642-00234-2.

[19] Richard O. Duda, Peter E. Hart, and David G. Stork. *Pattern Classification*. 2nd ed. Wiley, 2000.

[20] Trevor Hastie, Robert Tibshirani, and Jerome Friedman. *The Elements of Statistical Learning: Data Mining, Inference, and Prediction*. 2nd ed. Springer, New York, 2009. DOI: 10.1007/978-0-387-84858-7.

[21] Gareth James et al. *An Introduction to Statistical Learning. With applications in R*. 2nd ed. Springer Texts in Statistics. Springer, July 2021.

[22]   Sergios Theodoridis and Konstantinos Koutroumbas. *Pattern Recognition.* 4th ed. Academic Press, 2008.

[23]   Ian Goodfellow, Yoshua Bengio, and Aaron Courville. *Deep Learning.* MIT Press, Nov. 2016. URL: http://www.deeplearningbook.org/.

[24]   Francois Chollet. *Deep Learning with Python.* 2nd ed. Manning Publications, Dec. 2021.

[25]   Francois Chollet. *Deep Learning with R.* 2nd ed. Manning Publications, July 2022.

[26]   Ulrike von Luxburg and Bernhard Schölkopf. "Statistical Learning Theory: Models, Concepts, and Results". In: *Handbook of the History of Logic.* Vol. 10. Amsterdam, Niederlande: Elsevier North Holland, May 2011, pp. 651–706. DOI: 10.1016/b978-0-444-52936-7.50016-1. arXiv:0810.4752.

[27]   Mehryar Mohri, Afshin Rostamizadeh, and Ameet Talwalkar. *Foundations of Machine Learning.* 2nd ed. MIT Press, 2018.

[28]   Edward R. Tufte. *The Visual Display of Quantitative Information.* 2nd ed. Cheshire, CT: Graphics Press, Jan. 2001.

[29]   Cole Nussbaumer Knaflic. *Storytelling with Data. A Data Visualization Guide for Business Professionals.* Nashville, TN: John Wiley & Sons, Oct. 2015.

[30]   Bernd Jähne. *Digital Image Processing.* 6th ed. Springer, 2005. DOI: 10. 1007/3-540-27563-0.

[31]   Jacob Eisenstein. *Introduction to Natural Language Processing.* MIT Press, 2019.

[32]   Bing Liu. *Sentiment Analysis.* 2nd ed. Studies in Natural Language Processing. Cambridge University Press, Oct. 2020.

[33]   Mark Newman. *Networks.* 2nd ed. Oxford University Press, July 2018.

[34]   Avrim Blum, John Hopcroft, and Ravi Kannan. *Foundations of Data Science.* Cambridge University Press, Jan. 2020. DOI: 10.1017/978110875 5528.

[35]   Grant Sanderson. *3Blue1Brown, Season 3: Neural networks.* Aug. 2018. URL: https://youtube.com/playlist?list=PLZHQObOWTQDNU6R1_ 67000Dx_ZCJB-3pi.

[36]   Grant Sanderson. *3Blue1Brown: Bayes theorem, the geometry of changing beliefs.* Dec. 2019. URL: https://youtu.be/HZGCoVF3YvM.

# Index

© Springer-Verlag GmbH Germany, part of Springer Nature 2023
M. Plaue, *Data Science*, https://doi.org/10.1007/978-3-662-67882-4

Printed in the United States
by Baker & Taylor Publisher Services